Lightning in the Age of Benjamin Franklin

Lightning in the Age of Benjamin Franklin

Facts and Fictions in Science, Religion, and Art

Jan Wim Buisman

Translated by
Lee Preedy

Leiden University Press

◀ A so-called *Parapluie paratonnerre*. This (highly dangerous!) portable lightning conductor was designed by the French physician and botanist Jacques Barbeu-Dubourg (1709-1779). Nineteenth-century engraving from L. Figuier, *Les merveilles de la science*. Paris: Furne, Jouvet et Cie, 1870

This publication was partly made possible by financial support of Sormanifonds, Thijssen-Schoute Fonds, De Gijselaar-Hintzenfonds, and Stichting Gilles Hondius Foundation

Original title: *Onweer. Een kleine cultuurgeschiedenis, 1752-1830.*

Authorised translation from the Dutch language edition published by Vantilt (The Netherlands)

ISBN Hardback 978 90 8728 387 2
ISBN Paperback 978 90 8728 512 8
e-ISBN 978 94 0060 433 9 (e-PDF)
https://doi.org/10.24415/9789087283872
NUR 694

Cover design: Andre Klijsen
Cover illustration: *Oak Struck by Lightning*, Maksim Vorobyov, 1842. The State Tretyakov Gallery, Moscow
Design and typesetting: Friedemann Vervoort
Printer: Scanlaser bv, Zaandam

Contact EU General Product Safety Regulation (GPSR): productsafety@lup.nl
Printed and bound by CPI Group (UK) Ltd, Croydon, CR0 4YY

At base, science is no more than an investigation of a miracle we can never explain, and art is an interpretation of that miracle.

Ray Bradbury, *Martian Chronicles*

Table of contents

Introduction: Lightning after Franklin

The invention of the lightning rod by Benjamin Franklin (1706-1790) was one of the first major scientific and technological contributions (with cultural and philosophical significance) to transfer from the Americas to Europe, rather than the other way round. This book broadly describes its European reception. And while the focus is on reactions in the Netherlands, it relates those to American, German, French, and British developments. More importantly, the responses to the invention of the lightning rod are used here as a window on cultural developments, among them the reception of science and technology, but also the diverse variants of religious life, from the traditional to the esoteric, and the experience of nature in the context of the Enlightenment and Romanticism. The result is an intriguing combination or rather a reconnection of such relatively dissimilar scientific disciplines as meteorology, disaster research, and the study of religious and cultural mentalities. In short, this book traces both the physical and creative thunderstorms that reverberated in the Western consciousness during the second half of the eighteenth century and the first decades of the nineteenth.

The history of the Low Countries is one of a millennia-long battle against the elements, particularly water. From the west, the North Sea posed a constant threat; from the east, the great rivers Rhine and Meuse all too often flooded the land. Between 1568 and 1648, man-made disasters were added to natural ones as the Netherlands fought an eighty-year-long war of independence from Spain. Only with the signing of the Peace of Westphalia in 1648 was the Republic of the Seven United Netherlands, commonly referred to as the Dutch Republic, formally recognised as a sovereign state. It was the political culmination of the Dutch Golden Age. It also made permanent the position of the Reformed Church as a privileged denomination. Calvinists predominated, but they were by no means the only Protestants. Lutherans, Mennonites, and Remonstrants formed a significant and often very culturally active minority, despite being treated as second-class citizens. And while taken as a whole these congregations resulted in a population that was for the most part Protestant, a sizeable Roman Catholic minority still remained – albeit subject to discrimination. However much the economic and political position of the Dutch Republic may have declined in the course of the eighteenth century, there was little change in the diverse nature of its religious landscape. That diversity will also be apparent when we look at the way in which, in the second half of the eighteenth and the early nineteenth centuries, pious folk of all kinds reacted to the challenges of nature, especially thunder and lightning.

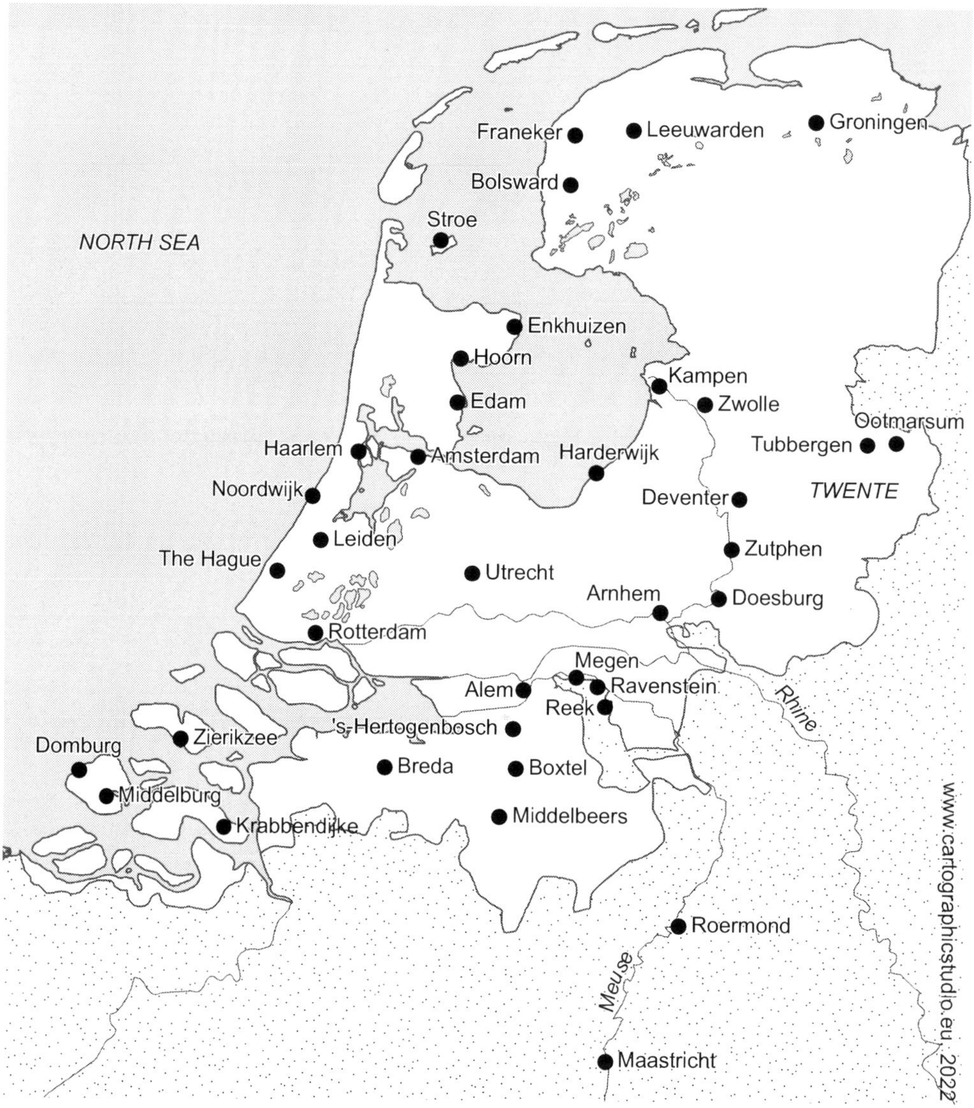

▲ Map of the Netherlands

Strikingly, those reactions were by no means always fearful. After all, the Dutch Republic was far from being a cultural backwater; it boasted a vibrant scholarly community, a comparatively tolerant spiritual milieu, and a lively press. In that confident climate, enlightened inquiry and fascination with the wonders of nature soon overcame any lingering superstitious dread, as is well illustrated by the reaction of a Calvinist minister, Johannes Florentius Martinet (1729-1795), who hailed from Zutphen in the east of the Republic, to a thunderstorm. "In all nature," he wrote in 1778, "no phenomenon is more striking to the eye, awesome to the ear, tremendous to the senses and useful to the world."[1] Moreover, he continued in lyrical vein, if a storm should arise while he was working, "I throw everything aside in order to hear and see it (...). Did not the holy poet also listen to it when he sang 'The God of Glory thunders' (Ps. 29:3), or were we not supposed to notice it? I never hear it and see it without feeling great reverence, deep amazement, and humble delight, devoid of foolish fear; unless it is very violent, and this blessing is turned into a judgment."[2]

◀ Portrait of Joannes Florentius Martinet, by Reinier Vinkeles, 1778

Martinet's liking for thunderstorms was not just a personal idiosyncrasy. His younger contemporary, the prolific man of letters Willem Emmery de Perponcher (1741-1819), was equally enthusiastic. Indeed, his biographer called the thunderstorm "an eminently Perponcherian theme",[3] though that overly credits this now-forgotten author, for as we shall see below, thunderstorms enjoyed a good deal of popularity in the culture of the time, not only in burgeoning science but also religion and art. The subject would become all the rage, in fact, particularly in the last quarter of the eighteenth century,[4] no doubt because in the declensionist mindset of that period it provided an obvious metaphor – storm clouds signifying moral, economic, and political decay gathering menacingly over the Republic and culminating in a transfer of power by revolutionary France in 1795.[5] But we must draw the circle wider still, for not only thunder and lightning but meteorological phenomena per se enjoyed a growing degree of popularity in this period of emerging Romanticism, in the Netherlands as well as the rest of Europe.[6]

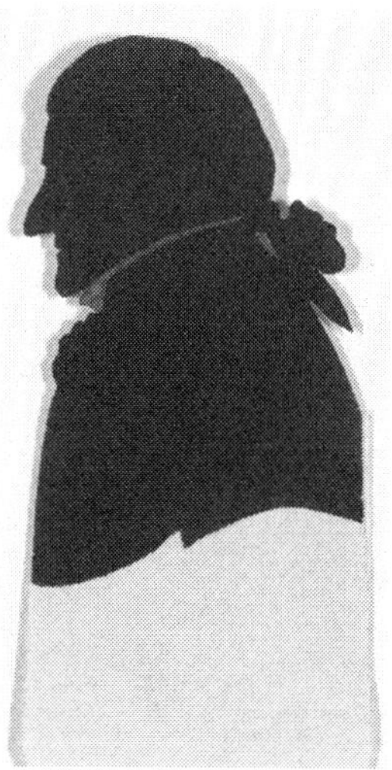

◀ Silhouette of Willem Emmery de Perponcher Sedlnitzky, by Casparis Haanen (1778-1849)

In the late spring of 1752 Benjamin Franklin invented the lightning rod, and interest in thunderstorms increased. Franklin, a self-taught polymath, was the first to prove what many scholars, including himself, had already suspected – that lightning was electrical in nature. Not content with theory, he put his invention to practical use, guiding the potentially destructive lightning via a simple metal wire to the ground, where it could do no harm. That same summer at his home in Philadelphia he installed the very first lightning rod, now no longer as an instrument of scientific research but a means of protecting the building from lightning strikes.[7] In the decades that followed, in the American colonies and Europe as well, especially at-risk edifices such as powder magazines, windmills, lighthouses, and church towers were fitted with lightning rods.

▲ Portrait of Benjamin Franklin, by Pierre Michel Alix, after Louis-Michel van Loo, 1793-1795

▲ The lightning rod installed by Franklin on the Philadelphia home of banker Benjamin West demonstrates its effectiveness during a thunderstorm in 1760, engraving, maker unknown

The long-term influence of the invention and introduction of the lightning rod, particularly among proponents of the Enlightenment, should certainly not be underestimated. Despite the initial scepticism among Franklin's contemporaries and the sometimes faulty installation of the rod, it was quite possibly the eighteenth century's most spectacular example of progress in controlling nature, a discovery with symbolic, even paradigmatic significance.[8] In retrospect that significance becomes all the greater when we consider that the lightning rod's lustre has faded simply because its utility now seems so self-evident.[9] But in the eighteenth century the fact that a natural phenomenon which had always been so unruly and unpredictable as the thunderstorm could apparently now be tamed must have been seen as a milestone. Enlightenment thinkers such as the German philosopher Immanuel Kant (1724-1804) described Franklin as a modern Prometheus, who had stolen fire from the gods.[10] And even though Kant's early characterisation might also have been meant as a caution against hubris – Prometheus was, after all, condemned to eternal punishment[11] – as time went by man's self-confidence certainly increased: more than sixty years later, for instance, the English poet Percy Bysshe Shelley (1792-1822), author of the lyrical drama *Prometheus Unbound* (1820), exulted that lightning was now man's slave.[12] Two years earlier his wife, Mary Shelley (1797-1851), had already gone a step further, likening her famous protagonist Victor

Frankenstein – who harnessed lightning, produced electricity, and with it created a sapient being – to a modern Prometheus.[13] Was it not the Titan Prometheus who in Greek mythology created man from clay? This opened up wide perspectives for those who had an eye for such things, for surely what man had accomplished in the matter of the lightning rod should be achievable in other areas too? In such fantasies of technological mastery there was, in other words, a certain suggestiveness that was not always easy to resist. Whatever the case, to enlightened minds it was obvious that a major step had been taken on the long path of progress. Moreover, Francis Bacon's (1561-1626) dictum that "we cannot command nature except by obeying her" had once again been proved valid: a conclusion that held great promise for the future.[14]

For others, however, this assertion was too bold. Conforming, consciously or not, to what would later become a somewhat caricatural Romantic perception of the Enlightenment,[15] they wondered – contemporaries as well as historians – whether the pursuit of a theoretical and above all practical control of nature had not made man so drunk with his own success that his sense of religious dependence was shaken. By examining the contemporary debate apropos the changing perception of the thunderstorm and its philosophical implications, we can therefore learn something about secularising trends in European culture in the late eighteenth and early nineteenth centuries. Some decades ago the German theologian Manfred Büttner even concluded that it was pre-eminently climatology – regarded by him as "the theory of weather events" – which in the eighteenth century "becomes the 'spokesman' in the discourse between theology and science."[16] And his fellow-countryman Richard van Dülmen also cited that same apparently insignificant phenomenon, the introduction of the lightning rod, as a striking example of the secularisation of everyday life.[17]

In Enlightenment eyes, what was being waged was nothing less than a war between "superstition" and "sensible" empirical knowledge. Considering the scarcity of sources documenting the views of lower social groups, earlier historiography has been rather too prone to giving literal credence to Enlightenment thought. Consequently, the lightning conductor has acted as a catalyst for an overly simple linear transition from "myth" to "science", resulting in a certain enlightened triumphalism.[18] Yet as we shall see, in reality the invention of the lightning rod – let alone its practical introduction – could hardly be described as an unqualified success. In the decades following its creation it was a constant topic of debate among scholars and it took a long time to become commonplace in society. Moreover, as mentioned briefly above, it was sometimes installed incorrectly and for quite some time its effectiveness in foiling lightning strikes was modest at best.

Enlightened minds, meanwhile, were not in the least daunted. As they perceived it only ignorance, superstition, and prejudice were to blame and they fought those enemies with total conviction and unwavering dedication. Was it not that very phenomenon, the thunderstorm, that throughout the ages had given rise to all kinds of magical fancies and irrational behaviours? In that respect the thunderstorm acted as such a point of crystallisation at the time that its study can still be used today as a conduit to understanding a number of interesting aspects of popular culture and how they were confronted and countered.

But the importance of such a study extends further still: it offers an insight into the eighteenth-century experience of nature and the ways in which that experience was then changing. Ever since Isaac Newton (1643-1727) had shown that the universe was arranged with order and system and could be understood by the human mind, the enlightened citizen could cast off fear and superstition. Indeed, with a sigh of relief or perhaps of "enlightenment" he could now more easily regard natural phenomena as so many proofs of the existence, goodness, and majesty of the providential God. The numerous adherents of that "physico-theology", a form of (mainly Protestant) natural theology that sought to comprehend the Creator from his creatures, basked in a harmonious universe. For those representatives of the Christian Enlightenment, the teaching from the book of nature was an instructive prelude to the teaching from the book of Revelation, which is to say the Bible. Moreover, the study of nature or the works of God was a source of knowledge and virtue, while also offering an excellent opportunity to counteract the influence of radically enlightened writings by all kinds of irreligious folk.

That old enlightened preference for uniformity, order, controllability, and computability had invariably led to a distaste for dramatic phenomena such as comets, volcanoes, gales, earthquakes, and thunderstorms. But Edmond Halley (1656-1742) had already discovered the periodicity in the appearance of his eponymous comet. And now yet another major step had been taken: since Franklin had shown by his invention of the lightning rod – the ultimate enlightened triumph – that even "wild" nature could be tamed by technology, that utterly uncontrollable natural phenomenon had been allotted a neatly defined place in the harmoniously enlightened universe. The German physicist Georg Christoph Lichtenberg (1742-1799) strikingly referred to Franklin's invention as a *Furchtableiter* – literally a "fear diverter".[19] That process had consequences not only in religious but also aesthetic terms, since there was now less reason for that distaste. Indeed, in the eyes of the cultural elite, fierce natural phenomena now fell into the aesthetic category of the sublime, and in that guise became the cynosure of artistic fascination. For in a broader sense it is not going too far to suggest that man's newly-gained (relative) mastery of lightning created the necessary precondition for being able to revel in incalculable, majestic nature. Precisely because lightning was not necessarily dangerous any longer – nay, had even become "an affair of man" to a certain extent,[20] – "sensitive" souls could afford to play with it in literature, music, and painting. As Johan Huizinga (1872-1945) put it, there was a transition from a scientific-intellectual to an aesthetic approach to nature,[21] whereby the former approach was, in my view, a precondition for the latter.

◀ Portrait of Georg Christoph Lichtenberg, by Friedrich Wilhelm Bollinger, after a painting by Johann Ludwig Strecker, 1818-1832

Thus, many nervous inhibitions disappeared and, especially when they were replaced by a hidebound bourgeois self-satisfaction, there was plenty of room and reason for "sensitive" shuddering at nature's caprices as a kind of emotional compensation. Do we catch here, in this sort of Hegelian "sublation" of the weak spot of the Enlightenment, one of its pivots towards the Age of Romanticism? For it was precisely in this cultural period – to the whole of which another meteorological phenomenon has not coincidentally given the name *Sturm und Drang*[22] – literally "Storm and Stress" – that such expressions of the sublime received so much attention. It is not the least ambition of this study, therefore, to shed some light on the transition from Neoclassicism to Romanticism – a problem "generally acknowledged to be one of the most perplexing and difficult to account for in literary history."[23] The fact that the study of wild weather "can help us to calibrate the transition from Classic to Romantic" is therefore an important argument.[24] In other words, the study of the thunderstorm also has its importance in the history of aesthetics, particularly the sublime and the *Unheimliche* or "uncanny".[25] Both are dealt with at length in Chapter 7. For although it had been a while developing since the early eighteenth century, in a sense the new interest in nature in the second half of that century became a link between the Enlightenment and Romanticism, with the Romantics celebrating the discoveries that enlightened natural philosophers had made.[26]

The key theme of this book is the reception of the lightning rod in the various layers of society in the Netherlands and elsewhere in the eighteenth and early nineteenth century. The subsequent and more specific aim, however, is to explore the changes which that invention brought about in the perception and interpretation of the thunderstorm in the broadest sense of the word. Ultimately, therefore, the focus is on the symbolic significance of both the lightning rod and the thunderstorm in the Western world.

This study has had the good fortune of being able to build on some pioneering works in the field. Foremost among those are *Playing with Fire, Histories of the Lightning Rod,* a collection of essays published in 2009 which are fundamental to the cultural history of the lightning rod, and the monograph by the German historian Oliver Hochadel, who published a broad study of the rod's introduction in Augsburg.[27] Even so, it is still the case that in the examination of changing perceptions of thunderstorms induced by the introduction of the lightning rod, literature, music, and painting are important but hitherto barely explored sources.[28]

If we broaden our artistic parameters, however, we find a splendid monograph on the reception of the lightning-bolt-hurling god of thunder in the history of poetry.[29] There are, moreover, several recent German, English, and French studies which establish an intriguing relationship between the more general theme of the burgeoning science of electricity – that enigmatic phenomenon which many scholars of the period could only describe through imagery and analogy – and developments in the cultural and particularly the literary field, though their authors are sometimes seemingly unaware of each other's research.[30] In Germany a compendium of literary meteorology was published recently, and in the US a magnificent study on severe weather and the theological reaction to it has appeared;[31] while in Paris a workshop on "Electricity and Imagination" held in 2012 led to a special issue of *Centaurus. Journal of the European Society for the History of Science, in* 2015.[32] Also highly relevant is the collection of essays on the cultural representation of climatic and meteorological phenomena, including thunderstorms, that was published a number of years ago in France, while a more recent English monograph offers a cultural history of volcanoes that is comparable in many ways.[33]

At this point, we come up against a curious paradox, however. For on the one hand it should be obvious that such a literary study must be multidisciplinary almost by definition, with the requisite data sometimes having to be gleaned from crumbling articles that have long been stashed away and seemingly irrelevant books whose dusty whereabouts are half forgotten. And even though we focus at first on the predominantly Protestant Netherlands between 1752 and around 1830 for our source material (mostly eighteenth-century printed matter such as sermons, almanacs, poems, and scientific treatises, plus objects like landscape paintings), the simple need to give all the various Christianities due attention forces us to look beyond the national borders to gauge Catholic reactions. A comparative European, indeed Western perspective is essential. And because the weather is a subject on which everyone has an opinion, our approach must also have a very broad social base. This implies that both the High Enlightenment and its subtle, popularised influences in the capillaries of society must be addressed, or in other words that we must consider not only the increasingly specialist scholarly discourse of Franklin and other leading figures, but also the demonstrators, sometimes barely distinguishable from fairground showmen, who entertained their audiences with their "mysterious" and sometimes spectacular electrical experiments.[34] After all, in the early study of electricity, everyone was an autodidact.

But on the other hand, however broad our approach may be and however heterogeneous the source material, the subject itself – the changing perception of the thunderstorm in the transition from Enlightenment to Romanticism – is very limited. Hence the best we can hope for is that this book resembles an hourglass: ideally, when all the research has passed through its narrow waist, a broad perspective opens out that incorporates numerous subjects and reveals the extent to which the experience and interpretation of the thunderstorm reflect the cultural changes of the late Enlightenment and early Romantic periods.[35]

▶ Stormy scene, painting from 1775 by Francesco Giuseppe Casanova (1727-1802). Terrified travellers are overtaken by a thunderstorm, while lightning strikes an oak tree

Science and Technology

◀ Benjamin Franklin draws electricity from the air by means of his "electric kite", painting by Benjamin West from ca. 1816

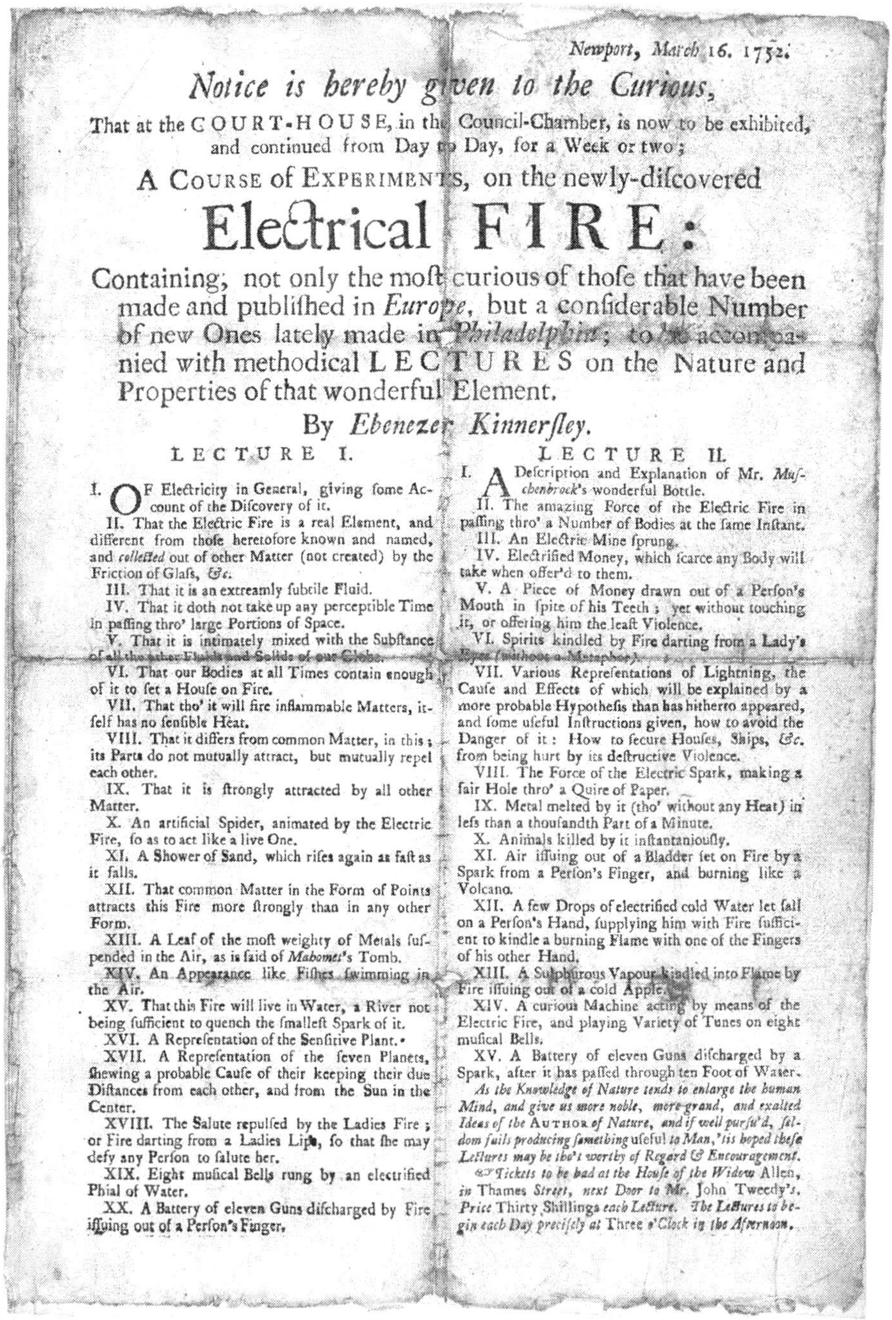

Newport, March 16. 1752.

Notice is hereby given to the Curious,

That at the COURT-HOUSE, in the Council-Chamber, is now to be exhibited, and continued from Day to Day, for a Week or two;

A COURSE of EXPERIMENTS, on the newly-diſcovered

Electrical FIRE:

Containing, not only the moſt curious of thoſe that have been made and publiſhed in *Europe*, but a conſiderable Number of new Ones lately made in *Philadelphia*; to be accompanied with methodical LECTURES on the Nature and Properties of that wonderful Element.

By *Ebenezer Kinnerſley.*

LECTURE I.

I. OF Electricity in General, giving ſome Account of the Diſcovery of it.

II. That the Electric Fire is a real Element, and different from thoſe heretofore known and named, and *collected* out of other Matter (not created) by the Friction of Glaſs, *&c.*

III. That it is an extreamly ſubtile Fluid.

IV. That it doth not take up any perceptible Time in paſſing thro' large Portions of Space.

V. That it is intimately mixed with the Subſtance of all the other Fluids and Solids of our Globe.

VI. That our Bodies at all Times contain enough of it to ſet a Houſe on Fire.

VII. That tho' it will fire inflammable Matters, itſelf has no ſenſible Heat.

VIII. That it differs from common Matter, in this; its Parts do not mutually attract, but mutually repel each other.

IX. That it is ſtrongly attracted by all other Matter.

X. An artificial Spider, animated by the Electric Fire, ſo as to act like a live One.

XI. A Shower of Sand, which riſes again as faſt as it falls.

XII. That common Matter in the Form of Points attracts this Fire more ſtrongly than in any other Form.

XIII. A Leaf of the moſt weighty of Metals ſuſpended in the Air, as is ſaid of *Mahomet*'s Tomb.

XIV. An Appearance like Fiſhes ſwimming in the Air.

XV. That this Fire will live in Water, a River not being ſufficient to quench the ſmalleſt Spark of it.

XVI. A Repreſentation of the Senſitive Plant.

XVII. A Repreſentation of the ſeven Planets, ſhewing a probable Cauſe of their keeping their due Diſtances from each other, and from the Sun in the Center.

XVIII. The Salute repulſed by the Ladies Fire; or Fire darting from a Ladies Lips, ſo that ſhe may defy any Perſon to ſalute her.

XIX. Eight muſical Bells rung by an electrified Phial of Water.

XX. A Battery of eleven Guns diſcharged by Fire iſſuing out of a Perſon's Finger.

LECTURE II.

I. A Deſcription and Explanation of Mr. *Muſchenbrock*'s wonderful Bottle.

II. The amazing Force of the Electric Fire in paſſing thro' a Number of Bodies at the ſame Inſtant.

III. An Electric Mine ſprung.

IV. Electrified Money, which ſcarce any Body will take when offer'd to them.

V. A Piece of Money drawn out of a Perſon's Mouth in ſpite of his Teeth; yet without touching it, or offering him the leaſt Violence.

VI. Spirits kindled by Fire darting from a Lady's Eyes (without a Metaphor).

VII. Various Repreſentations of Lightning, the Cauſe and Effects of which will be explained by a more probable Hypotheſis than has hitherto appeared, and ſome uſeful Inſtructions given, how to avoid the Danger of it: How to ſecure Houſes, Ships, *&c.* from being hurt by its deſtructive Violence.

VIII. The Force of the Electric Spark, making a fair Hole thro' a Quire of Paper.

IX. Metal melted by it (tho' without any Heat) in leſs than a thouſandth Part of a Minute.

X. Animals killed by it inſtantaniouſly.

XI. Air iſſuing out of a Bladder ſet on Fire by a Spark from a Perſon's Finger, and burning like a Volcano.

XII. A few Drops of electrified cold Water let fall on a Perſon's Hand, ſupplying him with Fire ſufficient to kindle a burning Flame with one of the Fingers of his other Hand.

XIII. A Sulphurous Vapour kindled into Flame by Fire iſſuing out of a cold Apple.

XIV. A curious Machine acting by means of the Electric Fire, and playing Variety of Tunes on eight muſical Bells.

XV. A Battery of eleven Guns diſcharged by a Spark, after it has paſſed through ten Foot of Water.

As the Knowledge of Nature tends to enlarge the human Mind, and give us more noble, more grand, and exalted Ideas of the AUTHOR *of Nature, and if well purſu'd, ſeldom fails producing ſomething* uſeful *to Man, 'tis hoped theſe Lectures may be tho't worthy of Regard & Encouragement.*

☞ *Tickets to be had at the Houſe of the Widow* Allen, *in* Thames *Street, next Door to Mr.* John Tweedy's, *Price* Thirty Shillings *each Lecture. The Lectures to begin each Day preciſely at* Three *o'Clock in the Afternoon.*

▲ Announcement of a course of experiments and lectures on the "newly-discovered Electrical Fire", to be given by Franklin's friend and co-experimenter Ebenezer Kinnersley, in Newport, Rhode Island, March 1752

1

A New Invention

When Petrus van Musschenbroek (1692-1761) invented the "Leyden jar" in 1746, it marked an important milestone in early electricity research. It was seen that when Leyden jars were charged, sparks closely resembling tiny bolts of lightning were produced. Many natural philosophers of the time found it easy to imagine that the two phenomena were similar – or even fundamentally identical. In 1749 Benjamin Franklin argued that artificially generated electricity and the natural discharge of lightning were one and the same, a theory he proved in 1752 when he carried out his kite experiment. In September of that year he installed the first non-experimental lightning rod on the chimney of his house in Philadelphia. Within a few years his invention was already being taken up in British America and – albeit with some initial hesitation – in Europe as well. Yet the lightning rod was far from being an overnight success: it encountered scepticism and hostility from other researchers, while disagreements among experts made its introduction a problematic process. Nevertheless, its repeated effectiveness was of decisive importance: success bred success.

Shall then a man on a wire of metal,
Catch and compel the celestial ray?[1]

Well before the invention of the lightning rod, the phenomenon of electricity, including atmospheric electricity, was already a subject of growing interest among natural philosophers – both scholars and amateurs alike. Their experimental, Newtonian brand of physics yielded practical results in the Netherlands and elsewhere. An important milestone was reached in 1746 when Petrus van Musschenbroek (1692-1761) invented the "Leyden jar".[2] This "condenser", or capacitor, made it possible to store electricity, more specifically the static electricity produced by an electrostatic generator. And when several jars were connected in series (which Benjamin Franklin did in 1748), they could even produce very large discharges similar to those of atmospheric electricity.

◀ Portrait of Petrus van Musschenbroek, print by Jacob Houbraken, after a painting by Jan Maurits Quinkhard, 1738-1740

It was seen that when the Leyden jars were charged they gave off sparks that closely resembled tiny bolts of lightning. Many natural philosophers of the time found it easy to imagine that those two phenomena were similar – or perhaps fundamentally identical. In 1749 the Bordeaux Academy of Arts and Sciences made the analogy the subject of its annual competition.[3] But how could such a theory, even one as plausible as that, be proved? For the problem of reliably reducing the phenomena of atmospheric electricity to proportions suited to a laboratory situation was still unsolved. It therefore took quite some time before it could be demonstrated that lightning, too, was electrical in character. In 1749 Benjamin Franklin used analogous argument to suggest that the artificially generated electricity to which he exposed his scale models and the natural discharge of lightning were basically identical. But it was not until he carried out his kite experiment in 1752 that he was able to prove that this analogical reasoning was correct: both artificial and natural electricity could charge Leyden jars, produce sparks, and give shocks.[4]

Prior to that, Franklin had already recorded the results of his early experiments in a pamphlet entitled *Experiments and Observations on Electricity, Made at Philadelphia in America*. Published in London and sold for two shillings and sixpence, it contained a series of letters he had written to Peter Collinson (1694-1768), a fellow of the Royal Society, in which he hypothesised that clouds are electrically charged and that lightning is caused by the sudden release of large amounts of "electric fluid". The theory could be tested, he proposed, by plucking the charge from the clouds with a pointed metal rod, though he was still unclear as to how pointed conductors could both "draw off" and "throw off" electricity at the same time.[5] His pamphlet also contained a specific proposal for such a test: an elaborated version of what is known as his "sentry box" experiment.

No sooner had scholars read Franklin's volume than they set to work. In the winter of 1751-1752, for example, Ebenezer Kinnersley (1711-1778), Franklin's co-experimenter, gave lectures in Philadelphia, Boston, Newport (Rhode Island), and elsewhere in the course of which he suggested ways "to preserve Houses, Ships, etc. from being ever Struck by Lightning."[6] In France, in a garden in Marly-la-Ville, just north of Paris, the botanist Thomas-François Dalibard (1709-1799) actually performed Franklin's proposed sentry box experiment (with significant technical assistance from a Mr Delor) as early as 10 May 1752. He did so at the instigation of none other than the renowned naturalist the Comte de Buffon (1707-1788), who had urged the botanist to prepare a French translation of Franklin's *Experiments and Observations* a few months earlier. Dalibard built a small enclosure – like a soldier's sentry box – containing a stand insulated from the ground and supporting a pointed metal rod over forty feet high. On 10 May storm clouds gathered above Marly-la-Ville. Inside the "sentry box" a military veteran, standing on the ground, held an earthed wire in an insulating handle close to the iron rod. A shower of sparks ensued – the ultimate proof, in the eyes of the experimenters, that (storm) clouds are electrically charged.[7] Dalibard also saw it as confirmation of Franklin's assumption that such earthed, pointed rods could protect buildings from lightning strikes.[8]

In the weeks and months that followed, Dalibard's experiment was successfully repeated by a number of electrical investigators, first in France and soon elsewhere as well. In Italy on 27 July 1752, for example, Giuseppe Veratti (1707-1793), a teacher of anatomy at the University of Bologna, erected the same sort of experimental conductor on the observatory tower of the Palazzo Poggi, where the Institute of Sciences was housed. Unfortunately, during this repetition of the Marly experiment two of Veratti's assistants received a violent shock, which led to such fierce protests from local Bolognans that the authorities were compelled to suspend Veratti's work and order the removal of the apparatus.[9]

Meanwhile, in the late spring of 1752, Franklin – who was apparently unaware of the Marly experiment – was pushing ahead with his research. After all, an experiment involving a sentry box was not very flexible and there must surely be something more manoeuvrable than a forty-feet long iron rod. The answer, he must have realised as the year wore on, was a kite. Not only was it easier to manipulate, it could also reach much greater heights. And so in June 1752, barely one month after Dalibard's experiment in France, Franklin conducted his famous kite experiment in Philadelphia. The kite he flew was made from a large silk handkerchief kept rigid by two light strips of cedar wood, with a pointed metal tip extending from the vertical strip. At the bottom of the kite string a key was attached with a silk ribbon tied to it. As silk is non-conducting, Franklin held the kite by this ribbon. When a storm cloud passed, sparks flew from the key to the knuckles of his hand. The kite also proved to be a suitable instrument for charging a Leyden jar. In short, Franklin had shown that there is electricity in the clouds.[10]

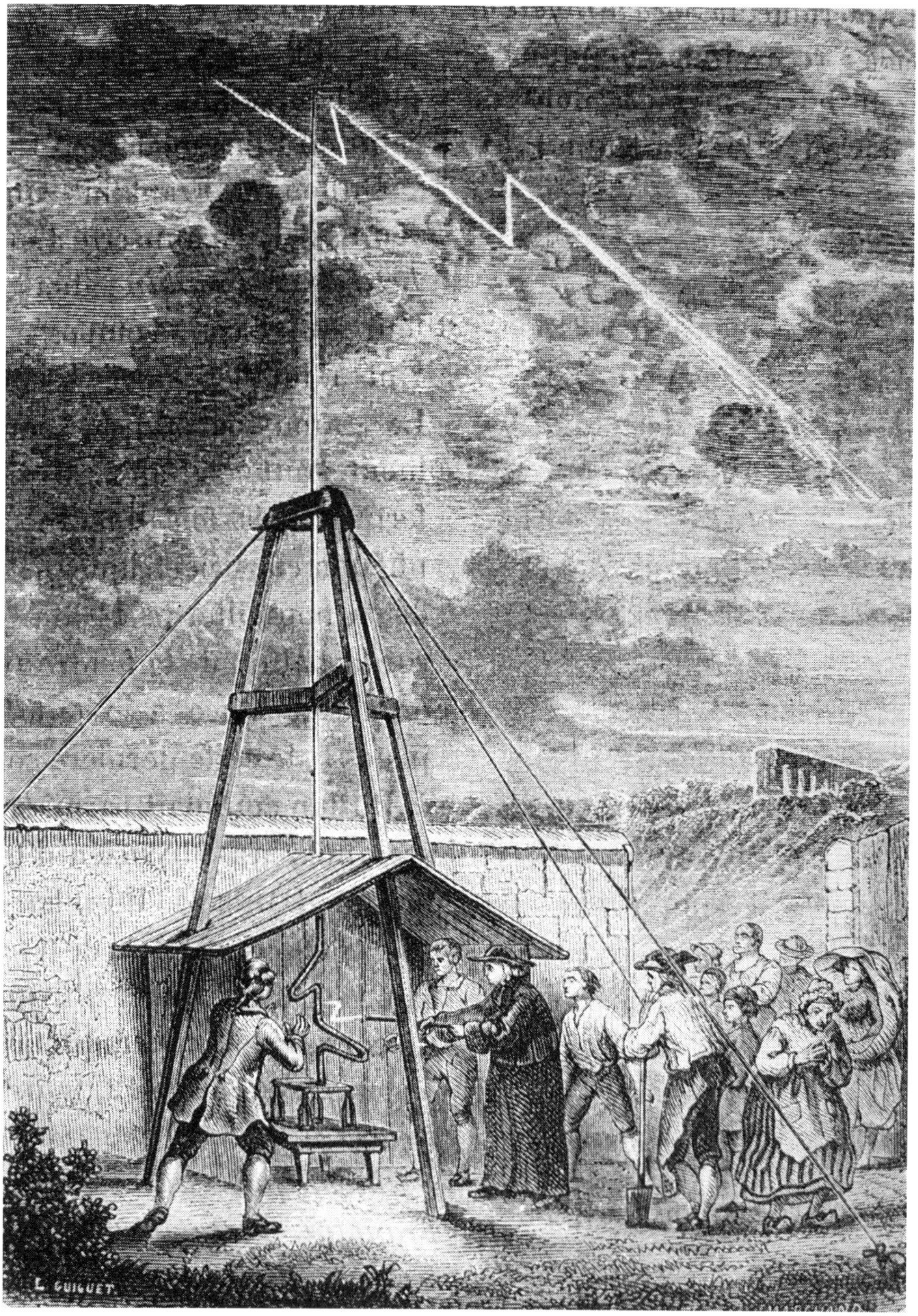

▲ Engraving, published in 1870, of the sentry box experiment conducted on 10 May 1752 by the botanist Thomas-François Dalibard, in Marly-la-Ville, just north of Paris

That experimenters were not always aware of the dangers of the newly discovered static electricity in the atmosphere was made clear by the shocking fate of Georg Wilhelm Richmann (1711-1753), a physicist of Baltic German descent. His attempt to conduct lightning to his laboratory in Saint Petersburg in order to measure its intensity cost him his life. The apparatus he used in his version of the sentry box experiment was not safely earthed.[11] As he leaned forward to read his electrometer he was struck dead by a side flash, even though he had made sure that the rest of his installation was earthed.[12]

Meanwhile, in September 1752 Franklin had installed the first non-experimental lightning rod on the chimney of his own house in Philadelphia. Before the decade ended his invention was already being taken up to a certain extent in British America, first in a number of towns across the country as well as on some plantations in the South. Europe soon followed, albeit with some initial hesitation. For although the "machina meteorologica" installed on 15 June 1754 in the town of Přímětice in Moravia was built differently from Franklin's conductor, there is little doubt that its idiosyncratic inventor, the Premonstratensian monk Procopius Diviš (1698-1765), would have known of the American's writings and experiments.[13]

◀ Portrait of Procopius Diviš, Premonstratensian monk and electrical pioneer, print by Jan Balzer (1746-1799)

Yet the lightning rod was not the resounding success it seems in hindsight. In the decades immediately following its invention its history and introduction went in fits and starts, without any certainty of the hoped-for result. For even historians who focus mainly on the technical operation of the innovation itself can hardly miss the fact that eighteenth-century experimenters were not always in agreement. And needless to say, those who prefer a description of the device's actual use can paint a much more fanciful picture of its beginnings, because for various reasons the lightning conductor encountered resistance.[14]

Although "natural philosophy" (including the study of electricity) underwent a major process of social integration in the second half of the eighteenth century[15] as science and technology became subjects of general public interest, Diviš's early lightning rod had only a short existence. In 1759 a great drought occurred which the local peasants attributed to the malign effect of the monk's sinister invention and after a short time they tore it down.[16] Before that, in 1755, Diviš's proposal to install a kind of lightning rod on the Hofburg (the imperial palace of the Habsburgs) in Vienna, had been rejected on the advice of mathematicians.[17] Here was another reason why the introduction of the lightning rod was sometimes a problematic process: disagreement among experts. The introduction of the lightning rod was held back in France, for instance, because for many years it was opposed by influential figures such as the learned Abbé Jean-Antoine Nollet (1700-1770). Franklin's invention and theories threatened to undermine the leading position that Nollet – whose demonstrations in Paris were renowned throughout Europe – had always held in French electrical research. He therefore condemned the use of metal rods outside a building as useless or dangerous, maintaining that they actually attracted lightning to a building that would otherwise be out of striking distance.[18] Nor did he believe that such a small piece of metal, no thicker than a finger, could safely conduct such a strong electrical charge to the ground.[19] Running through this was the personal animosity that existed between the Comte de Buffon and his seconds Dalibard and Delor on one hand and Nollet on the other,[20] as well as their broader disagreement over a "teleological" or "mechanical" approach to nature.[21] Only after Nollet's death did the tide start to turn. In 1772, Franklin was elected as an associate member of the French Academy of Sciences.[22] A year later, France got its first lightning rod.[23] Even then the introduction of the device was still a slow process – by 1781 the total number of lightning rods in the whole of France was a measly eleven.[24]

◀ Portrait of Jean-Antoine Nollet, print by Pascual Pedro Moles, after a painting by Maurice-Quentin de La Tour, 1771

Elsewhere, in England for example, with its close ties to the New World, there was less resistance to the lightning rod. The first was installed in 1760 on the Eddystone Lighthouse in Plymouth. It was the third lighthouse on the site, built in 1759 after its wooden predecessor had burned down. Subsequently, St Paul's Cathedral in London and the government's new powder magazines at Purfleet on the Thames were also fitted with lightning rods, in 1769 and 1772 respectively.[25] In both cases, Franklin himself served on the advisory committee.[26]

▲ Eddystone Lighthouse, Plymouth, on which the first lightning rod in England was installed in 1760; painting by Anton Melbye, 1846

Mainland Europe did not lag behind. Gradually, the conductor began to be accepted there too, in both Catholic and Protestant areas. Though in those early years, or at any rate before 1780, its introduction can hardly be described as lightning fast,[27] let alone as a triumphal march of Enlightenment thinking. It was often an accident that prompted installation, as when lightning struck St Mark's campanile in Venice in 1761 and again 1762, for example, causing extensive damage: a lightning rod was installed in 1766.[28] In Brescia in Lombardy the reason was an even greater disaster. The Church of St Nazarius was being used as a powder magazine. In August 1769 it was struck by lightning; the resulting explosion killed certainly hundreds and possibly as many as 3,000 people. That tragedy was not only sufficient reason to provide the new depot with a lightning rod but also encouraged the installation of conductors at similar depots elsewhere in Europe.[29] For instance, in 1772 the Dutch physician Jan Ingenhousz (1730-1799) and artillery commander Baron Leopold von Unterberger (1734-1818) were commissioned by Joseph II (1741-1790) to provide every powder magazine in the Austrian territories with a lightning rod.[30] And in that same year, Joseph's brother Leopold I (1747-1792), Grand Duke of Tuscany, also ordered something similar for his realm.[31] Enlightened absolutists took pride in promoting their subjects' welfare in this way. It is why, in 1776, the Elector Palatine had lightning rods placed on all the powder towers and castles in his lands, while in Prussia all the powder magazines were likewise protected by 1784 – over the protests of Frederick the Great (1712-1786) himself, who is said not to have wanted one on his own Sanssouci palace.[32]

In the long term the repeated effectiveness of the lightning rod proved to be of decisive importance: success bred success. In that respect, private initiatives were extremely helpful. In Germany in 1770, for example, the tower of St James's Church in Hamburg became the first to receive a lightning rod thanks to the efforts of physician Johann Reimarus (1729-1814), son of the well-known deist Hermann Samuel Reimarus (1694-1768). Reimarus was a keen advocate of the new invention; by 1794 he saw his efforts rewarded with no fewer than 130 lightning conductors in Hamburg alone.[33] Things were stirring in the German countryside, too: many castles and noble manor houses were equipped with lightning rods – a trend that started in the 1770s and rapidly accelerated from the early 1780s. The numerous thunderstorms in the hot summer of 1783 provided a particularly important incentive.[34]

◀ Portrait of Johann Albert Heinrich Reimarus, physician and installer of lightning rods, painting by Heinrich Gerhardt Fette, 1818

▲ Lightning rod next to the powder magazine at the Königstein fortress in Saxon Switzerland, ca. 1810, aquatint, maker unknown. There is still a *Blitzeiche* at the so-called *Königsnase*, the same prominent location in that large castle complex as depicted here, though at the time the tree was probably intended as a natural lightning rod

GUILIELMUS IACOBUS 'S GRAVESANDE
PHILOSOPHIAE ET MATHESEOS
PROFESSOR LUGD. BATAV.

2

The Introduction of the Lightning Rod in the Netherlands

Well before Franklin came up with his lightning rod in 1752 there had already been a good deal of electrical experimentation in the Netherlands, at universities as well as in learned societies. There was often no great distinction between the amateur and the scholar, some of whom reacted remarkably quickly when the results of Franklin's experiments became internationally known. It was not until the 1770s, however, that favourable publicity for the lightning rod began to become more widespread in the Dutch Republic, partly in works of popular science. The installation of a rod (not only in the Netherlands but elsewhere too) was often the reaction to an accident caused by a lightning strike. In the Republic, moreover, early installations tended to be the result of private initiative, on the part of both the installers and their clients. By around 1780 the lightning rod was becoming a fairly established fixture on both buildings and tall ships and installations by private individuals were starting to be more common. That was entirely in line with developments elsewhere in Europe – historiography speaks of a "tipping point" around 1780.

Early Experiments and Publications (1745-1779)

The story of the lightning rod's introduction in the Netherlands should be read as a transnational one, given that country's openness to influences from abroad and the frequent exchange of scientific ideas. As we shall see below, Dutch electrical experimenters often had close relationships with their French, English, and German counterparts. Benjamin Franklin himself visited the Dutch Republic on two occasions.

◀ Portrait of Willem Jacob 's Gravesande, professor of astronomy, mathematics, and philosophy at Leiden University, print by Jacob Houbraken, after Jan Wandelaar, mid-eighteenth century

▲ The huge electrostatic generator complete with Leyden jars, built in 1784 by John Cuthbertson at the request of museum director Martinus van Marum for display in Teylers Museum

Well before Franklin had come up with his invention there had already been many experiments with electricity in the Netherlands. As Lissa Roberts's seminal article (1999) on the early history of the electrostatic generator in the Dutch Republic makes clear, static electricity had been sparking great interest at Leiden University since the 1720s, due not least to the demonstrations and courses held by the well-known physicist Willem Jacob 's Gravesande (1688-1742).[1] Utrecht University had also acquired its first electrical machine before 1740.[2] That there was an interest in experiments with frictional electricity in the Dutch Republic in those early days is also apparent from the translations of the works of Johann Heinrich Winkler (1703-1770), a professor at Leipzig University who, although initially appointed to teach philosophy and philology, became increasingly drawn to natural philosophy.[3]

The investigation of electricity was not restricted to the universities; numerous more or less educated laymen were also active – instrument makers and "electrical showmen" for example – especially in countries like the Dutch Republic and the German principalities, where the culture of the Enlightenment had spread to urban centres. In 1745, for example, Lieve Kaas (1717-1799), a Rotterdam silversmith, autodidact, and co-founder of a society called "Diversity and Harmony" (*Verscheidenheid en overeenstemming*), succeeded in generating an electric spark with the help of an electrostatic generator. This accomplishment was certainly not unique; Kaas remarked on that occasion that a number of itinerant amateur experimenters in Amsterdam and Rotterdam were also conducting similar tests. According to the *Amsterdamsche Courant* of 25 October and 1 November 1746, for instance, a Polish "electriciser" would be giving electrical demonstrations in town before proceeding all the way to Venice. Such travelling showmen played a significant part in the popularisation of scientific ideas, even if their aim was primarily to profit financially from the increasing interest in electrical phenomena and the growing market for electrical products.[4] Ideally experiments such as those with the Leyden jar produced in spectators a simultaneous sensation of both pleasure and fear, the sort of delicious shudder typical of the experience of controlled danger.[5]

At a time when there was relatively little scientific specialisation, particularly when it came to pioneering experiments, there was often no great distinction between the amateur (as Franklin had been originally) and the scholar attached to a university. This also explains why, in spite of their entertainment value, amateur experiments frequently served as serious tests of existing theories or led to the creation of new ones – intentionally or otherwise.[6] It was only in the last decades of the eighteenth century that the gap between "amateurs" and "experts" grew as the process of professionalisation and specialisation gained momentum.[7] At the same time, the integration of what we now label "science" into social life continued. For in addition to the often successful "official", university-based scholars there were also their lesser-known brethren, their associates and followers, a long procession of dilettantes and minor characters. There was, in short, a diverse and broadly interested audience, representative of an emerging universal high culture,[8] that was steadily becoming better informed through the appearance of wide-ranging, general-interest magazines.

▲ Portrait of Jan and Petrus van Musschenbroek, painting by Hiëronymus van der Mij, 1715. Jan van Musschenbroek, Petrus's brother, was a renowned instrument maker

One early undisputed expert was the eminent scholar Petrus van Musschenbroek, whose consequential invention of the Leyden jar in 1746 was mentioned in the previous chapter: Winkler worked on improvements to it in that same year.[9] After Franklin had conducted his kite experiment in 1752, it was also Van Musschenbroek who replicated it – one of the first Dutch scholars to do so – even before a Dutch translation of Franklin's letter describing the experiment made it accessible to a wider audience.[10] On 16 September 1756 in Warmond (near Leiden) and 14 and 20 July 1757 near Noordwijk, for example, Van Musschenbroek was already flying kites during thunderstorms.[11] He may have received support for his experiments from the well-to-do Remonstrant amateur Abraham Edens (ca. 1690-1765), whom he had known since 1736.[12] At his death, Edens, who lived in Warmond, left an extensive collection of scientific instruments including at least one kite for electrical experiments and an electrostatic generator.[13] Obviously, Van Musschenbroek's reputation had long been firmly established, and when Benjamin Franklin and his son William made a tour through the Austrian Netherlands and the Dutch Republic in August and September 1761 they made a point of visiting the great scholar in Leiden. As William wrote to his sister Sarah, Van Musschenbroek in turn "was extreemly Glad to see my Father, and told him he was about Publishing a Work in which he should make Considerable Use of a Leter [*sic*] he had Received from him [from] America". Their visit was only just in time, for as William went on to say, "poor Gentleman though he appeared in witts and Health he died a Forginight [*sic*] after."[14]

Meanwhile, in that same early period, in the years 1757-1765, previous experiments with frictional electricity were being continued by another Dutch scholar. The Catholic physician Jan Ingenhousz, a gifted pupil of the famous Herman Boerhaave (1668-1738) who was then a general practitioner in Breda, worked day and night on experiments aiming at improving the electrostatic generator.[15] Was his curiosity aroused primarily by the potential medical uses of electricity, perhaps in connection with a large-scale public debate on the subject in England in the 1740s? Many authors there had recommended electricity as a treatment for all sorts of ailments. The Methodist John Wesley (1703-1791) had also argued eloquently for the application of what he saw as cheap and effective electrotherapy for his often impecunious crowd of followers and proselytes.[16] However much is still unknown about Ingenhousz's research in his Breda years, however much he was distrusted for it – his nocturnal investigations had earned him a reputation as a necromancer[17] – it must have been in that very period that he laid the foundations for his work as an installer of lightning rods in the Austria of Maria Theresa (1717-1780) and Joseph II, as mentioned in the previous chapter.

▲ Portrait of Jan Ingenhousz, biologist, physician, and electrical pioneer, print by Domenico Cunego, 1769

▲ Portrait of Martinus van Marum, print by Willem van Senus, after Charles Howard Hodges, 1830-1832

In fact, the relative predominance of medical men in early Dutch electrical research is striking. For instance, the Haarlem doctor Jan Engelman (1710-1782) published a treatise on the nature and operation of electricity as early as 1754, still without mentioning Franklin's invention.[18] And later, too, there were numerous physicians among the early implementers and promoters of experiments with electricity, such as the promising young doctor Martinus van Marum (1750-1837), who was already advocating the installation of lightning rods in 1776.[19] We shall meet Martinus and other well-known physicians such as Johan Rudolph Deiman (1743-1808) and Cornelis Rudolphus Theodorus Kraijenhoff (1758-1840) later on.

Was it mere coincidence that when Johannes Theodorus Rossyn (1744-1817) defended his doctoral thesis in Franeker (Friesland) in 1762, becoming one of the very first Dutch scholars to focus on the scientific assimilation of Franklin's discovery, his supervisor was another physician, professor Christiaan Everhard de Lille (ca. 1725-1795)?[20] Be that as it may, the eighteen-year-old doctoral student, who would eventually become a professor of philosophy in Utrecht, was not so much concerned with emulating electrical experiments as with theoretical reflections on them. For all that, he was really well-informed. His thesis addressed the Abbé Nollet's objections to the lightning rod, for example, and made an explicit case for the new invention.[21] What we would call nowadays theoretical physics had found a practical legitimacy.

Rossyn's early advocacy is all the more emphatic considering how few publications on that subject appeared in the Republic in the 1760s. Certainly, general-interest magazines such as the *Vaderlandsche letter-oefeningen* did their best to keep a wider readership relatively quickly informed via translations, and no doubt reinforced confidence in the lightning rod's effectiveness by doing so. In 1766 the journal published a letter from Franklin,[22] and a few years later it included a translation of one by Harvard professor John Winthrop (1714-1779), once again underlining the usefulness of the lightning rod.[23] Information about the experiments in France also became available to a broader public fairly quickly, for instance through the translation of Jacques-Christophe Valmont de Bomare's (1731-1807) multi-volume *Dictionnaire raisonné universel d'histoire naturelle*, first published in Paris in 1764. Yet such publications were exceptions in the Dutch Republic in the 1760s, and often their tone was rather critical as well. For although, partly thanks to Valmont's encyclopaedia, Dutch readers could learn about developments such as the sentry box experiment carried out at Marly-la-Ville, they would also be aware of French scepticism about the lightning rod, since Valmont repeated Nollet's objection that the disproportion "between the effect and the cause" was too great to allow of high expectations of the new invention.[24] Typically, there was not a single direct reference to Franklin. The first volume (published in 1769) of a comparable Dutch publication, Johannes le Francq van Berkhey's (1729-1812) four-part *Natuurlyke historie van Holland*, did refer to the experiments of Franklin, Nollet, and others but was oddly silent about the invention of the lightning rod itself, let alone about its potential usefulness.[25]

It was not until the 1770s that the lightning rod began to receive more widespread favourable publicity in the Dutch Republic, some of it in works of popular science. And even then, opinions were initially divided. Apparently, what the historian of technology Wiebe Bijker has called the "closure" of an invention – the generally accepted, stable interpretation of its usefulness – had not yet occurred.[26] Take the Dutch translation, published in 1773, of a work by Nollet. Entirely in the spirit of the abbé himself, the editor argued that although the "points" of the rods might attract part of the "electrical matter", that would still not prevent the electricity "from also spreading and communicating itself to all other bodies closer to the ground, and of quite a different shape."[27] And in 1778, when the second edition of the Dutch version of the *Dictionnaire oeconomique contenant l'art de faire valoir les terres, et de mettre à profit les endroits les plus stériles* by French cleric Noël Chomel (1633-1712) was published, its translator and editor, Jacques Alexandre de Chalmot (1734-1801), was, to put it mildly, equally ambivalent. In his new version of the entry on lightning the information he provided was fairly outdated, while in the entry on thunder he seems to have made a half-hearted attempt to compensate for that by at least mentioning Franklin's kite experiment but without drawing the firm conclusion that thunderstorms should be regarded as electrical phenomena. At the crucial point in the discussion, De Chalmot referred not to Franklin but to the views of Winkler, Nollet, and others. Not until a decade later, in the *Vervolg* or "supplement" (1787) was it acknowledged that Franklin was correct.[28] The Delft instrument maker Johannes Rechter (1730-1801) was more up to date: in 1779 he coupled an enthusiastic assessment of the lightning rod with a warning against reckless experimentation. In doing so, he not only showed a thorough understanding of the kite experiments carried out by Franklin in America and Jacques de Romas in France, but also of the dangers those experiments entailed. Accordingly, he advised against "keeping the Kite in the Air when a thunderstorm is actually raging nearby, because (...) a large quantity of lightning matter might run down the kite string and render fruitless all means of protection, however well chosen."[29]

Such a warning was by no means unnecessary for there was a good deal of experimentation going on with replicas of Franklin's kite, known in the Netherlands as "thunder dragons". In a sense, electricity had become a science vogue. In Amsterdam the man at the centre of practical electrical research was the English instrument maker John Cuthbertson (1743-1821), who had settled there in 1768.[30] He and his brother Jonathan (1744-1806) had taken private lessons in Amsterdam with physics teacher Dr Benjamin Bosma (1724-1795?).[31] John Cuthbertson, later best known as the maker of the gigantic electrostatic generator on display in Teylers Museum in Haarlem,[32] manufactured all kinds of electrical apparatus and devices, including "electric kites". These he offered for sale as standard products in his workshop in Sint Luciasteeg. Anyone who had the wherewithal in 1773 could become the proud owner of one for the princely sum of 80 guilders.[33] Cuthbertson also regularly conducted his own experiments with them, assisted by friends such as the merchant Adriaan Paets van Troostwijk (1752-1837), the aforementioned physician Johan Rudolf Deiman and the mathematician and physicist Dr Henricus Aeneae (1743-1810).[34] In those years he also constructed an entire kite cabinet, a peculiar combination of the static sentry box and more flexible kite experiments.[35] Another friend of Cuthbertson's, the bookseller Martinus Magerus, who worked as such in Amsterdam from 1757 to 1774[36] and co-translated with Cuthbertson a famous work by the natural philosopher Joseph Priestley (1733-1804),[37] had already conducted the kite experiment a number of times in the summer of 1772.[38] And, to cite a last example, Prince Dimitri Alexeievich Gallitzin (1734-1803), Russian envoy to The Hague and friend of philosophers such as Voltaire (1694-1778) and Denis Diderot (1713-1784), went so far as to install a complete kite belvedere in which to carry out experiments. He reported his findings to Saint Petersburg's Academy of Sciences, of which he had been a member since 1767.[39] All these individuals were part of an newly emerging audience of educated laymen whose curiosity, fashionable consumerism, sociability, and typical Enlightenment desire to disseminate knowledge and skill, joined to the endeavours of university-trained scholars, went far to overcome the initial suspicion of the lightning rod – in the Republic as well as elsewhere.

▲ "Electric kite cabinet" built by the Amsterdam-based instrument maker John Cuthbertson; illustration in J. Cuthbertson, *Practical Electricity and Galvanism*, London: J. Callow, 1807

Amsterdam and The Hague were not the only places where experiments were being carried out. Rotterdam was humming with electrical activity too. In June 1766, when Benjamin Franklin passed through the Dutch Republic again, this time on his way to the German spa town of Pyrmont, he visited Rotterdam and met the city physician Salomon de Monchy (1716-1794) and his son and successor François Willem (1749-1796). Those contacts were not without result. For three years later, when Salomon became a director of the newly-founded (and still-existing) Batavian Society of Experimental Philosophy, Franklin was offered a corresponding membership, an honour he graciously accepted and one he shared with scholars such as Ingenhousz, for example.[40] In 1779 the Batavian Society's pioneering role became less symbolic and more tangible. On 9 August of that year it announced an academic competition whose entrants were required "To show by experiments which atmospheric phenomena depend on the operation of natural electricity; how they are produced by it; and what are the most suitable means of safeguarding our houses, ships, and persons from their harmful influence." It was the aforementioned Martinus van Marum whose entry won the gold medal on 14 August 1780.[41]

Apart from academic competitions, the lectures and transactions of various learned societies sometimes included arguments in favour of the lightning rod. With their activities and publications, those prestigious institutes undeniably stimulated that invention's "closure". In 1770, for instance, Captain William May, son of the English shipwright John May (1694-1779) who worked in Amsterdam, was already using the transactions of the Holland Society of Sciences as a vehicle for communicating his concerns about safety aboard naval vessels. In 1749 he had been on a ship struck by lightning and observed that the bolt had seriously interfered with the operation of the magnetic compass needles. Now that Franklin's invention was available it should be possible to avoid not only that hazard but also the risk that lightning would ignite the gunpowder kept on board. May, who oversaw the outfitting of the Amsterdam Admiralty's vessels, therefore recommended that warships be equipped with lightning rods.[42] One thing that his treatise for the Holland Society might have mentioned but did not was the dreadful disaster that had befallen a ship of the VOC – the Dutch East India Company – which showed how justified his warning was. On 24 February 1765 the *Pylswaard*, sailing from Enkhuizen and bound for Batavia, was struck by lightning just before it reached its destination. Fire broke out, the ship's gunpowder exploded, 183 men perished, including the captain and first mate.[43] Even before that catastrophe another Englishman, the physician William Watson (1715-1787), had also taken up his pen to urge the installation of lightning rods on ships. His recommendation found its way, in translation, into the *Vaderlandsche letter-oefeningen* in 1764.[44]

As we shall see below, these calls do seem gradually to have been heeded. Yet in 1787 they had to be repeated by another member of a learned society, the Middelburg regent and Zeeland VOC director Daniel Radermacher (1722-1803). He quoted from a ship's log which recorded the lightning strike on an East Indiaman on 20 February 1785 that left three dead and eleven injured. To Radermacher the conclusion was obvious: "Such events are (...) very compelling reasons for the use of the well-known lightning conductors on our sea-going ships." The installation of such conductors, for example on British vessels, had in his opinion demonstrated their usefulness.[45] It was not long before the advice was followed.

The First Installations

In the meantime, the Amsterdam Chamber of the Dutch East India Company seems to have moved faster in this respect than its Zeeland counterpart. Could their alacrity have been motivated by an incident in October 1770, when a Dutch East Indiaman anchored in the Batavia roads was very badly damaged by lightning due to an iron spindle at the top of the mainmast, while Captain James Cook's *Endeavour*, lying nearby, suffered no ill effects from the same lightning strike, being equipped with a lightning conductor? As Cook put it in his journal on 12 October 1770:

About 9 o'clock in the Evening we had much rain, with some very heavy Claps of Thunder, one of which carried away a Dutch Indiaman's Main Mast by the Deck, and split it, the Main top mast and Top gallant mast all to shivers. She had had an Iron Spindle at the Main top gallant Mast head which had first attracted the Lightning. The ship lay about 2 Cable lengths from us, and we were struck with the Thunder at the same time, and in all probability we should have shared the same fate as the Dutchman, had it not been for the Electrical Chain which we had but just before got up; this carried the Lightning or Electrical matter over the side clear of the Ship. The Shock was so great as to shake the whole ship very sencibly [sic]. *This instance alone is sufficient to recommend these Chains to all Ships whatever, and that of the Dutchman ought to Caution people from having Iron Spindles at their Mast heads.*[46]

▲ On 12 October 1770 Captain James Cook's ship *Endeavour* was struck by lightning while lying in the Batavia roads but suffered no damage thanks to a lightning rod, whereas a nearby Dutch East Indiaman unprotected by a rod was badly damaged

Although there are no indications that the contrasting fates of the ships led to immediate discussion, let alone direct action,[47] the directors of the Dutch East India Company, the *Heren XVII* or "Gentlemen Seventeen", must gradually have become aware that installing lightning rods on their vessels was a worthwhile investment. At any rate, in

1780 the "prudent and exemplary" decision of the Amsterdam directors to equip their ships with lightning conductors is mentioned.[48] And was it any wonder that they called upon the expertise available in Amsterdam by awarding the contract for the job to John Cuthbertson?[49] For much as he enjoyed conducting useful and pleasing experiments with friends, learned or otherwise – collaboration between scholars and educated laymen was characteristic of learned societies – the commercial services of instrument makers like Cuthbertson were as essential in the actual installation of lightning conductors as they were in the practical organisation of scientific experiments. Between 1775 and 1783 Cuthbertson fitted no fewer than 56 vessels with lightning rods in the Dutch Republic[50] and no doubt many of them were Dutch East India Company ships.[51] Evidently developments elsewhere were somewhat slower. In 1787 Zeeland was still lagging behind – witness Daniel Radermacher's argument – while as late as 1819 Martinus van Marum was forced to conclude that the lightning rod was not common on Dutch ships.[52]

It was not only ships that needed lightning rods, however. Increasingly they were being installed on edifices on land such as windmills, churches, and country houses. One telling fact is that in 1779, the very year of the Batavian Society's original academic competition, the Republic's first commercial lightning rod installation company was founded: the firm of Schaap in Deventer, which still exists today.[53] One of the first buildings to be fitted with such a device was the tower of Rosendael Castle near Arnhem. Its owner, Baron Assueer Jan Torck (1733-1793), had already had a lightning rod installed in

▲ Print of Rosendael Castle near Arnhem, by Abraham Rademaker, 1728-1733

1778, for the sum of around 50 guilders, by the aforementioned Russian envoy Dimitri Alexeievich von Gallitzin, who lived opposite Torck's winter residence on the Voorhout in The Hague and was thus able to put his experimentally-acquired knowledge and skill into practice. On 6 July 1778 Prince Gallitzin sent a written record of the event to the Academy of Sciences in Brussels, of which he had become a member on 3 April of that year. At the meeting held on 9 November his letter, which was accompanied by a memorandum on the best form of lightning rod, was read out.[54] A few years later, in 1783, the prince resigned his diplomatic post and devoted himself entirely to the study of electricity while enjoying the baron's hospitality at Rosendael Castle. By that time he even had his own rooms there.[55]

While it might seem that much of the theoretical experimenting with electricity was taking place in the urban centres of the province of Holland, that was certainly not the case when it came to the practical installation of lightning rods. For on 24 December 1782, not long after Torck's tower acquired its conductor, the spire of St Martin's Church in Doesburg in Gelderland was also fitted with one, thanks to the handy efforts of Cornelis Kraijenhoff (1758-1840), a promising young student of philosophy and medicine at Harderwijk University. The following year he published a detailed account of this deed in an appendix to his adaptation of a French book.[56] The Doesburg bell tower had apparently been struck by lightning so often in the past that on 25 February 1782 the town council, prompted by the burgomaster, J.H. van Leenhof de Lespierre (1731-1807), agreed to have a lightning rod installed. The need for such a device was underscored when lightning struck again – allegedly for the tenth time since 1737 – on 26 July of that year, even before the rod was put in place.[57] Kraijenhoff later claimed that this was the first lightning conductor to be installed on any [Dutch] public building.[58] And while the physicist Gerard Moll (1785-1838) did not directly second that assertion his admiration for Kraijenhoff's pioneering role was clear when he spoke of him as "justly boasting of the first specimen of his vigour."[59] Whatever the truth of the matter, the Doesburg lightning rod was evidently such a novelty in 1791 that it attracted tourists who requested special permission to see it from the presiding burgomaster.[60]

▶ Illustration of the lightning rod on the spire of St Martin's Church in Doesburg, installed on 24 December 1782 by Cornelis Rudolphus Theodorus Kraijenhoff, from L.S. Jacquet, *Proef eener elektrische natuurkunde*, pl. X

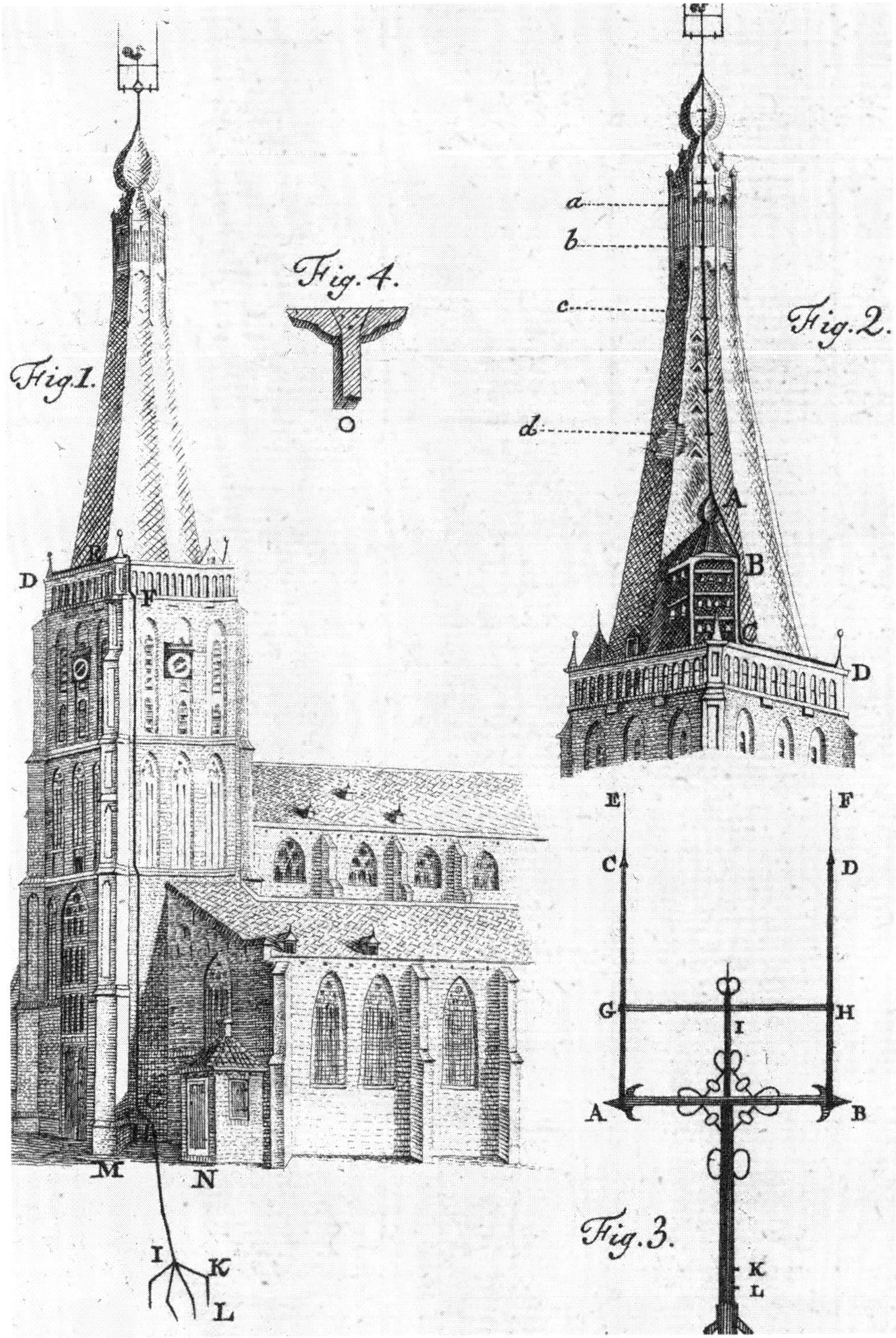
Fig.1.
Fig.4.
O
Fig.2.
a
b
c
d
A
B
C
D
E
F
M
N
I
K
L
Fig.3.
E
F
C
D
G
H
I
A
B
K
L

◀ Portrait of Lieutenant General Cornelis Kraijenhoff, pastel drawing by Hermanus J. Slothouwer, 1838

Some Tentative Conclusions

On balance, the picture of the early installation of lightning rods in the Dutch Republic is still rather fragmentary. And for now the exact nature of the link between experiments and commerce is also somewhat unclear. What is certain, however, is that here too, at an early stage, a group of demonstrators formed – sometimes perhaps little more than travelling showmen, but more often educated laymen. Included in that group were estimable and sedentary members of learned societies such as Lieve Kaas in Rotterdam in the 1740s or, over half a century later, the self-taught Hendrik Lugt (1747-1815), active in Zaandam and Haarlem, who tried to earn money by giving lessons and doing experiments.[61] There will certainly have been a number of lightning rod installers among them. Foremost was undoubtedly John Cuthbertson in Amsterdam, but there must have been technicians doing similar things in the province, too, such as the tinsmith and instrument maker Cornelis Bayens (1751-1823) from Megen in North Brabant, who was already active installing lightning rods before May 1783.[62] Around that time a market must have opened up for technicians like him, for the founding in 1779 of the first commercial lightning rod installation company – the aforementioned firm of Schaap in Deventer – can be regarded as a milestone in that respect. For a long time the number of those interested will have been greater than the number of customers since the actual number of lightning conductors must still have been limited. Nevertheless, a growing public was becoming aware of the rod's utility through the transactions of learned societies, for instance, or general-interest journals such as the *Vaderlandsche letter-oefeningen* (founded in 1760). Thus a new audience with an above-average education was reached, readers who not only absorbed the information about the new invention passively but in many cases were prompted by it to experiment themselves. In doing so they also played an important part in combating the vernacular meteorological superstitions they ascribed to the lower classes.[63]

One thing, meanwhile, is crystal clear: the introduction of the lightning rod in the Netherlands can only be understood as part of an aggregate of international developments, one that also includes a colonial dimension. The departure of Dutch East India Company ships, the comings and goings of scholars and technicians, as well as the free exchange of ideas, are so many proofs that telling a purely Dutch story is not enough.

It is also clear that the installation of a lightning rod (in the Dutch Republic as well as elsewhere) was often in direct reaction to an accident. The introduction of the device was often less straightforward than might be expected, for sometimes it took a while for the realisation to dawn that in the long run a lightning conductor saved more than it cost. The early installations were therefore not the predictable outcome of a linear process of public acceptance but rather the reverse; it appears that in the Republic they were mainly the result of private initiative, on the part of both the installers and their clients.[64] The latter were invariably wealthy people – fitting a lighting conductor was a fairly costly business – around 50 or 60 guilders – notwithstanding Prince Gallitzin's contention that its price was quite reasonable.[65] The lightning rod would remain an elite phenomenon until the advent of mass production in the second half of the nineteenth century.[66]

Placing lightning rods on towers and churches still cost considerably more than fitting them on houses. In 1836 Kraijenhoff cited prices of 243 guilders for a tower around 63 metres high and 309 guilders for one reaching approximately 94 metres. Nonetheless, he once again reiterated his belief that fitting a lightning rod would more than pay for itself when the accidents it prevented were taken into account. Not to install one for reasons of cost was therefore a false economy.[67]

Even when orders were received from local or provincial authorities in that early period, it was almost always as a result of the actions of a few individuals. It was the very absence of a strong national authority that led, for example, to the country's powder magazines being without protection until 1797 – which is to say after the French had introduced a powerful central government in the Netherlands. And even then it was Kraijenhoff who, as director of fortifications for the province of Holland, had decided to do something about it himself. Holland's example was soon followed elsewhere in the country "at the express orders of the government." As late as 1819, Kraijenhoff had a pamphlet on the subject published.[68]

In 1836, perhaps from a sense of wounded national pride, Kraijenhoff concluded that his country was lagging behind.[69] He may have been right as far as the actual number of lightning conductors is concerned: in countries with a more centralised government their installation might eventually have proceeded with more vigour than it did in the Netherlands. But when we view the matter from different perspectives and focus on more specific periods of time the answer is less ambiguous. For it is precisely because of its international orientation that the idea that the mid-eighteenth-century Dutch intellectual world was in some way retrograde fails to hold water. Some of the members of that world (including a conspicuous number of physicians[70]) reacted remarkably quickly and with very definite interest when the results of Franklin's experiments became known worldwide, with Van Musschenbroek and Rossyn leading the way.

Understandably the second phase, the wider introduction and implementation of the new ideas, took more time and did not truly gain momentum until the late 1770s. At the heart of this development were the learned societies, an enlightened setting in which theoretical reflection (sometimes inspired by the translated works of physicists in other countries) always went hand in hand with the carrying out of useful and pleasant experiments, some of which involved scale models known as "thunder houses".[71] By formulating questions to be answered via competitions, organising lectures, and publishing scholarly writings, learned societies thenceforth made an essential contribution to propagating the new invention and to its eventual success. At almost the same time, around 1780, the first non-experimental lightning rods were installed in the Dutch Republic. It is probably going a step too far to call this a real breakthrough, since as late as 1804 a ferocious thunderstorm claimed victims here and there and caused tremendous damage,[72] as did a lightning strike on the St Martin's Tower in Groningen in 1822[73] – there were evidently hiccoughs in the practical process of introduction. But what can safely be said is that from around 1780 the lightning rod became an established phenomenon on land and at sea: the process of installation by private individuals definitely began to accelerate, sometimes even outside the circles of the learned societies.[74] And that conclusion is again entirely in line with developments elsewhere in Europe – historiography even speaks of a "tipping point" around 1780,[75] because shortly afterwards the extraordinary number of thunderstorms that occurred during the hot summer of 1783 led enlightened absolutists in Germany to launch an energetic programme of installing lightning rods, with the support of a major media campaign.[76]

▶ Title page from Dominikus Beck, *Kurzer Entwurf der Lehre von der Elektricität*, Salzburg: Waisenhausbuchhandlung, 1787. The inset illustration aims to make clear the electrical nature of lightning: the discharge of a Leyden jar by a human hand causes fire in the same way as the lightning strike on the church on the left. Some maintained that this was man playing God, but in the eyes of many others he was merely showing the workings of God's creation and how to guard against the damaging effects of lightning

Kurzer Entwurf
der
Lehre
von der
Elektricität.

Verfaßt zum
Gebrauche
seiner Zuhörer
von
Dominikus Beck,
Hochfürstl. Rathe, der Mathematik und Experimentalphysik öffentlichen Lehrer in Salzburg.

Mit VIII. Kupfertafeln.

Salzburg
im Verlage der Hochfürstl. akad. Waisenhausbuchhandlung
1787.

▲ The French physicist Pierre Bertholon thought that with his *électro-végéto-mètre* the "sulphurous" electrical "fluid" could be extracted from the air and then reused to fertilise crops by means of electric "rays", anonymous print, eighteenth century

3

Eighteenth-Century Physical Theories on Thunderstorms

Drawing on ancient Greek tradition, many eighteenth-century scholars made a connection between volcanic activity, earthquakes, and thunderstorms. Subterranean sulphuric vapours were believed to be the main cause of both eruptions and earthquakes, while thunderstorms were thought result from the spontaneous combustion of sulphurous and nitrous vapours in the air. In the chemical mindset of the time, electricity itself was usually seen as sulphurous in character, which made the lightning rod a potentially perilous apparatus. For drawing electricity from the air and sending it into the ground increased the amount of electricity in the earth, which could trigger an earthquake. One of the principal physics-based objections to the lightning rod was that the only way it could render lightning harmless was by attracting it. But the rod's proponents replied that its action was at most receptive: ultimately, arguments to the contrary failed to stand up. The discourse about sulphur was international, but by the standards of the time it can be hard to determine where "science" ended and "superstition" began.

The International Discourse

The introduction of the lightning rod or at least the discussion of it was sometimes helped by an extreme weather event. The summer of 1783 certainly saw one of those. On 8 June, in the south-east of Iceland, the Laki volcanic fissure erupted. It was the start of an eight-month-long series of violent explosions which threw up such enormous clouds of ash that much of the northern hemisphere was affected. In large parts of north-western Europe the haze was anxiously observed as dry mist, dry fog or *Höherauch*. Many scholars regarded these evil vapours as the main cause of the extreme summer heat and the subsequent extraordinary frequency of thunderstorms that year.[1]

Some also believed that there was a more direct relationship between those two natural phenomena. According to an early modern theory that can be traced back in part to Aristotle (384-322 BCE) and Heraclitus (ca. 500 BCE),[2] thunderstorms were caused by the spontaneous combustion of sulphurous and nitrous vapours in the air. It was a tenacious concept that in the second half of the eighteenth century still had a wide group of

▲ *The Destruction of Sodom and Gomorrah*, painting by John Martin from 1852. The biblical story (Gen. 19) tells how God executed his judgement on the wicked cities with a rain of burning sulphur. Sulphur had long been associated – and still was by many in the eighteenth century – with hell's lake of fire and brimstone and the sulphur-like particles of which lightning was thought to consist

supporters, including the internationally renowned Petrus van Musschenbroek himself, though he based his defence of it on the Bible rather than the classics – after all, it was also by a rain of burning sulphur that the godless city of Sodom perished.[3]

The dry fog, which no amount of sun could dissipate, was not the only phenomenon that seemed to prove supporters of the sulphur theory right. Some believed that the earthquakes in Calabria that began on 5 February of that same year, 1783, and continued off and on until 1786, were the result of exploding sulphurous and nitrous fumes – the result, and also the cause. For as had happened in Iceland, when those subterranean explosions combined with subsequent hypogeal fire they caused volcanic eruptions: if some of the vapours remained undischarged, earthquakes were the inevitable result. In both cases, however, such large quantities of fumes containing such a high level of sulphur were released into the atmosphere that they constituted "a kind of aerial gunpowder."[4] Whether the vapours then exploded immediately or first led to the formation of dry fog,[5] the result was the same – frequent and violent thunderstorms. And importantly, however much scholars may have differed about the effects of and interaction between those natural phenomena, many of them made a connection between volcanic activity, earthquakes, and thunderstorms – yet another age-old idea, for according to classical mythology was it not Vulcan who forged lightning for Jupiter in the bowels of Etna? Then there was Pliny the Elder (first century CE), for instance, who interpreted earthquakes as the result of subterranean storms.[6] And in early modern times, too, earthquakes were seen as subterranean thunderstorms and thunderstorms as superterrestrial earthquakes.[7] It was that supposition that in 1737 caused Franklin to endorse the opinion that thunder, lightning, and earthquakes were materially related.[8] So it is entirely typical of the persisting popularity of the sulphur discourse in the eighteenth century that when Georg Philipp Telemann (1681-1767) composed a cantata in response to the Lisbon earthquake in 1756 he called it *Die Donner-Ode* – "The Thunder Ode".

But earthquakes were not always accompanied by vapour or smoke. It was that observation which by 1750 had already led the English clergyman and scholar William Stukeley (1687-1765), a man of wide-ranging interests, to conjecture that earthquakes might also be electrical phenomena, the literally "shocking" consequences of a discharge between a highly electrical atmosphere and an approaching "unelectric" cloud, felt on the earth's surface. The cleric and professor of experimental physics at Turin, Giovanni Battista Beccaria (1716-1781), saw further evidence for that theory in the hitherto inexplicable velocity with which earthquakes could travel over immense distances.[9]

Accepting that hypothesis did not necessarily mean rejecting the sulphur combustion theory. The French natural philosopher Pierre Bertholon de Saint-Lazare (1741-1800) combined both with the view that electricity was the principal cause of earthquakes for it ignited subterranean fires – producing earthquakes as well as volcanic eruptions.[10] So in a sense, according to this erudite abbé, earthquakes literally "fell out of the sky." But was it certain that lightning only travelled in one direction, from above to below? Or did it sometimes follow the reverse route, discharging sulphuric exhalations upward from the ground into the clouds? In either case, Bertholon declared, there was often a disturbance of the "electrical equilibrium" between atmospheric and terrestrial electricity. The balance could be restored, he believed, by installing a new variety of "ascending" lightning rod. He suggested that certain areas in France should be provided with a *para-tremblement de terre* or "para-earthquake" – a very long iron rod with various points and wires at both ends – that was to be driven as deeply as possible into the ground while still leaving the upper end exposed. This would not only prevent the damaging effects of thunderstorms but would draw the excess electricity out of the earth as well, and so preclude earthquakes.[11] Bertholon's reasoning had undeniably gained in plausibility since Alessandro Volta (1745-1827) discovered the marsh gas methane in 1776 and was able to ignite it with an electric spark.[12]

But not everyone who posited a relationship between thunderstorms and earthquakes was convinced of the usefulness of the lightning rod, certainly not at first. The Reverend Thomas Prince (1687-1758) of Boston seemed quite certain that Franklin's invention was actually dangerous, an opinion in which he was far from alone. Prince and his partisans shared the view, widely held in the chemical mindset of the time, that electricity itself had a sulphurous character, and they feared that lightning rods would drain so much "lightning matter" from the air to the ground that massive earthquakes would occur – a line of reasoning by which the pastor explained the earthquake in Massachusetts on 18 November 1755 and at the same time justified his opposition

Reverend Thomas Prince of Boston, opponent of the lightning rod in part because he believed it increased the risk of earthquakes. Mezzotint, by Peter Pelham, after John Greenwood, 1750

(also religiously motivated) to lightning rods, of which Boston already had quite a few.[13] Still in 1783, an anonymous German sceptic also argued that by drawing electricity from the air into the ground the lightning rod affected the amount of electricity in the earth and could therefore encourage earthquakes.[14] This was met with the rejoinder that there had not been a single lightning rod at the scenes of colossal disasters such as Lisbon and Messina, destroyed by earthquakes in 1755 and 1783 respectively. Others tried to put a positive spin on the matter: earthquakes could actually be prevented by installing enough lightning rods.[15] One author, the Jena physicist Johann Ernst Basilius Wiedeburg (1733-1789), went so far as to assert that the reason why there were no earthquakes in Egypt – a fact already noted by Herodotus, incidentally – was that the pyramids had been specifically built as conductors of both atmospheric and subterranean thunder.[16]

At that time, therefore, the notion that thunderstorms and earthquakes were closely related was fuelled by opinions about the nature of (atmospheric) electricity, usually regarded as a collection of sulphurous and nitrous vapours or "lightning matter". Often, electricity or "electric fire" was more or less identified with phlogiston,[17] the invisible "fire substance" supposed by natural philosophers to exist in all combustible bodies and to be released during combustion. The air was often full of it, and its supposed flammability predisposed it to unite with the aforementioned vapours.[18] Then again, electricity was also increasingly being seen as a mysterious "electric fluid", without the two interpretations being mutually exclusive.[19] Franklin had been a frontrunner with his theory of a single electric fluid, an abundance of which led to a positive charge and a shortage to a negative charge – a theory that was quite widely accepted at the time, because it seemed to explain the working of the Leyden jar; only much later was it shown to be incorrect.[20] This makes it all the more understandable that Bertholon built his lightning rod in such a way that it could be used to discharge (what he considered to be) an excess of that single electrical fluid to both the air and the earth. Moreover, it was the relative simplicity (as it initially seemed) of Franklin's theory that made it more attractive than the two-fluid concept which other scholars (including Nollet, up to a point) were convinced was correct.

▶ Title page from Athanasius Kircher, *Sphinx mystagoga, sive diatribe hieroglyphica, qua Mumiae, ex Memphiticis Pyramidum adytis erutae, et non ita pridem in Galliam transmissae*, Amsterdam: Johannes Janssonius van Waesberghe 1676 (http://diglib.hab.de/wdb.php?dir=drucke/15-9-quod-2f&distype=thumbs, 00010). Many western artists depicted Egyptian pyramids (which they had never seen) as more pointed than they really were. This led the German physicist Johann Wiedeburg to argue that the Egyptians had purposely built the pyramids that way so as to drain away both atmospheric and "subterranean" electricity

Whatever the practitioners of the science of the time thought about the nature of this natural phenomenon, most of them believed that atmospheric electricity had a strong influence on all kinds of living beings, including plants.[21] Which is why the introduction of the lightning rod was such a cause for concern, even if the objections were sometimes rather far-fetched, especially seen from our modern perspective. For instance, it was suggested that since the new invention drew electricity from the air the "sulphurous vaporisation" in the air was no longer bound to anything, leaving a residue of "dry fog" floating in the atmosphere, with all the malign consequences that were presumed to follow. Then there those who argued that the lightning rod drew so much electricity from the air that it stopped the rain falling.[22] That was the reasoning that led Moravian peasants to blame a period of severe drought on the lightning conductor or "machina meteorologica" installed by the Premonstratensian monk Procopius Diviš and to tear it down.[23] In short, the fertility of the fields suffered from the lack of electricity in the atmosphere – a lack induced by the evil lightning rod. Sometimes opponents of the device seemed convinced that the best way to guarantee the supply of electricity to the fields was by a direct lightning strike,[24] believing that this was the only way that enough of the beneficial "nitrous" fertiliser saltpetre could be ensured.

Countering that was the argument adduced by the lightning rod's supporters, who maintained that the new invention promoted the fertility of the land. It was all a question of getting the balance right. Bertholon therefore designed an *électro-végéto-mètre*, a curious instrument with which he thought the electrical "fluid" (interpreted as sulphurous) could be removed from the atmosphere and then reused to fertilise crops by means of electric "rays". With a few adjustments, his invention could also serve as a lightning rod, he thought.[25] Others followed a less complicated line of reasoning, simply stating that the lightning rod did not cause drought but instead made it rain, as it attracted showers.

At that point, however, the discussion took a new turn, for one of the main physical objections to the lightning rod was that the only way it could render lightning harmless was by attracting it. Advocates of the rod went to great lengths to nullify this objection. The problem was that scholars knew very well that Franklin's invention worked but had different opinions about exactly how. The technology seemed to be ahead of the science.[26] Did the conductor avert lightning by dissipating atmospheric electricity or did it actually attract it?[27] Franklin's answer to that question was somewhat equivocal[28] and gave his adversaries more than a little justification for their scepticism or rejection purely on principle. For example, Nollet and many others were convinced that the thunderstorm more or less "hovered" in one place due to the lightning rod's power of attraction, whereas without such an apparatus that same place would probably have been spared from thunderstorms altogether.[29] That was the logic that held the rods installed in Düsseldorf in 1782 responsible for the numerous thunderstorms the following year, for instance.[30] There was also the fear that unprotected buildings in the area often suffered needless damage. But proponents of the new invention replied that the lightning rod did not actively attract lightning: its action was at most receptive and aimed only at preventing the lightning's natural path via a chain of metal objects from being interrupted by non-conductive objects. Even if there were some specific attraction, it should be considered a positive thing because of the beneficial rains it induced (as we have just seen). Other, unprotected buildings had nothing to fear from the presence of such a newfangled device, as lightning would be diverted via the protected building, provided it was the highest in the area.

Here, in fact, was another bone of contention. Some scholars believed that the lightning conductor's potential tractive action was greatly dependent on its design and favoured rods that ended in a knob rather than a point. In England one of the most determined champions of the blunt conductor was the portrait painter and experimental philosopher Benjamin Wilson (1721-1788), who supposed that pointed lightning rods worked too efficiently, so to speak. So great was their power of attraction that stronger discharges were the result; indeed, they sometimes occasioned discharges that would otherwise never have occurred.[31] In 1777 the Royal Navy's powder magazines at Purfleet, which (on the previous advice of Franklin and others) were equipped with pointed conductors, suffered some minor lightning damage – a result of the rods being improperly earthed. To Wilson, however, it was the vindication of his knob theory: obviously the problems were caused by use of pointed rods, which at the time had been fitted against

Self-portrait of the experimental philosopher and painter Benjamin Wilson, etching 1749

his warning. As an eminent portraitist, Wilson had friends in very high places. When he shared his misgivings about the lightning rods – also pointed – that had been installed on Buckingham Palace his words were heard by the most illustrious ears: George III ordered that Wilson was to be given whatever resources he needed for new experiments. And while the knob or point controversy was primarily a learned discourse within the British scientific community, such royal interference and perhaps the envy it caused, inadvertently allowed a modest political element to intrude.

At a time when the American War of Independence was raging, overly-suspicious folk discerned in Franklin's arguments for pointed rods a plot to harm the king – branded a "tyrant" by the Americans – in his own palace. To George III, Wilson's scientific findings, even though they were disputed by most of the scholarly world at the time, were clear enough in themselves, and he gave orders for his pointed lightning rods to be replaced by blunt-ended ones. But the enlightened French finance minister Anne-Robert-Jacques Turgot (1727-1781) – an ally of Britain's rebellious colonies in America – gleefully added fuel to the political fire with his now famous epigram that celebrated the heroic Franklin not only as a lightning-fighting modern Prometheus but also as the conqueror of tyrannical rule: *Eripuit caelo fulmen sceptrumque tyrannis* – "He wrested lightning from heaven and the sceptre from tyrants."[32]

The great majority of contemporary scholars, including Ingenhousz,[33] concurred with Franklin in his preference for pointed lightning rods. Although later research established that a rod's height is much more critical than its shape,[34] at the time the dogmatic certainty with which some authors defended the pointed version seems really remarkable. Evidently, those often enlightened individuals believed that even a neutral stance in the matter of points or knobs was tantamount to accepting the "unreasonable" fear that conductors would attract lightning. Conversely, in the contemporary context there was no clearer way to express confidence in the conductivity of Franklin's invention than to opt for pointed rods.[35]

Although, in the first decades after 1752, there was certainly no consensus among scholars, it was obvious to the new invention's numerous supporters that physics-based objections ultimately failed to stand up. Not only was the fear that the conductor would attract lightning unfounded, they maintained, but other physical arguments were also groundless. There was no reason to believe that the lightning rod caused earthquakes, for instance, any more than to fancy that it adversely affected the fertility of the soil. The final objection (advanced by Nollet and Valmont de Bomare, for example), that there was an extremely dangerous disparity between the vast amount of electricity in the atmosphere and the simple metal wire that was supposed to render it harmless, was equally unjustified in their view.[36] Of course, care had to be taken that the device was earthed and that enough of the right material was used: after investigating the conducting power of different metals, experts concluded that copper was best suited for the purpose as it produced the fewest sideflashes.[37]

The Netherlands

In the Dutch Republic the dry fog or "sulphurous fumes" of 1783 were also observed with some concern.[38] There too such phenomena were sometimes interpreted in an electrical context, all the more so since just a few years earlier, in August 1779, with interest in electrostatic phenomena at its height, the Batavian Society of Experimental Philosophy had held a competition to explain them.[39] And in 1786 the merchant Paets van Troostwijk, encountered in the previous chapter, gave a lecture on discoveries in electrical research that led to the "atmospheric phenomena" being seen as electrical in character.[40] But although opinions in the Netherlands hardly differed from those elsewhere and the summer of 1783 was exceptionally warm there as well,[41] evil vapours or frequent thunderstorms seem not to have generated the sort of massive wave of publications and lightning rod installations that occurred in Germany.[42]

◀ Humidor made in 1787 by silversmith Johannes Schiotling to commemorate the award on 26 January 1787 by the Académie Royale et Patriotique of Valence of an honorary prize to Adriaan Paets van Troostwijk and Cornelis Kraijenhoff. The prize was for their treatise *De l'application de l'électricité à la physique et à la medicine*, written in response to an academic competition of 1786 and published in Amsterdam by D.J. Changuion, in 1788

Certainly, when the summer of 1783 arrived, a lightning conductor had already been fitted on Doesburg's much-struck church tower. And needless to say, on that occasion and even before it, on 28 May 1783, many agreed that the rod had proved its usefulness, as its installer Cornelis Kraijenhoff did not fail to point out.[43] But the lecture and article in which he did so, published in 1785 (together with translations of writings by Ingenhousz and others and with learned society papers such as Radermacher's)[44] can at most be regarded as a distant and fairly solitary echo of the far greater number of publications popularising the lightning rod in Germany. Nevertheless, Kraijenhoff went to great lengths to refute the physical objections to the lightning rod. His essay reprised almost the entire international discourse up to that point, from the objections based on the disparity between atmospheric electricity and the wire used in the conductor or the rod's supposed potential to attract rather than conduct electricity, to its effect on the fertility of the soil and the risk of it causing earthquakes.[45]

As to that last point, Kraijenhoff – proceeding from Franklin's theory of a single electric fluid seeking equilibrium, and apparently also inspired by Bertholon's ideas – assumed that the lightning rod was able to "suck in" or "drive out" lightning matter from or to the clouds, thereby implying that it could both render thunderstorms harmless and, if it did not prevent earthquakes, could at least reduce their intensity.[46] We also find that two-way motion idea in the writings of the Mennonite preacher and installer of lightning rods Gerrit Hesselink (1755-1811). In 1804 – and undoubtedly influenced in part by Bertholon – he defended the idea that during earthquakes sulphurous substances accumulated beneath the ground and that these, if they escaped in the form of fumes, caused thunderstorms.[47]

As we have seen above, in 1783 seismic and volcanic activity and frequent thunderstorms jointly lent a degree of plausibility to the sulphur discourse as an old causation theory. Conversely, the fact that those phenomena sometimes occurred simultaneously gave rise to the idea that earthquakes also had an electrical character. Dutch researchers sometimes proved to be just as convinced of the latter as their foreign fellows. In that regard, the Dutch scholar Martinus van Marum, for example, shared the view of his English colleague Joseph Priestley, who, like himself, had conducted extensive experimental research on the subject.[48] And wholly in line with the contemporary chemical way of thinking, the dilettante bookseller Martinus Magerus believed that he could remove "lightning matter" from the air with his "electric kite" in the same manner as Priestley, who also used a kite to try to restore the balance between atmospheric and terrestrial electricity. We have already seen that Magerus had translated a book by that great scholar.[49] All in all, therefore, in the Dutch Republic too it was invariably "lightning matter" that was held responsible for a whole miscellany of miseries.

Perhaps the best evidence for the similarity between Dutch and international thinking is thus to be found in the frequency and persistence with which vapours of all kinds were the subject of debate. In 1769, for example, the Leiden lecturer Johannes le Francq van Berkhey already suspected that lightning matter, which drifted from hot countries to the Netherlands, could be regarded as one of the electrical atmospheric phenomena – finding support for that conclusion in both Franklin and Nollet. Berkhey, who like most of his contemporaries considered fire a substance, thought it possible that "lightning or the electric, or perhaps phosphoric fire-element (...) is also fed inside our earth, and is revealed even in wintertime." The release of those vapours, he said, resulted in a fireball which burst into pieces in the air, thus causing thunder and lightning.[50] Other authors put more emphasis on the part played by sulphur. Jacques Alexandre de Chalmot, the Dutch translator and editor of Noël Chomel's *Dictionnaire oeconomique*, wrote most clearly on the topic: "Lightning (...) arises from collections of sulphurous and saltpetrous vapours, which gradually rise in the air, gather there, and by a light breeze or rotations of the air are so compacted and compressed that all the damp parts of the air are driven out, and the sulphurous and saltpetrous parts rub together so much that they become hot, eventually catch fire, and burst." Death by lightning occurred in one of three ways, he informed the reader: "either from suffocation (...) or from a cessation of their fluids caused by the lightning's acid sulphur, or finally from bodily injuries inflicted by lightning."[51] A practical example was provided – up to a point, at least – on 21 August 1768, when lighting struck the church and tower of the hamlet of Alem while a service was in progress. Fortunately there were no fatalities, but the officiating minister Samuel Keuchenius (1737-1824) did note an oppressive sulphurous smell that filled the church for more than two hours.[52] Some years later, his cousin and fellow-minister Johannes Florentius Martinet published a detailed account of the events that Keuchenius had recorded at the time.[53] The sulphurous stench would not have surprised him, for in his well-known *Katechismus der natuur* he too explained lightning as resulting from the electrification of risen oil, sulphur, and saltpetre vapours.[54]

Even though some were sure that the precipitation of those various vapours encouraged plant growth,[55] in the short term it was important to keep the sulphur content in the air as low as possible, especially indoors. As we have just seen, Chomel's dictionary cited suffocation as a leading cause of death from thunderstorms. To avert that danger, whenever a storm occurred it was essential that the room one was in had a door that could be opened at any moment to escape the stifling air. For the same reason, if it stormed during the night it was wise to get out of bed, so Kraijenhoff advised.[56] Even in 1853, people were still being warned that during a thunderstorm they should be sure to keep the air in the house fresh by opening a window.[57] According to Chomel's dictionary the only door that should be kept securely closed was the one to the cellar in order to protect perishable foodstuffs, for the "electric vapours" quickly soured butter, cream, and milk. Sulphurous air was also very bad for anything involving fermentation, such as bread, beer, and wine, as it held up the process.[58] People, on the other hand, should not shut themselves in their cellars, Martinet stressed,[59] but should go outside with their children, at least while the storm was rising.[60]

Conclusion: "Science" or "Superstition"?

In short, the villain of the piece turned out to be sulphur every time. On this point too, the discourse in the Dutch Republic was entirely consistent with what was being said elsewhere in Europe. In other countries, especially Germany, there were also warnings against sultry air in the house. During a thunderstorm, one should do as little hard work as possible because all the sweat that was produced would create extra sulphurous vapour, not only increasing the risk of suffocation but also the danger of explosion.[61] And it was not merely exertion that should be avoided: overanxiety also made people sweat and the vapours thus caused made them far more likely to attract lightning than if they had kept calm, argued the Lutheran minister Johann August Ephraim Goeze (1731-1793) from Quedlinburg in eastern Germany. His warning could still be read in 1801 in a Dutch journal.[62] Some even advised holding one's breath during a thunderclap.[63] In line with the foregoing, it was also considered prudent to stay out of the kitchen during a thunderstorm and to keep as far away from the chimney as possible, for the sulphur in the smoke could attract lightning.[64] For the same reason, the fire should be extinguished[65] and all tobacco smoking should stop immediately.[66] Finally, one should stay away from "vaporous" places such as marshes, middens, cattle sheds, tanneries, hospitals, and privies.[67]

◀ Portrait of the Lutheran minister Johann August Ephraim Goeze from Quedlinburg, artist unknown, ca. 1780. From the 1770s Goeze devoted himself increasingly to the study of the natural sciences

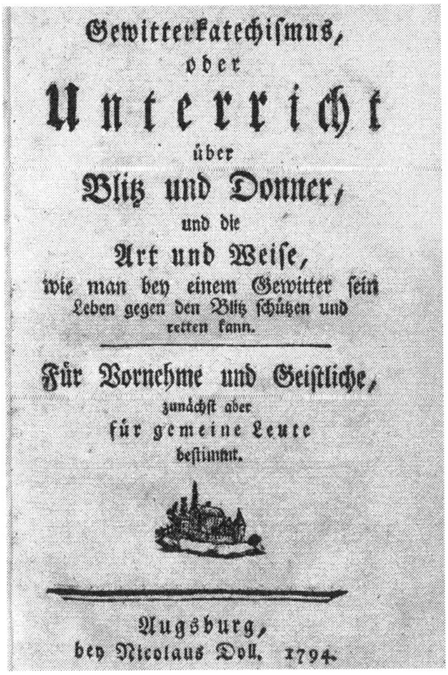
Gewitterkatechismus,
oder
Unterricht
über
Blitz und Donner,
und die
Art und Weise,
wie man bey einem Gewitter sein
Leben gegen den Blitz schützen und
retten kann.

Für Vornehme und Geistliche,
zunächst aber
für gemeine Leute
bestimmt.

Augsburg,
bey Nicolaus Doll, 1794.

◀ Title page from Joseph Kraus, *Gewitterkatechismus*, Augsburg: N. Doll, 1794

The discourse about sulphur may have a completely international character but, according to the standards of the time, it can be hard to determine where "science" ended and "superstition" began. For reasoning along such lines, trying to thin out the evil vapours must have seemed the obvious thing to do, not only inside, by opening windows and doors, but also outside in the open air, and thus by more powerful means. Outdoors they could be detonated in controlled explosions, for instance. Which is why thunderstorms were sometimes bombarded with cannon fire, particularly in the Alps. Was there not a certain logic in that custom, despite the fact that such *Wetterschießen* or "firing at thunderclouds" came in for considerable criticism from Enlightenment thinkers both at the time and later? Even without a combined explosion, it was thought that the enormous sound waves would at least disperse the sulphurous vapours. That was also the reason behind another old custom that was still more controversial than firing cannonballs at thunderclouds – ringing the church bells when a storm approached. Intellectuals like the Potsdam doctor Friedrich Leberecht Supprian (1723-1789) in 1746 deemed this so-called "weather-ringing" a suitable means of diluting the vapours.[68] And in the following years, even the *Encyclopédie* of Diderot and d'Alembert recommended it in the entry under "Tonnerre" (thunder), which was written by none other than the inventor of the Leyden jar, Petrus van Musschenbroek.[69] In Chapter 5 we shall see how "weather-ringing" gradually became the target of comprehensive and vehement protest from those for whom it was as much the textbook example of "superstition" as the lightning rod was the prototypical emblem of the Enlightenment. First, however, the attitude of "official" religion to thunderstorms and lightning rods demands our attention.

Religion

◀ Prometheus brings fire to mankind, painting by Heinrich Friedrich Füger, ca. 1817

4

Official Religion[1]

In the religious mindset of the sixteenth-, seventeenth- and eighteenth-century West it was generally accepted that God addressed people directly by means of thunder and lightning. These were empyrean signs, usually betokening divine wrath and the coming Day of Judgement. Until the mid-eighteenth century, many, both Protestant and Catholic, regarded that as self-evident. Preaching was therefore intended to evoke penitence and repentance. In the second half of the eighteenth century, however, the newly-invented lightning rod deprived God of his supposed instrument of punishment and thus the nature of man's relationship with him was changed. Fatalism, though still endorsed by many, was gradually replaced by personal responsibility, which enabled man to utilise the rational preventive device that God had placed in his hands. Nature no longer posed a threat but was the result of God's marvellous work of creation, and could be understood through reason and increasingly also through feeling. God, man, and nature all worked in unison. So-called physico-theology stressed the goodness, wisdom, and omnipotence of the Creator evinced in the impressive natural phenomenon of the thunderstorm.

God's Wrath

In practically every cultural tradition, thunder and lightning are regarded not as mere natural phenomena but as direct manifestations of higher powers. In the Andaman archipelago, thunder is the voice of the pre-eminent deity. In Mesoamerica lightning has powerful associations with primary deities; in Mongolia it represents the displeasure of the creator of balance in the world. From the Baltic to Bolivia, Iceland to India, Jamaica to Japan, thunderstorms are interpreted as signs from the Demiurge. Up to a point, there is a parallel with the monotheistic Judeo-Christian tradition, which conceives of Yahweh as an invisible supreme being whose speech emanates from the storm.[2]

◀ Moses receives the Ten Commandments on Mount Sinai (Ex. 19 and 20), painting by Benjamin West, 1784. A bolt of lightning is faintly visible at top right

That God addresses people directly in thunder and lightning was also a generally accepted idea in the religious mentality of the sixteenth-, seventeenth- and eighteenth-century West. "During the Baroque era thunderstorms were simply considered the appropriate language of the Godhead," as German historian Christian Begemann pointed out some decades ago.[3] Christian preachers were supported in that belief by the Bible and had long been used to citing passages from the Psalms, such as "the voice of the Lord is over the waters; the God of glory thunders" (29:3) and "the waters take flight at the sound of God's thunder" (104:7) or from Job, "he thunders with his majestic voice and he does not restrain the lightnings" (37:4-5), to show how that sublime natural phenomenon was also identified with divine utterance. In an age of biblical literalism texts like these convinced many Christians that the thunderstorm was not just a metaphor for the voice of God but the direct expression of it.

That notion of the immediate and clearly audible presence of the Almighty could have extreme consequences. It led some devout souls to put the word "*onweer*" (bad weather, thunderstorm) consistently in quotation marks.[4] In certain circles it came to be regarded as disrespectful to eat, work, or even speak during thunderstorms – to interrupt God, as it were.[5] The strictest believers so revered this natural phenomenon that they would neither look nor even point at it and forbade their children from doing so.[6] In the Lüneburg region in northern Germany, such behaviour was actually considered a sin, for which the punishment would be a crooked finger,[7] while others feared that "sinners" of that kind would be turned to stone.[8] And in Normandy, France, it was thought that such unwise folk were likely to be struck by lightning.[9] It goes without saying, therefore, that in the eyes of these often pietistic people, installing a lightning rod was the very height of irreverence: not only an arrogant and futile effort to improve God's creation, but worse, a presumptuous and pointless attempt to evade his judgement.

Everything should be geared to listening attentively to the voice of God and correctly interpreting what was said. For however directly the Lord spoke to mankind, as when handing down the Ten Commandments on Mount Sinai (Ex. 19 and 20), his outer manifestation invariably served the higher purpose of the revelation of his message. Like every other catastrophic natural phenomenon, therefore, thunder and lightning were first and foremost signs, more specifically of God's wrath and the coming Day of Judgement. Up until the second half of the eighteenth century that was regarded as self-evident, in the Dutch Republic as elsewhere (Germany and Great Britain, for example), among both Protestants[10] and Catholics and in both high and low social groups. Lightning was God's weapon. In painting, lightning bolts were typically depicted with arrowheads, particularly from the early seventeenth century.[11] Between the sixteenth and part of the eighteenth century, the preaching of both great confessions was mostly still aimed at allowing thunder and lightning to act on the churchgoer's "heart", the seat of his or her conscience. Preachers of the fire and brimstone type had no problem whatever in coupling a religious and moral storm to the meteorological sort: when Christ sent out

▲ Fresco in the Domed Hall of Bruchsal Palace, by Johann Zick and his son Januarius, mid-eighteenth century. Notably, the bolt of lightning has an arrowhead – an expression of the ancient view of lightning as God's weapon

▲ Porcelain dish decorated with various objects and insects, as well as arrow-headed shafts of lightning, maker unknown, Brussels ca. 1800

▶ Playing card from lottery game, with arrow-headed streaks of lightning, second half of the eighteenth century

his apostles to preach did he not call James and John "sons of thunder"? (Mk. 3: 17). One upshot of that was that in the early modern period the thunderstorm sometimes became a metaphor for preaching.[12] That thundering word – necessarily uttered fortissimo in a time without microphones – was intended to evoke penitence and repentance in a subtle interaction between nature and revelation. For the thunderstorm was used as a means of emphasising the need for humility, as the Lord first did in the Holy Scriptures. Ignore that instruction, those preachers warned, and the Almighty would be compelled to speak more forcefully and directly from the book of nature by making his punitive hand felt in the thunderstorm.

Catholic preachers, too, often saw the thunderstorm as retribution and so emphasised the need for repentance before ritual or even "superstitious" devotions were resorted to.[13] According to the French bishop Jacques-Bénigne Bossuet (1627-1704), a sermon should speak to men's consciences; preachers should seek "flashes of light that pierce, thunder that moves, a bolt of lightning that sunders hearts."[14] But inter-denominational as such a "pedagogy of fear" may have been, there were differences in approach. To Protestants – and *a fortiori* the pietists among them – repentance was an eminently internal matter, much more than it was to Catholics. Among Protestants, the thunderstorm did not occasion public forms of penance such as processions, but gave rise – in Germany and elsewhere – to a whole tradition of so-called *Wetterpredigte* or "thunder sermons", which called for the serious examination of the conscience. Public worship was the proper place for effecting this process of repentance, whereby the sinner was shocked out of his sinful slumber by the sound of thunder. The heart-sundering would then be followed by sincere belief, the fruit of which was heartfelt prayer. Not that this so-called *Wettergebet* or "thunder prayer" should be limited to public worship. After all, for pietists, the family was like a church in microcosm. Hence, the *Wettergebet* also had a legitimate place in the *Hausandacht*, the practice of meditative piety in the home, in which the domestic servants might also participate. In some homes furniture was specially placed for that purpose, the so-called *Gewitterbank* or "thunder bench".[15] In addition, of course, there was the time-honoured private reading of devotional books in response to the thunderstorm.[16] And sometimes penitential hymns were sung. Traditionally these had both an edifying and a motivating purpose, while the end of the thunderstorm was often followed by hymns of thanksgiving.[17]

Thunder hymns could be heard elsewhere, too. In England, for instance, a sudden storm on 20 August 1697 inspired the non-conformist preacher Isaac Watts (1674-1748), the "godfather of English hymnody", to compose a hymn incorporating meteorological notions of divine power: God appears as "the Thunderer, who makes the clouds his throne."[18]

The *Wetterpredigte* and its equivalents in other countries (such as Puritan England or – as we shall soon see – the Dutch Republic) often had a fixed structure. The judicial element was almost always paramount: the Lord was a stern judge who was an-

▲ In Catholic parts of Germany a *Herrgottswinkel* with a crucifix and other religious artefacts was often installed in a corner, where people prayed and lit a *Wetterkerze* (candle) during thunderstorms

gered by sin and punished it with the thunderstorm. In the strictest or most traditional sermons a whole catalogue of sins was cited; in a few cases the preacher believed he knew exactly what had drawn down the divine wrath and what the appropriate punishment would be:[19] perjury, church theft, or the profanation of the Sabbath, for instance, would assuredly result in death by lightning.[20] There was also the age-old belief that "thunderstones" (objects later identified by science as belemnite fossils) were the instruments the Lord used to strike such sinners directly.[21] But besides these special providences or more general judicial elements, such sermons usually also contained an important instructive or cautionary component: the violence of nature was intended to lead each individual sinner to repentance, to a salutary spiritual "storm" in his inner life, and at the same time to warn him against more serious calamities, even of eternal judgement itself[22] should such a fundamental change in lifestyle be unforthcoming.[23] After all, the really evil exhalations were the wicked deeds of men![24] On the other hand, God's wrath would be appeased if he saw a sinner's heart sundered.[25] Only then would the *Wettergebet* really be effective.[26]

◀ A (Catholic) rural family pray in their living room during a thunderstorm, ca. 1833, painting by Lorenzo Quaglio the Younger (1793-1869)

In colonial America, preaching and writing were much the same. Witness the well-known story of the Reverend Cotton Mather (1663-1728). On 12 September 1694, while praying with his congregation at the start of a service, Mather had a premonition that prompted him to lay aside the sermon he had prepared and to extemporise on Psalm 29:3. The loud claps of thunder that almost drowned out his discourse were, he told his parishioners, God's own glorious voice. Even when interrupted by a messenger with the news that his own house had been badly damaged by lightning, Mather continued his sermon. For as he explained to the congregation, the material things of life were far less important than contemplating God's call. The following year he published that sermon and the one he gave on the Sunday after, under the title *Brontologia sacra*, an edifying work which, "while useful for all men at all times" was "especially intended for an entertainment in the hours of thunder," according to the title page. Repentance was necessary, but at the same time, Mather stressed, the true believer should not be afraid for nothing happened without God's will.[27] The great preacher Jonathan Edwards (1703-1758) agreed with him, describing thunder and lightning as part of "God's artillery", the perfect instruments of divine wrath.[28]

In the Dutch Republic a similar tone was adopted in sermons and writings. Think, for instance, of the enduring penitent reactions to the lightning strike and subsequent fire that destroyed Edam's Grote Kerk on 24 February 1602 and which led to an annual local service of thanksgiving and prayer that continued to be held well into the eighteenth century[29] and, according to some, even into the late nineteenth.[30] Or how, when the tower of Middelburg Abbey was struck by lightning on 1 August 1712, the strict Calvinist preacher Carolus Tuinman (1659-1728) saw in that devastating natural force God's punishment for the lax way in which urban and regional authorities combated heresies of every kind. That people showed themselves obstinate and unrepentant under the punitive hand of the Lord in the storm, was also a source of concern to the Catholic burgomaster's valet Johannes Janssen (ca. 1700-1780) of Aachen.[31] And his fellow believer, the educated Limburg farmer, alderman, and chronicler, Winand Mengels (1733-1778), agreed with him. In his history of the village of Kanne near Maastricht he recorded a case of very real and immediate divine retribution, the sad tale of four women who, on the afternoon of 25 July 1777, in Lixhe just north-west of Visé, were struck by lightning as they sat in the fields: two of them were killed, a third was seriously injured, while the fourth, a girl of about fifteen, was unharmed; for as the storm approached she had started to pray, whereas her three companions had laughed mockingly and used foul language.[32] In a similar way, in the early nineteenth century, a preacher used the fear of thunderstorms to combat cursing.[33]

▶ Portrait of Pieter Boddaert, print by Jacob Houbraken, after a painting by Philip van Dijk, mid-eighteenth century

The call to repentance is so central to some adherents of the punishment theory that one might think they saw the storm as an occasion for edifying reflections rather than a natural phenomenon, as is the case, for example, in the poem written by the pious poet Pieter Corneliz Boddaert (1694-1760) published in 1726:

How fearfully howls and crashes and rumbles the thunder,
(...) As if it were a prelude to the world's youngest day.

(...) Consider, sinner, if this can make your heart tremble,
As God's patient forbearance keeps you alive,
What will it be if his everlasting vengeance should find you?

Concur with the rod he has appointed, by repenting in time
in order that he may think of you for the good.
Delay, tarry not, lest his lightning might devour you![34]

That old mode of thought with its associated image of a stern God was still current at the end of the century. As late as 1799, a recent storm prompted the poet Willem de Perponcher to call for humiliation, confession of sin, and repentance. Reflecting in retrospect, he asked his readers: "Who amongst us (...) did not consider, with fear and contrition, his guilt and transgressions; seeing above his head his Father, so often provoked by disobedience, in the attitude of a wrathful Judge, enthroned upon the dark clouds, as on an awful Judgement seat, with the terrible bolt of lightning in his hand and ready to crush his enemies?"[35]

Yet at the same time, De Perponcher's work suggests that the religious mindset had undergone some changes in the course of the eighteenth century – often subtle changes that had not only a Dutch but a general Western character. For however culpable man might be in the poet's eyes, it was owing to the heavenly Father's forbearance that he was nonetheless spared. Even those who had been harmed by the thunderstorm might remember that it was a Father who punished, chastised, or tried them, and whose only motives were mercy and love. On the other hand, the storm also brought blessings which likewise, in a gentle way, meant nothing less than to bring man back to God.[36]

A Change in Mentality

However much some thought that thunder theology had been one of the most powerful foundations of Christian doctrine from earliest times,[37] it gradually became less obvious to think of punishment in that connection. Although, in the (early) modern era, virtually no one doubted the workings of Providence, such a punitive interpretation of the storm had lost a lot of credibility, particularly since the second half of the eighteenth century.[38]

▲ Jupiter with his thunderbolts before a palace or temple, pen and ink wash drawing by A. Egel, eighteenth century

Yet this change in mentality already had a modest history. For even though it was initially represented by only a small handful of individuals, some Baroque thinkers such as Scipion Dupleix (1569-1661) had already tried to counteract the fear that they saw as underpinning the punishment theory, citing authors from classical Antiquity since they often emphasised the natural causes of phenomena such as thunderstorms.[39] Somewhat later, in the late seventeenth century, and in a slightly wider circle, the theory of punishment had been further devalued when the mechanisation of the world view meant that God was no longer regarded as the *causa proxima* ("proximate cause") but at most the *causa remota* ("remote cause") of the thunderstorm.[40]

This critical undercurrent rumbled on into the eighteenth century. In 1744, for instance, an article in the *American Magazine* warning against superstitious fears of things like thunderstorms stated that we should not "measure the Work of God by our scanty Capacities." Consequently, we should not claim heaven's approval of our own undertakings, nor should we resort to supernatural explanations for things of whose natural cause we are ignorant.[41] A few years later, back in Europe, the same line of thought could be found in an article in the journal *L'Observateur hollandois*, probably written by Jacques-André Courtonne (1711-1773), minister of the Walloon church of Leeuwarden. In that case too, criticism was levelled at people who took thunder to be God's judgement and misused it to suit their own ends or interests. For as a result, "we make Heaven speak a very obscure language," which all too often led to a superstitious religion of fear. Courtonne turned to Seneca to refute the thunderstorm-equals-punishment theory: thunder and lightning were simply natural phenomena, nothing more.[42]

In the second half of the eighteenth century, however, the newly-invented lightning rod also contributed to this more distant view of God as it deprived him of his supposed instrument of punishment. And for some, Antiquity was once again a source of inspiration. Taking the classical writer Lucian (ca. 125-ca. 180) as his exemplar, the German writer Christoph Martin Wieland (1733-1813) humorously expressed the diminution of divine power in his *Göttergespräche* ("Talks Between the Gods") (1789-1793). Positing an Olympian conversation between the elderly Jupiter and Juno, he has the wise supreme god say to his wife that given the waning of religious belief, the time for divine intervention was really over: "would I make people more reasonable with lightning and thunderbolts?," he asks her scornfully.[43] In other words, Jupiter was enlightened enough to accommodate himself to the new order and to avoid making himself ridiculous by futile displays of power now that Franklin had invented the lightning rod, thereby proving that it was not man's moral failings that affected the direction of lightning but its natural attraction to things made of metal and tall objects like trees and towers.[44] Thus Wieland distanced himself from the traditional view – expressed earlier by Jonathan Edwards, for example – that God employed lightning to punish the pride implicit in lofty man-made structures.[45]

▶ Portrait of Christoph Martin Wieland, painting by Ferdinand Carl Christian Jagemann, 1805

To Wieland, the image of a punitive God was obsolete. But he was not alone in thinking so. To socio-cultural layers below the enlightened elite as well, the thunderstorm was no longer the prelude to terror of sin and examination of conscience. On the contrary, the enlightened citizen could now face thunder without trembling for the very reason that his conscience was clear and his right-doing was not merely the product of a dread of punishment. Conversely, those who were still afraid of thunder and lightning were soon accused of harbouring a false idea of God. For it was no longer God's storm-induced repentance, without which all precautions would be futile, that was the main thing, but the use of (ultimately God-given) effective devices like the lightning rod. In a certain sense, punishment therefore became mercy. The result of this conceptual about-face was that lightning thus tamed became an integral and salutary component of nature.[46]

The Fatalism Controversy

Yet this was by no means a linear development, as if the invention and adoption of the lightning rod – that icon of the American Enlightenment – had caused "superstition" and fear of lightning's destructive power to melt away like snow in the sun.[47] Enlightened eighteenth-century polemicists may have liked to think so, setting Franklin's rod in diametric opposition to God's beating birch, but a more nuanced historiography avoids such easy triumphalism, if only because Franklin's own ideas about Providence had undergone some modification: as a young man, he had been opposed to inoculation, for instance, an attitude he regretted afterwards when his four-year-old son died of smallpox. It was that loss which subsequently motivated him to find a preventive form of protection against needless harm from lightning.[48]

In a certain sense, however, the lightning rod created a new anxiety in the secular-minded. For if one was nonetheless hurt by a thunderstorm, the old familiar path of serious self-examination in order to discover the causes of God's wrath was now an unlikely one. In needing to take adequate preventive measures one was left to fend for oneself against a nature whose workings were increasingly perceived as entirely random.[49] Moreover, the phenomenon of electricity (atmospheric or otherwise) was still surrounded by all kinds of unsolved humility-inducing scientific riddles, while on the other hand not all religious folk were opposed to the introduction of the lightning rod. It was, after all, a Moravian monk, Procopius Diviš, who had built one of the earliest devices.[50] Another cleric, the Padua physics professor Abbé Giuseppe Toaldo (1719-1797), introduced the lightning conductor to the Venetian Republic and supervised the installation of a rod on Siena Cathedral,[51] while his colleague, the aforementioned Turin professor of experimental physics Father Giovanni Baptista Beccaria, also publicised Franklin's ideas.[52] The lightning rod won papal approval, too: Benedict XIV and Pius VI expressly recommended its use,[53] while on the Protestant side the Quedlinburg minister and naturalist Johann Goeze was likewise an advocate of the new invention.[54] The lightning rod's proponents, even when they were not members of the clergy themselves, were very often devout individuals who believed in the divine providential guidance of creation. God governed natural phenomena according to laws that were immutable but also intelligible; their operation was therefore predictable. Thus the benevolent Deity had enabled rational man to invent the lightning rod. Hence the thunderstorm's technical "disenchantment" could be seen as contributing to the enlightened purging of religion, which should, more than ever, become an internal matter.[55]

But even if there was less conflict between science and religion in the Age of Reason than is generally thought, the extremes were sometimes very far apart. On the one hand, the most traditional champions of the thunder-as-punishment theory did indeed draw fatalistic consequences from their interpretation and had no hesitation in saying so in the strongest terms: it was, they reckoned, as impious as it was pointless to try to resist the "artillery of heaven"[56] with this newfangled *Teufelszeug*[57] or "hellish stuff" since no one could escape the wrath of an omnipotent Providence. Indeed, according to some of the residents of Nuremberg in 1788 it was tantamount to "pure blasphemy."[58] Was it not illusion and presumption to imagine that God's wrath could be evaded by conducting electricity into the earth? Would it not cause earthquakes instead, due to what was believed to be the natural accumulation of sulphurous vapours?

Reverend Thomas Prince of Boston certainly thought so. And as we have already seen, his point was proved – so he maintained – by the severe earthquake that shook the city on 18 November 1755, undoubtedly as a result of the many lightning rods that were installed there. That sparked a controversial polemic with Professor John Winthrop of Harvard over whether or not the lightning rod was a tangible expression of man's presumption.[59] In these years there were some who said that even the late Abbé Nollet would not have shied away from extreme fatalistic utterances: according to a statement attributed to him, attempting to ward off God's lightning was as impious as for a child to resist the chastening rod of the father.[60]

A decade or so after Nollet's death, between 1780 and 1784, the question of man's supposed presumption was in the limelight again in the northern French town of Saint-Omer, where a certain Charles Dominique de Vissery de Bois-Valé (d. 1784) had installed a lightning rod on his house, only for his neighbours to institute legal proceedings to force him to remove the offending object. De Vissery lost his case then appealed and won, thanks in part to the closing argument of his young lawyer Maximilien Robespierre (1758-1794), who poured scorn on the notion that installing such a device amounted to defying the Lord's wrath or depriving him of the means of punishing man's crimes. Robespierre had his plea printed and lost no time in sending off a copy to Franklin.[61]

In the religious battle over the lightning rod, as in many other polemical contests, there was a sedulous search for scapegoats. All kinds of dividing lines became or were made visible. In Catholic Italy, for instance, lightning conductors were known as "heretical rods".[62] Elsewhere, national rather than confessional differences were deployed. In Switzerland for example, a farmer in the canton of Aargau believed that it was "the God-denying French" who invented the lightning rod in order to make God jealous.[63] And thirdly, social differences were also often involved, though it would be incorrect to see – as the enlightened elite were inclined to do – fatalistic resistance to the lightning rod as occurring exclusively in the lower socio-cultural classes. When Pierre Bertholon had the honour of installing the first two lightning rods in Paris, around 1780, opposition came from theologians as well: to interfere with lightning demonstrated an egregious lack of reverence, they protested, for it was the manifestation of divine wrath.[64]

At the other extreme was the sometimes equally oversimplified assertion of man's autonomy. In the febrile atmosphere of those days, Franklin himself often became the focus of the controversy, not as a scapegoat now but rather the consummate hero who had boldly penetrated the realm of God's dominion. That process of heroisation had begun in the 1750s when Kant described Franklin as a modern Prometheus,[65] although (as we have seen) at the time he doubtless also intended the comparison as a warning against man's hubris and was not entirely convinced, at least initially, of the lightning rod's prophylactic effect.[66] Nevertheless, as a typical *Selbsthelfer*[67] ("self-helper")

▲ Riot in Saint-Omer over the installation of a lightning rod, 1780-1784

Franklin, and his invention, had increased man's confidence in his own powers to such an extent that the more radical spirits soon endowed his figure with almost mythic proportions.[68] For how royally rewarded was the rebellion of sovereign man against the punitive instruments of divine omnipotence! And what breathtaking perspectives it opened up for the future! Inventing the lightning rod was, Robespierre argued in 1783, "the boldest and most surprising idea mankind ever came up with."[69]

A few years later, the notion of stealing lightning from the gods was equally pleasing to the Marquis de Sade (1740-1814), if only because it put an end to the old topos that turned lightning into an instrument of an atheist's execution. But de Sade did not stop there. The most important result of Franklin's invention, in his view, was that every libertine could now take his fate into his own hands and could, metaphorically speaking, pluck arrows from his own quiver and audaciously shoot thunderbolts back. He declared with sardonic pleasure that (without the mitigating effect of the lightning rod) unleashed nature seemed to punish virtue and reward vice, like a somewhat indifferent criminal. Thus the whole idea of a deity became little more than a chimera.[70]

▲ Portrait of the Marquis de Sade, by Charles-Amédée-Philippe van Loo, 1760

But even if conclusions of an atheistic kind were thought to be too radical, the secularising implications of such a clarion call of man's self-assertion against the divine world were clear enough. In the eighteenth century the real significance of the lightning rod lay not in the safety actually achieved, but in the fact that it made it possible on a large scale for the first time.[71] That paradigmatic meaning emanated a strongly suggestive effect, transforming mortal man from sorcerer's apprentice to sorcerer, a champion who challenged or even dethroned the gods. This metaphysical revolt is particularly typical of the zeitgeist of *Sturm und Drang*,[72] although it by no means disappeared after that. In fact, had man not already succeeded in playing God by imitating lightning in scale-model experiments and freely manipulating it at that micro-level?[73] And although this did not yet work in full-scale reality, an inscription next to the lightning rod on the tower of St Margaret's Church in Gotha in eastern Germany triumphantly announced to posterity: "We have prescribed the lightning its path."[74] In fact, as mentioned earlier,[75] lightning had become man's slave, or so the poet Shelley concluded in *Prometheus Unbound* (1820). His wife, Mary Shelley, clearly concurred, at any rate in *Frankenstein: Or the Modern Prometheus* (1818). Professor Waldman of Ingolstadt speaks with evident satisfaction of the "new philosophers" who "ascend into the heavens (...). They have acquired new and almost unlimited powers; they can command the thunders of heaven, mimic the earthquake, and even mock the invisible world with its own shadows."[76]

Almost imperceptibly, the existential sense of the Enlightenment gave way to the mood of *Sturm und Drang* and Romanticism. Later on we shall take a closer look at the major aesthetic aspects of that cultural transformation.[77] Suffice it to say here that man, whether ablaze with enlightened-scientific or Romantic-artistic self-confidence, saw in the thunderstorm above all a challenge to confront the elements armed with nothing but his own strength.

After all, everyone was the maker of their own happiness. To autonomous man that not only implied a metaphysical revolt but at the same time a political rebellion, for Jupiter's thunderstones often symbolised the instruments of absolutist sovereignty.[78] It would be going too far to expand on that here, to explain how in the political arena – that pre-eminent field of the makeable – Franklin also met the requirements of Promethean symbolism wonderfully well, according to his admirers. For whether he took up arms against harmful lightning or the "tyrannical" power of George III he was equally undaunted. Suffice it simply to recall in this regard that his triumph over both of those supposedly invincible powers inspired Turgot's well-known epigram for Houdon's 1778 bust of Franklin: *Eripuit caelo fulmen sceptrumque tyrannis* – "He wrested lightning from heaven and the sceptre from tyrants."[79]

For some, especially in the American Enlightenment,[80] the metaphysical revolt definitely went too far. They continued to doubt whether that unrestrained and at times chilling emphasis on human responsibility might not lead to consequences that were too audacious for religion. After all, it was not necessary to embrace "superstitious" fatalism vis-à-vis the lightning rod before wondering whether the Enlightenment motto, *sapere aude* ("dare to know"), did not sometimes verge on hubris – on a desire to control nature for reasons that seemingly went beyond pure self-preservation and could therefore only be described as self-overestimation. Rather, they invoked the fate of King Salmoneus of Greek mythology, who paid with his life for his attempt to imitate Zeus's thunder and lightning. In short, how drunk was mortal man with his own success? Was there no middle way?

If so, it seemed hard to find. Abbé Nollet, for example, who saw Prometheus as an awful warning precisely because of his hubris and who could already hear the eagle pecking at the Titan's liver in the Caucasian distance, so to speak, rather compromised his own objections by apparently combining them with a form of fatalism.[81] Nor did it help that in the other camp people like the German philosopher Kant somewhat categorically posited a simple dichotomy between Enlightenment and "superstition", thus literally and by definition excluding any intermediate solution: "Emancipation from superstition is called Enlightenment."[82]

▲ Zeus's eagle devours Prometheus's ever-regenerating liver, painting attributed to Pieter van Portengen, 1643

More useful, indeed worthy of consideration, was what Kant had written elsewhere in the same *Critique of the Power of Judgement* about the self-assertion of reasonable man in the face of overwhelming nature.[83] Although it was a discourse on aesthetics, it evinced a more general mentality that appealed much more to this moderate public, who could be described as supporters of the Christian Enlightenment. For as the renowned Genevan naturalist and Mont Blanc mountaineer Horace-Bénédict de Saussure (1740-1799) likewise pointed out, man derived new strength and dignity from battling the elements.[84] Personal responsibility could be taken here by gratefully utilising the rational preventive device that the Lord had placed in man's hands. De Saussure himself had set the example in 1771 by installing a lightning rod right next to his house – the first in Switzerland – and penning a pamphlet about it.[85] Did not Proverbs (22:3) say that "the clever see danger and hide; but the simple go on, and suffer for it"?[86] Viewed in that way, the fatalistic resistance to the lightning conductor was nothing less than culpable negligence. Moreover, whoever defended this gift of God could never be accused of human presumption, because the lightning rod did not work *against* creation, but actually *with* it.[87]

▲ Descent of Mont Blanc by Horace-Bénédict de Saussure in August 1787, coloured copper engraving by Christian von Mechel, 1790

Protestant Enlightenment and Moderate Orthodoxy in the Netherlands

For a long time, as far as actual preaching was concerned, that change in mindset had often been a hesitant and subtle process. Yet even in places that the most anodyne version of the Enlightenment had bypassed, something had altered, though often the difference was barely discernible. Admittedly, many preachers, like some poets and editors of journals, continued to insist that natural phenomena did not happen by chance but really meant something, and that the thunderstorm should therefore be the occasion for personal devout meditation, indeed, for repentance – a "subject-centredness in the experience of nature" that characterised the entire early modern period.[88] Yet there was now a diminishing stress on the inevitable sinfulness and concomitant punishment of all men, while the distinction was increasingly made between the effects which the storm had or ought to have on the various categories of audience: what was intended as punishment for one did not necessarily mean punishment for the other. In part that individualisation was consistent with the acute personal emphasis that had always characterised preaching in Protestantism in general and Pietism in particular.[89] At the same time, such a diversification naturally gave wide ramifications to the deeply rooted moralism of the time.

The Dutch pastor Jan van Eijk (1757-1822), from Loosduinen near The Hague, saw the thunderstorm as a source of terror, especially for the "wicked". His work contained various reminiscences of the so-called "Second" or "Further" Reformation – a post-Reformation pietistic movement in Dutch church history. In a dialogue on lightning in poetic form between "Thomas" and "Philippus" it ran thus:

He who mocks God and religion,
That wretch may tremble before such a God!
But he who lives for the Lord,
And has Jesus for his Advocate,
May always be of good cheer,
For God's omnipotence protects him.[90]

Other authors (young and old) agreed with him. The thunderstorm might be a punishment for the wicked, but even in the most fearful danger pious folk trusted in God. The Lord himself directed the frequent downpours and floods "to punish his enemies and benefit his people," so said the Reverend Samuel Johannes van de Wijnpersse (1759-1842) as late as 5 August 1792 as he delivered a genuine *Wetterpredigt* with unmistakably pietistic overtones in Amsterdam's Oude Kerk.[91] His fellow-clergyman Jacob van Loo (1754-1797), who held a living in Ootmarsum, had given a sermon to the same effect a few years earlier, and he too was not the only one.[92] The Almighty was only terrible

▶ Portrait of the learned Amsterdam pastor, Reverend Samuel Johannes van de Wijnpersse, print by Jan Willem Pieneman (draughtsman) and Frederik Christiaan Bierweiler (engraver), first half nineteenth century

to the sinner who misjudged God,[93] for, as De Perponcher likewise argued, the Lord "sought to compel the recalcitrant into submission by fear, as well as eliciting the more willing hearts by his favour."[94]

This kind of "separating preaching" by pastors and poets could not infrequently be joined up with the old topos of the hardened libertine who was led to repent by a severe weather event.[95] For example, in his pious early years, the poet Gerrit Paape (1752-1803) challenged the *esprit fort* ("freethinker") thus:

Where is your courage now, Unbeliever!
Behold, see him hide himself!
He who boldly disputes the Almighty's being
Creeps faint-heartedly into deep pits.
Come, strong spirit! plead your case,
As the thunder rattles!
Deny Heaven's favour and vengeance here and now! (...)
But what is this! you are silent (...) no wonder![96]

This was still all fairly traditional stuff. Profligates, libertines, and similar dissolute persons were often enlisted for little more than the sake of contrast; for most other folk, wheeling out such "big guns" in a sermon was unnecessary. Things became more lively, however, when the usual good-bad dichotomy was replaced by a triad of types. The Reverend Van de Wijnpersse, for instance, structured his whole sermon around the distinction between three different groups: callous characters whose "heathen indifference" caused them to see the thunderstorm merely as a purely natural phenomenon; terror-stricken souls who were trapped in cringing fear of what they regarded as the "ignited rays of God's wrath"; and the god-fearing who repented and found shelter in "the throne of eternal love."[97] Even if those in the last group, the true believers, were struck down by the thunderstorm, such a sorrowful path still led to their salvation. Hence Van de Wijnpersse was manifestly opposed to the punishment theory held by those who, out of "most woeful oppression," saw in thunder and lightning only the proofs of God's vengeance.[98] Clearly, given that kind of diversification of his intended audience, the Amsterdam preacher had insensibly drifted away from the old orthodoxy, even if he can hardly be called enlightened yet. Conversely, from then on the development that he and his kind experienced unintentionally classed the unchanging remainder as ultra-orthodox.

That process of individualisation was often closely connected with a similar process of subjectification. After all, believers with a clear conscience could face the thunderstorm without fear; only the hearts of the "wicked" trembled, and quite rightly: "every clap of thunder makes him (...) shudder; he hears in it the voice of the Almighty, who calls him to stern judgement; and in every flash of lightning he considers an ever-burning fire that shall be the reward of his misdeeds."[99] Those preachers and poets who were less inclined to talk of divine judgements dwelt all the more keenly on the gnawing pangs of a troubled conscience.[100] It was not so much thunder that frightened people as their own sin, opined the likes of Hamburg preacher Christian Samuel Ulber (1714-1776). Thus, though he was anything but enlightened yet, Ulber interpreted the traditional theory of punishment as a process in which the sinner projected his fear onto a terrifying natural phenomenon and then wrongly regarded it as a retributory agent.[101] His book was also translated into Dutch.[102] And even if those preachers and poets were content to preserve the idea of punishment, what true believers readily heard or read in their words was something in the nature of a paternal chastisement or trial. That was clear in the case of the poet Rhijnvis Feith (1753-1824), who believed that those who trusted in God heard a Father speak in every clap of thunder and therefore dared to say:

◀ Thunder and lightning above the village of Stroe on the island of Wieringen, 1778, print by Theodorus de Roode, after Pieter Jan van Cuyck

The disasters He sends are bonds of love;
The punishment is a blessing in his hands:
Heavenly joy is its fruit.[103]

God was like a good paterfamilias, who might sometimes scold his children but was really only being cruel to be kind, so thought the editor of a general-interest magazine.[104] Of course, such benign, amending chastisement was no longer likely to arouse cringing fear but childlike obedience instead, posited an almanac poet a few years later, for the Lord was not a tyrant meting out vengeance, but rather the contrary: "A beneficent, mighty God is my faithful Father."[105]

All in all, therefore, the punishment theory had become less self-evident than before, even among orthodox sections of Dutch society. Obviously, that change of mentality was part of a more general autonomous process,[106] but in this connection it was still to some extent a consequence of the invention of the lightning rod. Yesterday's harsh image of God, in which love and mercy were not entirely absent but were not much emphasised compared to the Almighty's ire, was also becoming an increasing ambivalent one in orthodox circles, though it was still far from the unequivocally mild image of God propounded by dissenters and other representatives of the Christian Enlightenment.

In turn, the lightning rod added a new argument to earlier enlightened scepticism of what was regarded as religiosity inspired by fear and therefore insincere. For as we saw above, *L'Observateur hollandois* had already jettisoned the punishment theory and the "superstitious" fear of thunderstorms in 1751 – before the lightning rod had even been invented.[107] Moreover, such critically-minded individuals could now point all the more forcefully to the ill consequences of what they conceived to be an erroneous image of God and man, for it fuelled fatalistic objections to the introduction of that really effective means of prevention, the lightning rod. As the journal *De menschenvriend*, a mouthpiece of the Protestant Enlightenment, argued: "God's love has (...) given man's ingenuity an opportunity to invent the most beneficial means by which we can protect ourselves against lightning." Therefore, "no tower, no tall building (...) should be without a lightning rod; every devastation wrought by lightning in churches or other buildings reproaches us with the most hateful ingratitude with which we scorn the divine goodness." And this was also the tenor of two discourses published by the Maatschappij tot Nut van 't Algemeen ("Society for Public Welfare").[108] In short, the thunderstorm should not lead to a penitential sermon but to the installation of that beneficial apparatus.

Stressing man's responsibility, as enlightened minds did, was entirely consistent with the ubiquitous topical moralism of the age. By spurning the gift of the new invention bestowed by a bountiful Creator upon his rational creature, they argued, fearful souls were guilty of gross negligence. Passive and burdened with prejudices, those all-too-pious people forgot that their prayers should go hand-in-hand with works. Opponents of the lightning rod were therefore increasingly held responsible for deaths caused by thunderstorms, certainly from around 1780.[109] But as early as 1762 Johannes Rossyn had also turned against Nollet and the fatalistic objections to the lightning rod.[110] And – to mention only one later case – at the start of the nineteenth century the Mennonite minister Pieter Wepkes Feenstra (1768-1858) likewise not only rejected the idea that the thunderstorm was God's punishment but also placed the responsibility for lightning strikes expressly on man.[111] To the moderately orthodox Reverend Van de Wijnpersse, that would have been going a tad too far, for although he had no qualms about the lightning rod, he had already warned against over-reliance on it in 1792, since it was unlikely that human forethought could prevent every disaster caused by lightning.[112]

Some of the criticism that representatives of the Christian Enlightenment aimed at the rigidly orthodox individuals mentioned above was that they were guilty of an exaggerated fear, as if it were "a sort of godliness to tremble at the sight of an approaching storm."[113] Perhaps implicitly dismissing Nollet's (supposed) view that warding off lightning by means of a conductor showed a lack of reverence, *De vaderlander* (a moral weekly) posed the rhetorical question: "should a child, an obedient child, sob and weep with fear when he hears the voice of his beneficent Father in Heaven?"[114] According to the enlightened pioneer of the Maatschappij tot Nut van 't Algemeen, Martinus Nieuwenhuyzen (1759-1793), that misplaced piety was often the result of injudicious educators deliberately inculcating the fear that thunder and lightning were expressions of God's wrath. Enlightened opinion often held mothers, grandmothers, and especially pietistic-superstitious servants to blame. In the latter case, social and religious distinctions frequently reinforced each other.[115] Unsound religious nurturing was not the only cause, however; a bad conscience and a weak physical constitution were significant factors as well.[116] But however different the causes, the effects were no less harmful: for man – nota bene the "king of nature"[117] – such a fear was not only unreasonable but also extremely impractical, for it often had such a paralysing effect that he was unable to provide adequate assistance.[118] And last but by no means least, that shrinking anxiety was also unchristian, because in the thunderstorm man did not apprehend in reverence and awe the greatness and goodness of the Creator but only trembled before that natural phenomenon[119] – a conclusion with which the more orthodox also wholeheartedly agreed, because they interpreted that fear as the consequence of a lack of trust in God.

▶ Faith fends off Fanaticism, print by Jan Punt, after Jean Michel Moreau, 1777. Faith pushes back the blindfolded Fanaticism from Charity. Fanaticism is represented as a male figure armed with a dagger, a flaming torch, and chains. Charity, on the other hand, is a defenceless woman who has fallen to the ground

It was their very condemnation of the rigidly orthodox that characterised the representatives of the Christian Enlightenment in the Netherlands. They were often more thoroughgoing in that criticism than they were in censuring the representatives of the more radical variants of the Enlightenment, most likely because the latter were simply less numerous in the Dutch Republic and perhaps also because their arguments were inherently more difficult to refute. The supporters of the Christian Enlightenment preferred to fight by indirect means: by accusing the rigidly orthodox of disgracing the true religion with their "superstition" and "prejudice", enlightened Christian authors hoped to stymie more extreme criticism of religion. In practice they left grappling with the more radical forms of Enlightenment to the representatives of (moderate) orthodoxy. As time went on this increasing polarisation was further encouraged by the ideas of the emerging Réveil, a religious revival and anti-modernist movement of the first half of the nineteenth century.

Finally, these patterns are well illustrated by the varying opinions about Franklin in the Netherlands. For instance, a versified tale about the "enemies of the light" ridicules a "zealot" who thinks it a sin to light a candle at nightfall and thus rebuke nature. Understandably, this inflexibly orthodox character has no truck with the invention of the "impudent Quaker orator" Franklin, who with his iron rods insolently argues with God.[120] Christian-enlightened writers, on the other hand, preferred to entrust resis-

◀ Portrait of Johannes Immerzeel, print by Philippus Velijn, after Hendrik Breukelaar, 1797-1828

tance to the secularising consequences of man's proclaimed autonomy to authors more orthodox than themselves. For example, no less a light than the poet Hieronymus van Alphen (1746-1803) dashed off a critical evaluation of Turgot's well-known epigram in his *Christelyke spectator*. That the famous American had wrested the lightning from heaven with his invention was, he wrote, "boasting, conceited boasting," all the more so since, in his opinion, the lightning rod's usefulness was still a matter of debate.[121] Rather than be guilty of such a metaphysical revolt, the poet preferred a thorough reformulation of the epigram, applying it to Christ, who had wrested the wretched from death and his sceptre from the devil.[122] Those who boasted of the makeability of life were soon at risk of secular conceit. In orthodox eyes this was clearly demonstrated by the rather categorical statement of a collective of prominent citizens in 1796. In response to the astonishing inventions of modern times, like lightning rods and hot air balloons, these people concluded that, "one will almost have to say man can do anything if he only wants to." Their opponents would have been delighted to note that those Christian-enlightened folk too were apparently so shocked by the secularising implications of their pertinent remark that they promptly censored themselves by adding, "or rather, God has enabled him to do anything, if he but chooses to make use of the powers granted to him."[123] Rhijnvis Feith also disapproved of what he saw as man's hubris. He was particularly annoyed by the profane drinking song that his fellow poet Johannes Immerzeel (1776-1841) had composed in loose imitation of the German writer Gotthold Ephraim Lessing:

It thunders! – Friends! empty your glasses;
Drink merrily! let the hurricanes rage!
We are safest taking shelter in wine
From the awesome vengeance of god Jupiter,
When his rough thunder claps:
Drink up – Why are you trembling like a reed!
Drink up – he blazes along the waters,
He does not blaze in his nectar.[124]

But in 1819, the year that this verse appeared, the separation of minds had largely taken place. A few years later, Abraham Capadose (1795-1874), who was part of the Réveil circle, revived old-fashioned, unadulterated fatalism when he remarked that the lightning rod had a diabolical character because, in his terrible pride, sinful man boasted of disarming God's ire with an iron wire.[125] It was not until 1869 that another, more moderate follower of the Réveil, the preacher-poet Jan Jakob Lodewijk ten Kate (1819-1889), made an attempt to bridge the gap that had arisen: without regard for secularising implications, the Dutch divine unconcernedly sang the praises of the "pious" Franklin, considering his invention the answer to a prayer and even giving the lightning rod a religious sanction by comparing it to the bronze serpent that Moses set up in the wilderness.[126] Was he aiming to eliminate objections to the lightning rod with a distant variant of what the anthropologist Clifford Geertz (1926-2006) has called "local knowledge"?[127] In any case, as far as Ten Kate himself was concerned the conductor was by now so well-established that in a certain sense he could give it a biblical legitimacy:

Yet that venture was no shameful delusion,
Feverishly arisen in the brain of a Titan.
Thinking and trying,
Study and struggle,
Sleepless nights,
Ripened it, prayers sanctified it.
Behold, there he is, the Discoverer! His eye
Is looking up:
"God of Nature! O crown my desire!
Thou who knowest, no selfish wish,
The salvation of Mankind,
Stimulates my striving." – And the deathly pallor of the cheeks
Is replaced by the burning glow of enthusiasm.
"Hither the rod!"
And there it appears – like the Brazen Serpent
For those who believe
Token of salvation – on the Temple above! (...)
There lives a God who rewards heroism:
F r a n k l i n ! your name is crowned forever.[128]

In Ten Kate, who, in that same poem, rejected the punishment theory and here and elsewhere always sought the ecclesiastical middle way, we can, despite the wholesale polarisation of the nineteenth century, finally recognise the centripetal tendency that had also been so typical of Van de Wijnpersse and moderate orthodoxy: each tried in his own way to adjust to the "spirit of the age". That was also true, mutatis mutandis, of the representatives of the Christian Enlightenment, who undertook a similar search for the right middle path, although the scale of their undertaking was considerably greater. It often led to a battle on two fronts – with the radical Enlightenment and ecclesiastical orthodoxy. Amid all the divisions, however, there was one movement in which birds of all feathers could apparently flock together almost without effort: physico-theology. That movement, which in the eighteenth century was also extremely popular among those who occupied themselves with thunderstorms, thus deserves separate attention.

Physico-Theology: The Netherlands

The aim of physico-theology was not to mock the thunderstorm but also not to be cowed by it. It provided a third path, so to speak, via reverence for the goodness, wisdom, and all-powerfulness of the Creator evinced in that impressive natural phenomenon.[129] Its most conservative representatives – first and foremost certain pastors and poets – were inclined to put a rather one-sided emphasis on God's omnipotence, and consequently to regard formidable natural phenomena primarily as evidence of the (sometimes even punitive) majesty of the Lord.[130] In their case this somewhat narrow view, and particularly the motivation from which it arose, gave physico-theology a rather different character. For many other adherents, however, physico-theological ideas involved a more munificent image of God (as will become clear). The inevitable result of all that was that physico-theology had something of a dual nature in both content and following.

While it might not be entirely accurate to describe the poet Hubert Kornelisz Poot (1689-1733) as a physico-theologian, to his receptive ear the thunderstorm pealed God's praises and testified to the indisputable power of the Most High.[131] The Lord was also a formidable force to the Reverend Engelberts of Hoorn, who insisted that man could never flee "the omnipresent Supreme Being" and his thunder, even though "the All-Governing [can] bring forth light from the darkness and turn the worst accidents into blessings. Thus does he instil awe and reverence for his formidable Majesty in the minds of reckless mortals."[132] Inculcating a trembling awe of God's splendour in the thunderstorm in the puny earthworm that was man – that was what those physico-theo-

▶ Portrait of the Reverend E.M. Engelberts of Hoorn, painting by Jacobus Buijs, 1768

logians were aiming at. The epistolatory novel *Het land*, for example, by the Dutch poet and author Elisabeth Maria Post (1755-1812) included a hymn whose entire emphasis was on God's majesty, thus stressing the need for constant recognition of the Lord's gracious mercy.[133] Many other poets also turned a hand to physico-theological verse, such as the anonymous bard whose piece was published in 1800 in the *Nieuwe vaderlandsche bibliotheek*: it lauded the Almighty as "Cloud-gatherer, welkin-darkener and earth-ruler."[134] The line between awe and fear was sometimes quite thin. To conservative physico-theologians of this kind, fear was by no means an improper response. At most, fear of God put fear of the thunderstorm in perspective, for it was his very omnipotence that enabled the Lord to bring a merciful halt to the violence of nature.

That it should be a freemason, Isaac van Haastert (1753-1834), who elaborated on that last point – in a poem on nature in which he professes that thunderclouds were also curbed at the all-encompassing command of our "faithful Father" – is no coincidence.[135] Obviously, the character the devout view of nature assumed was invariably dependent on the image of God with which it was associated. As has already been mentioned, to the traditional believer, God's omnipotence often functioned as an instrument of punishment for the inveterate sinner who should quake at the thought of the Last Judgement.[136] But for those who knew that during a thunderstorm they were in the gentle hands of the Almighty, the case was decidedly different: fear was superseded by faith in God. In a meditative exposition on clouds written in 1772, *De denker*, a moral weekly, sought to see the omnipotence, wisdom, and goodness of the Lord as pertaining to the same entity:

> *What is (...) more wonderful than to see a hundred blazing Arrows of lightning streak from the dark clouds, now up, then down, or to the sides, in the darkened celestial-blue sky, and instantaneously disappear into it? (...) Lift up thine eyes on high and behold, who hath created these things (...) what can be more pleasing than to see at work in my presence Him whose Wisdom and Goodness to me is as limitless as his Omnipotence?*[137]

This was the path followed by the representatives of mainstream physico-theology in the Netherlands and elsewhere: the Reverend Martinet, for example, counselled against being "insensible" but neither should one be "cringingly" fearful.[138] And the aforementioned Quedlinburg minister Johann Goeze, whose reflections found their way into a Dutch journal, also had this explicit recommendation: "Let us take the middle way," he proposed, "being not childishly afraid during thunderstorms, but also not thoughtlessly mocking."[139] As *De menschenvriend* pointed out in 1791, we might be seized by a godly trembling when we saw a thunderstorm burst upon us, and a voice within us might cry with the psalmist "the God of glory thunders!" Yet still God was love, and he remained so even in the midst of thunder and lightning.[140]

It was not only God's omnipotence and goodness that were evinced in that natural phenomenon, however, but also his wisdom. In the optimistic, harmonious worldview of physico-theology, everything had purpose and utility – the meanwhile amenable thunderstorm included. And in any case, rational man was able to discover the wise intentions of the Creator by an attentive study of nature. To the physico-theologian it was precisely this teleology that formed a "reasonable" (for which, read accessible and powerful) argument for the existence and providence of the Creator when faced with unbelieving "mockers". For if physico-theology readily distanced itself from the old orthodoxy through its image of man and God, its apologetic tendencies also kept more radical enlightened minds at arms length.

The thunderstorm was useful, decidedly useful. It was a mantra that physico-theologians never tired of repeating. To start with, there was its cleansing action in the atmosphere, purifying the human circulation[141] and killing germs. God's superlative wisdom and goodness radiated in and through it, a moral weekly affirmed in 1775 for instance.[142] Another journal from the same year also assured its readers that as a result of a thunderstorm "all those unhealthy sulphurous and other vapours that we inhale are removed from the air to some extent," thus preventing the outbreak of all kinds of deadly diseases.[143] For the earth, on the other hand, the huge quantity of "oily, saline, sulphurous particles" that were dissolved in the rain was beneficial for it acted as a fertiliser, the more so since thunder had first made the earth receptive by turning it up with its tremors. "Are these not benefactions," *De menschenvriend* enquired rhetorically, "Is this not matter for which to glorify him who blesses us with thunder and lightning?"[144] Conversely, a reasoned, apologetic argument cautioning man against his ungrateful meddling was equally apropos, according to one poet:

May the fuel barn so full of matter garnered by the Sun's fire
Not then winnow those strange mixtures,
Not refine, purify, and prepare,
To pour them like a fertile liquid upon this Earth?[145]

As the soil was being fertilised, and in addition to the cleansing already mentioned, the air was also refreshed,[146] a relief which became a favourite theme in many poems of the time (as we shall see later on). In the refreshed air, observed Hieronymus van Alphen, the birds begin to sing again and fields and fruits gleam with a new brilliance.[147] The consequences were beneficial not only for nature but also for man, another poet noted, thus underlining the often anthropocentric character of physico-theology: "Oh how the herbs smell now, and the flowers in the field! It seems to be delicate airs full of incense, the scent of balsam, ordered for me."[148] In the freshly budding foliage and the new flowers that opened in the fields, it was clear that even in its most terrible manifestations, nature was ever benevolent and "always the work of him who is perfect goodness and wisdom."[149] In short, praise to the Lord resounded in everything.[150]

N° 23.
Doopsgezinde Kamper-Eilanders, op weg naar de Vermaning, by een opkomend onweder.
"De donder bromt Gods naam ter eere!"
Mennonites de l'île de Campen, Se rendant à l'office religieux, à l'approche d'un orage.
"Il publie aux mortels la gloire du Très-Haut."
Amsterdam, by E. Maaskamp, Nieuwendyk N° 251.

◀ Mennonites from the island of Kampen, on their way to the *vermaning* or religious meetinghouse, as a storm approaches, engraving by P.H.L. van der Meulen, after L. Seidel, 1804-1809: "The thunder rumbles in honour of God's name!"

In fourth and final place, lightning was also beneficial because it killed the larvae of all kinds of harmful insects – the great Van Musschenbroek himself had voiced this surmise.[151] The Reverend Martinet was a tad more assertive, concluding that, "We know (...) that nothing better rids our gardens and fields of harmful caterpillars, which, it seems, cannot bear its flicker (...) or better restores the whole of Nature (...) than the thunderstorm."[152] In other words, rain, wind, thunder, lightning, and so forth were "so many fine examples of the Creator's unfathomed Wisdom, his incalculable Goodness, and his astonishing Omnipotence, and Order."[153] In the end, the thunderstorm's advantages far outweighed its disadvantages, the reverend assured his readers.[154] On balance, it resulted not in destruction but in re-creation, so too stressed the poet Willem de Perponcher, and never did nature bestow more benefits on mankind than in this phenomenon, which was so unjustly, nay, so ungratefully called "bad" weather.[155] Thanks to the munificent powers that the provident Creator had built into nature for its recovery,[156] the thunderstorm became an unequivocal benefit for man and animal: indeed, exulted Rhijnvis Feith, everything dripped with blessings.[157] Could this peaceful and harmonious result of seemingly contradictory tendencies be anything but the work of one and the same divine hand?[158] The conclusion, therefore, was that both the doubt of freethinkers and the fears of those of little faith must finally give way to pure praise:

> *Such abject cringing! Worship the Lord;*
> *Thank Him for this blessed weather,*
> *Which cools the hot air;*
> *Washes the impurity out of the atmosphere;*
> *Furnishes refreshment to man and beast,*
> *And the thirsty earth slakes and drenches.*[159]

In the discourse of this kind of popular Enlightenment (*Volksaufklärung*), nature no longer posed a threat but was the result of God's marvellous work of creation, which could be understood with reason and increasingly also with feeling, for the study of it could only prompt admiration and amazement in puny man. To a certain extent the invention of the lightning rod not only helped to controvert the theory of divine punishment but also furthered the development of physico-theology. It is why the editor of *De menschenvriend* hastened to reassure his readers that the use of the rod did not detract from the beneficial effects of lightning.[160] God, man, and nature all worked in unison. Anything that threatened to disrupt that Christian-enlightened harmony could count on formidable resistance.

Physico-Theology: Europe

The physico-theological ideas in circulation in the Dutch Republic were shared by many other authors elsewhere. Here, a few examples will suffice. For instance, in his well-known book *Physico-Theology* William Derham (1657-1735), an English clergyman and natural philosopher who was always ready for teleological arguments, cited the Swedish Catholic scholar Olaus Magnus (1490-1557) as pointing out that God sometimes shattered rocks by means of lightning strikes to expose the veins of silver in them.[161] In that same early period, in 1698, the Swiss physicist Salomon Hottinger (1649-1713) even published a complete *Brontologia physico-sacra*,[162] paving the way for what, in the eighteenth century, would become a special branch of physico-theology known as "brontotheology", the theology of thunder.[163] Its best-known representative is undoubtedly Peter Ahlwardt (1710-1791), professor of philosophy at the University of Greifswald, who needed more than four hundred pages to outline God's wise intentions with the thunderstorm and whose work reflects the aforementioned change in mentality from the traditional *Wetterpredigt* to Christian-enlightened physico-theology.[164]

◀ Portrait of Peter Ahlwardt, professor of philosophy at the University of Greifswald, engraving by Johann Martin Bernigeroth, 1730-1747

It was not only theologians who experienced this kind of process; poets did too. For many of them, such as the German poet and physico-theologian Barthold Heinrich Brockes (1680-1747), according to his "Gesang zur Zeit des Ungewitters",[165] the thunderstorm was no longer an expression of God's wrath. To be fearful of it was therefore fundamentally wrong and easily led to all kinds of superstition instead of the recognition of the Creator's wisdom, as the Swiss physician and poet Albrecht von Haller (1708-1777) agreed.[166] In the second half of the eighteenth century, thunder and lightning often still gave rise to pious reflections, as in the case of the great Friedrich Gottlieb Klopstock (1724-1803). In his hymn *Die Frühlingsfeier*, composed in 1759, it was not punishment he saw approaching but the heavenly Father's love and goodness. Despite the depiction of God's formidable omnipotence in the storm, the threat it once posed had largely disappeared.[167] Poets inspired by physico-theology likewise rarely dealt with eschatological matters; if they did, it was usually not in reference to the Last Judgement but to the future glory of man and the world that present nature might already foreshadow.[168] Apologetics was a genre more suited to that kind of Christian-enlightened author, although towards the turn of the century, feeling was starting to be regarded as more important than reason. That tendency can be observed in Goethe too, for although he admitted that teleological proof of God no longer stood up to critical reason, things were different, he thought, for the religious soul. Might we not see in the storm the imminence of a mighty Majesty, and then, in the perfume of flowers and the soft soughing of the wind, the coming of a loving Being?[169] Perhaps the rationalistic defence of religion, which at times seemed aimed at exorcising the demon of doubt with stock physico-theological arguments, had meanwhile overplayed its hand.

▲ Portrait of Barthold Heinrich Brockes, poet and physico-theologian, painting by Dominicus van der Smissen, second quarter of the eighteenth century

▲ Portrait of the poet Friedrich Gottlieb Klopstock, painting by Jens Juel, ca. 1779

Whatever the case may be, although physico-theology was still a primarily Protestant affair, eighteenth-century Catholics also had something to say on the topic.[170] One of the most popular works of the time was by the French priest and teacher Noël-Antoine Pluche (1688-1761), whose well-known *Spectacle de la nature* was intended as a tribute to the Creator and his wise providence. For thunder and lightning were no less the work of God than flowers and fruit, and violent storms obeyed the Lord just as much as gentle breezes did. However diverse in nature, everything in creation sang his praises. Furthermore, the thunderstorm was beneficial both morally and physically. For not only did it admonish those who abused God's mercies, it also dispelled the suffocating heat of a scorching summer's day. At the same time it got rid of an excess of insects and put an end to a potential drought.[171] If, thanks to the Abbé Pluche, the "irregular" had thus gained a tidy place in a well-regulated universe, that was still more the case after the invention of the lightning rod. Now there was even less cause for fear. As the learned cleric Giovanni Battista Beccaria often stressed, the forces of nature were also subject to the power, goodness, and wisdom of the Lord. He therefore assumed that a properly educated priest would reassure his parishioners and teach them to "admire the immensity of nature's operations and its Creator's omnipotence."[172] In short, reverence, wonder, and utilitarianism went hand in hand with a "sensible" and "pleasurable" scientific practice.[173] The idea that the introduction of the lightning rod would have led to the erosion of the physico-theological link between religion and science is therefore incorrect.[174]

◀ Portrait of the French priest and teacher Noël-Antoine Pluche, print by Louis Jacques Cathelin, after Nicholas Blakey, ca. 1772

However, (as mentioned earlier) physico-theology was often coupled with the pursuit of *Volksaufklärung* or popular Enlightenment, which had an equally interdenominational character. Enlightened Protestants and Catholics both wanted to liberate simple folk from ignorance, fear, and "superstition" by good instruction – an aim that categorically targeted children too.[175] For in enlightened eyes, fear of the thunderstorm was not only evidence of a wrong understanding of God but also the chief cause of the origin and proliferation of all kinds of "superstitious" ideas and practices. We shall see below just what a help they thought the lightning rod would prove.

▲ The English missionary Boniface (ca. 672–754) fells the "Donar oak" near Geismar around 723, to demonstrate to the Germanic tribes the powerlessness of the thunder god to whom the tree was dedicated. German school poster, ca. 1900. Boniface probably delegated the actual tree-felling. According to tradition, its wood was used to build a church

5

Marginal and Marginalised Religious Reactions

Enlightened Protestants and Catholics condemned the old (and perilous!) custom of ringing consecrated church bells to avert an approaching thunderstorm. If the lightning rod was the icon of the Enlightenment, they asserted, apotropaic bell-ringing was its unenlightened antithesis. In the Middle Ages it was believed that thunderstorms were caused by devils. The Church's thunderstorm demonology merged with the ancient pagan conviction that spirits could be repelled by loud noises – by what would be described today as sympathetic magic. In the Netherlands and parts of Germany and Austria no figure was more popular than the "thunder saint", Donatus. But the way his devotional material was used sometimes went well beyond what was ecclesiastically proper. On the other hand, in the century of Enlightenment age-old ritualism not infrequently gave way to processes of allegorisation and spiritualisation. Popular beliefs flourished in the countryside, where demons were kept at bay by safeguarding the threshold and hearth with sharp metal objects and lightning was repelled by particular plants, birds' nests, and carcasses.

Thunderstorms: The Devil's Work?[1]

Weather Prophet

Three clouds in the sky,
What should that mean?
The sexton should go home,
He should ring the weather bell![2]

◀ Allegory on the consequences of superstition and hypocrisy, print by Reinier Vinkeles, after a design by Cornelis Ploos van Amstel, 1770. The personifications of Nature and Reason are thrust aside by blindfolded Superstition, who brandishes a dagger and torch. Behind him and pushing him on is Hypocrisy, hiding her face behind a mask

It will be evident by now that in the first instance the focus of this book is on the reception of the lightning rod in the Netherlands, where the Reformation found fertile ground. Almost inevitably, therefore, the previous pages have had a rather Protestant slant. But that bias is chiefly due to the asymmetrical availability of sources. At the time, the Dutch Reformed Church was the only officially permitted religious institution in the Republic. Catholicism was tolerated practically and with difficulty at best, hence the Dutch Catholic community, albeit a large minority, was fairly marginalised. Catholic reactions to the new invention – even those endorsed by Rome – thus found very little outlet in the Republic. Nevertheless they are certainly important, so if we want to look at Catholic responses in general we must turn to sources and literature produced elsewhere. This we shall do a little later on. Matters become yet more complex because even after the stricter decrees issued by the Council of Trent the Roman Catholic Church still took a more lenient attitude to popular devotion, which often derived from a belief in the devil or the veneration of local saints. The Protestant denominations, on the other hand, categorically condemned popish "insolences" such as processions and consecration rites. As will become apparent, for all their personal reservations even enlightened Catholics tended to incorporate popular belief in a somewhat modified form rather than reject it outright on the basis of a rigid Protestant zeal for truth.

However, that distinction no way detracts from the interconfessional character of the pursuit of popular Enlightenment. In many respects enlightened Protestants and Catholics held similar views and took equal issue with what they regarded as "superstition" among their less educated co-religionists. In this chapter we shall examine three themes that are not always clearly distinguishable: the almost ubiquitous belief in the devil, the veneration of saints, and, finally, nature as the hook on which popular beliefs were hung.

Enlightened representatives of both denominations unanimously and steadfastly condemned the old custom of ringing consecrated church bells to avert an approaching storm, a particular source of irritation because of its "superstitious" nature. If the lightning rod was the icon of the Enlightenment, then *Wetterläuten* or apotropaic bell-ringing was, in their eyes, its unenlightened antithesis.[3]

▶ As the farmers of Lapeyre in the department of Aveyron trembled for their harvest during a heavy thunderstorm, a certain Pierre Boudes climbed the bell tower. Shortly afterwards the weather-ringing suddenly stopped: Boudes had been struck by lightning, illustration from *Le Petit Journal, Supplément illustré*, 11 September 1910

Le Petit Journal

ADMINISTRATION
61, RUE LAFAYETTE, 61

Les manuscrits ne sont pas rendus

On s'abonne sans frais
dans tous les bureaux de poste

5 CENT. SUPPLÉMENT ILLUSTRÉ 5 CENT.

21me Année — Numéro 1.074

DIMANCHE 11 SEPTÈMBRE 1910

ABONNEMENTS

SEINE et SEINE-ET-OISE.. 2 fr. 3 fr. 50
DÉPARTEMENTS.......... 2 fr. 4 fr. »
ÉTRANGER.............. 2 50 5 fr. »

UN SONNEUR DE CLOCHES FOUDROYÉ

The generalisation of such a simple dichotomy, or, as the influential liberal Protestant historian Andrew Dickson White (1832-1918) famously phrased it, "the warfare of science with theology in Christendom," would certainly have gone too far for the representatives of the Christian Enlightenment, however. For while there are indications that the Church itself instituted the custom of bell-ringing[4] and in any case certainly approved of it,[5] enlightened Catholics of the eighteenth century could point to the fact that from an early date the Church had combated the abuses associated with the tradition and had tried hard to make the distinction between belief and superstition. Indeed, in this respect representatives of the Catholic Enlightenment could refer to the writings of the Italian Jesuit Robert Bellarmine (1542-1621), a leading Counter-Reformation theologian and one of the few Catholics of his generation to question the utility of ringing consecrated bells to ward off thunderstorms.[6] In short, the dividing line between belief and "superstition" was somewhat erratic.

On the other hand, White's "conflict thesis" (sometimes represented as more irreconcilable than its author intended it to be[7]) no longer commands much support in current historiography. A scientific approach simply requires more distance and nuance. In these pages, therefore, "superstition" can be read simply as "elements of popular religion unsanctioned by the Church."[8] We shall see below that the multifaceted phenomenon of bell-ringing was widespread in medieval and early modern Europe, even though it is not always possible to tell from the available (and often somewhat older) literature where and when the actual examples were from.

One thing, at any rate, seems clear: ultimately the origin of the custom goes back to Paul's Letter to the Ephesians, in which allusion is made to "the ruler of the power of the air" (Eph. 2:2) and "the spiritual forces of evil in the heavenly places" (Eph. 6:12) against which the believer must contend.[9] Later – according to some at least since the time of Charlemagne[10] but more obviously since Thomas Aquinas,[11] the late Middle Ages, and the Catholic Reformation with its elaborate demonology and angelology – the idea gained ground that it was often devils who (with God's permission) caused a thunderstorm.[12] Notwithstanding all their opposition to bell-ringing, that was also the view of the leaders of the Protestant Reformation.[13] And although the controversial minister Balthasar Bekker (1634-1698) had stoutly defended the view that the devil was no longer able to carry out his works on earth since he was safely locked up in hell, even at the end of the enlightened century many people – Protestant and Catholic, learned and unlearned – were by no means convinced of that. The Amsterdam clergyman and professor Jacob van Nuys Klinkenberg (1744-1817), for example, in his major apologetic work *Onderwys in den godsdienst*, explicitly challenged Bekker's ideas. He was prepared to admit that devils could not do miracles, yet he also persisted in his belief in an age-old tradition: "They seem to influence the regulation of the weather," he wrote in 1788, though "not without God's permission, and subordinate to his infinite supremacy."[14] Many others had fewer mental reservations and were adamant that thunder was the work of the devil. That view was still prevalent (at least according to an enlightened Calvinist minister) around 1800 in the region around 's-Hertogenbosch in North Brabant, for example.[15]

▶ Portrait of Jacob van Nuys Klinkenberg, Amsterdam clergyman and professor, mezzotint by Frederik Christiaan Bierweiler, early nineteenth century

In the Roman Catholic tradition, thunderstorm demonology had become associated with the ancient pagan idea that spirits could be repelled and driven out by loud noise, a case of what a later generation would describe as sympathetic or imitative magic, which is to say that what is desired is attained not by propitiating a deity or demon but by carrying out rituals. These practices use objects or actions resembling or symbolically associated with the event or being over which influence is sought.[16] After all, clamour has long been credited with a deterrent or even an invocative effect.[17] Was it an expression of the primordial urge to create a din as a means of drowning out one's own fear?[18] The deafening racket made by dairy farmers in Styria during thunderstorms certainly suggests that: by yelling and roaring and making as much noise as possible with cow bells, chains, and pans, they tried to chase away the malignant powers.[19] In France and elsewhere, pots and pans were beaten, wine casks were pounded, and sickles were clattered to avert storms.[20] Likewise in Großarl in Salzburgerland, whose inhabitants cracked whips and drummed as hard as possible on milk churns with consecrated sticks during thunderstorms besides ringing all kinds of bells.[21] Around Aachen the preference was for earthenware horns that were blown loudly during thunderstorms. During the great septennial Marian pilgrimages to the Imperial city when the cathedral's most sacred relics were displayed, so many pilgrims' horns were blown with such deafening volume that "one would not have heard thunder the dear God," as the saying went. Those self-same instruments were later repurposed as *Wetterhörner* ("thunder horns") and blown to drive out the demons.[22]

For sheer volume of sound in any town or village nothing rivalled the church bell. In many European countries Catholics gradually came to believe that the ringing of such a sacred bronze colossus could deter the demons responsible for thunder and lightning. The French priest Jean-Baptiste Thiers (1636-1703), for example, was convinced of the salutary effect of the *trompettes du Roi éternel*.[23] The bell was considered to be the New Testament perpetuation of the Old Testament trumpets.[24] To be effective as a repellent the bell had to have been properly consecrated: to ring it was to call on the power its hallowing had conferred on it for the protection of hearth and home.[25]

During their consecration the bells were not only blessed but also exorcised with powerful incantations and psalms, then washed ("baptised") with holy water and salt and anointed with holy oil – a procedure described step-by-step in the *Pontificale Romanum* published in Brussels in 1735.[26] That kind of bell-blessing was also performed in pre-Reformation England.[27] In France bells sometimes received a double blessing, first of the metal prior to casting, then of the finished bell, before it was hung in the bell tower.[28] That final consecration ritual was described in particular detail and was a lengthy business: it included an exorcism and a fine selection of prayers – the optimal method of discouraging demons – as well as making the sign of the cross and reading Bible passages, particularly from Luke's Gospel and the Psalms.[29] Bell-blessings often resembled a christening ceremony, which according to Catholic doctrine also drives out devils.[30] The suggestion of baptism was reinforced by the naming of the bell, often after the patron saint of the church in which it was to hang, and sometimes it also received an animal nickname based on its timbre.[31] Occasionally the parallel was taken so far that in parts of France[32] and Germany[33] the baptised bell even had godparents!

The official ritual was performed by the bishop or his representative. But either before or after the service, various elements of popular religious culture unapproved by the Church were added. Some seem to be little more than legendary. For instance, thanks in part to the German novelist and poet Theodor Fontane (1819-1898) we know of tales from several places in the March of Brandenburg that tell how a snake – that satanic serpent of Eden – was cast into the metal being melted down for the bell, so that its ringing would not only banish thunderstorm demons as far as its sound reached but snakes as well.[34] Given the relatively low risk of a deadly snakebite in northern and central Europe, however, it seems more plausible that the zigzagging reptile should be interpreted as a symbol of lightning itself – an ancient and widely held view. In Upper

▶ *The Fall of the Rebel Angels* (cf. Revelation 12:9), watercolour by Edward Dayes, 1798

Lusatia and Bohemia, for instance, *Blitzschlangen* ("lightning snakes") were attached to the outside of houses to act as a repellent. And, to cite quite another example, the Hopi of North America also symbolised lightning in the form of a snake.[35] Which leads one to wonder whether Jan Jakob Lodewijk ten Kate's aforementioned comparison between the lightning rod and the bronze serpent might not incorporate a (sub)conscious reminiscence of that old popular belief?[36]

A *Blitzschlange* or "lightning snake" attached to the gable of a house in Upper Lusatia and Bohemia to prevent lightning strikes. The lightning snake pictured here is on the front of the house at number 4 Röntgenstraße in Ebersbach (Saxony)

Be that as it may, the reports that relics or crosses were also sometimes added to the molten metal from which the bell was cast seem less apocryphal.[37] Equipped with such sacred amplification the bell was considered highly efficacious against thunderstorms in Bavaria, for example.[38] Such bells were soon referred to in popular parlance as *hochgeweiht* ("highly consecrated").[39] Indeed, the best bells began to ring by themselves during thunderstorms, some thought; such stories found willing credence in popular belief.[40] Even its size could give a bell high status, for the bigger the bell and the louder its peal, the greater its capacity for demon-dispelling, so it was said.[41] We find similar thinking in the French village of Sennely, to the south-east of Orléans, where such large bells were installed that they threatened to bring the belfry down.[42] But the demons dispersed by church bells had to go somewhere. If the place with the biggest bell suffered least from thunderstorms by chasing them into neighbouring villages, it is hardly surprising that conflicts arose – in Bavaria and elsewhere.[43] In an often fierce competition, each village tried to acquire the largest bell in the area, having it cast locally if possible, of course, but more often by buying or, if necessary, stealing it.[44] In the end the solution was often an agreement to ring all the bells in the area at same time during thunderstorms, so that the demons fled the entire region.[45]

It will be clear by now that for many believers not all church bells were equal, either in sacral terms or in their supposed level of effectiveness in driving away thunderstorm demons. In the latter respect, moreover, a second distinction was often made between "ordinary" priests and priests who were especially adept at warding off thunder and lightning. Such special clergy – often assistant curates[46] – were popularly known as *wettergerecht* (which is to say particularly capable of exorcising thunderstorm demons), regardless of whether they were born with that gift or had acquired it through prayer or study. The title was granted to both secular and regular clergy (particularly Capuchins) and there could be many advantages to securing their services. For although the Church refused to acknowledge any such distinction, parishioners continued to insist on the appointment of *wettergerechte* priests, believing that they could give the village lasting protection from thunderstorms.[47] Sometimes, therefore, villages would not only compete for the best bell but also for the appointment of a *wettergerechte* priest, although there was a risk that if every village in the region had such a cleric, the thunderstorm would have nowhere to go.[48]

The importance of *wettergerechte* clerics was all the greater because it related to a third, more general distinction (which has been continuously assumed in the foregoing) between thunderstorms unleashed (with God's permission) by demons – by far the greatest and most dangerous category – and thunderstorms directly ordained by God. The special spiritual powers attributed to *wettergerechte* clerics made them pre-eminently suited to exorcising the *gemachte Wetter* – the thunderstorms caused by demonic malice. Such diabolic works demanded the deployment of all the supernatural means available, ecclesiastically sanctioned or not. It was in this case that bell-ringing was appropriate,[49] for on no account should those thunderstorms be allowed to run their course.[50]

But demons did not bear all the responsibility for thunderstorms or other kinds of meteorological misery. They could also be caused by evil people – even malevolent clerics.[51] In such cases folk tradition whispered of *Hexenwetter* ("witches' thunderstorms"),[52] since given the sheer speed with which these tempests arose they could only be the result of dealings with demons. More than any other evil, this sort of sorcery must be routed by the ringing of a consecrated bell,[53] a practice known as *Hexenläuten* ("witch-ringing"). In 1623, the archbishop of Salzburg, Paris von Lodron (1586-1653), had instituted the *Johannisläuten* or St John's peal, a loud relay of the city's church bells that lasted throughout St John's Eve – the night of 23 to 24 June. According to the archbishop, on St John's Eve – which with only a couple of days difference harks back to the night of the summer solstice, dedicated to Donar, the god of thunder and lightning in pre-Christian times – all kinds of occult practices occurred, resulting in thunderstorms and other troubles. That form of continuous *Hexenläuten* was still going strong in 1746, well over a century later – both in Salzburg[54] and the aforementioned village of Sennely in central France.[55] Just how effective the bell-ringing could be was evidenced by the missing pieces around the bells' rims, bitten off by enraged witches[56] – damage that later generations would attribute to the effects of prolonged or excessively enthusiastic ringing. Witches tried in all sorts of ways to rob the consecrated church bells of their sound and thus of their power, disabling them by spiking, for example, or undermining the bell tower.[57]

Since the late thirteenth century the devil-exorcising action attributed to consecrated bells had been proclaimed by their inscriptions: *Vivos voco, mortuos plango, fulgura frango* ("I call the living, I mourn the dead, I repel the lightnings") for instance; the words running around a bell cast in 1486 for the minster church in Schaffhausen and made famous through Friedrich Schiller's (1759-1805) *Lied von der Glocke* of 1799.[58] But other inscriptions also testified to the exorcism these instruments were thought to perform. A bell in the small Hessian town of Butzbach unadornedly states: *Est mea vox bambam, potens repellere Satan* ("My voice is bambam, powerful in repelling Satan").[59] In Seekirchen, just north of Salzburg, the oldest bell (1671) is inscribed in the vernacular: *Zu Ehren Gottes und des heiligen Apostels, Johannes und Jakobus, der Söhne des Donners* ("In honour of God and the Holy Apostles, St John and St James, the sons of thunder").[60] Sometimes, in another instance of sympathetic magic thinking, images of lightning were included, as in the case of a bell from Biberach in southern Germany, cast in 1777.[61] In some places the addition of devil-dispelling inscriptions continued for a long time: the latest known example – a new bell made for the church of Höchenschwand in the south of the Black Forest which bears the *Vivos voco* inscription – actually dates from 1825.[62] Even in the second half of the nineteenth century, bells were cast in France that still had inscriptions emphasising their apotropaic function.[63] That the inscriptions were often in Latin, the language of the Church, does indicate that the views on demon-dispelling bell-ringing should not simply be dismissed as popular belief.[64]

In spite of all the ritual precautions, however, there was still any number of ways in which things could go wrong. For while many might believe that the sound of a properly consecrated church bell could put the demonic legions to flight and even force them to hide deep in the mountains,[65] bells that had not been sanctified were vulnerable, and easy prey for the devil.[66] And what of the bell-ringers? Most certainly they too were vulnerable. Believing themselves to be the devil's particular target, some of them wore sacred clothing: a surplice with a priest's stole over it.[67] It was even more important that the sexton and the other bell-ringers carried out their duties with precision. And, in the case of *gemachte Wetter*, that they started in time, preferably while the sun was still shining.[68] In 1729 in Asselt near Roermond, for example, complaints of negligence in the face of an approaching thunderstorm were laid against the bell-ringers.[69] Even in the twentieth century in Hungary, the mere hint of a storm caused the sexton to be scapegoated: why had he let it get this far?[70] If, despite all the clanging, lightning did strike, it was not the effectiveness of the custom itself that was questioned but the quality of the bell-ringing; it started too late, the aggrieved population would conclude, or maybe it had not been loud enough![71] So it was also essential to have enough well-paid bell-ringers to produce the loudest sound possible, louder than on other occasions.[72] And in the third place, finally, it came down to a question of the bell-ringers' competence: according to an old belief, the skill lay in swinging the bell to its fullest extent for such a long time that the demon was stifled by it.[73]

The expulsion of demons by ringing the church bell cleared the air in another respect too. For not only did it curb the rampaging of demonic spirits in the heavenly places, which prevented believers from praying,[74] but also encouraged the free circulation of angels. In the spiritual dogfight that ensued, the sound of the bells was meant to invite the angels to join in the prayers of the faithful, especially those of the parish priest – a practice that was still viewed as a form of exorcism in France in the nineteenth century.[75] Father Augustine de Angelis (1606-1681), rector of the Clementine College at Rome, had already argued that the physical effect of ringing church bells (the dispers-

▲ Caspar David Friedrich, *Engel in Anbetung*, from: "Die Lebensalter", Blatt 7, sepia, ca. 1826. In the thunderstorm discourse of the eighteenth and nineteenth centuries, the idea gained ground that the sound of the bells invited the angels to join their prayers with those of the faithful

ing of sulphur vapours) should go hand-in-hand with a moral or spiritual effect, "because by the sound the faithful are stirred to pour forth their prayers, by which they win from God the turning away of the thunderbolt."[76] The Protestant Irenäus Montanus of Saxony, pseudonym of the poet Gottfried Benjamin Hancke (1695-1750), was evidently of the same mind, for as he stated in 1726, the best weather bell was pious prayer.[77] As time went on, this more or less allegorical interpretation gained strength: bell-ringing was mainly about warning the villagers and a summoning to prayer.[78] Although the idea that ringing church bells could have a magical or medicinal effect was certainly tenacious,[79] there was a growing emphasis on the need to back up the bell's purification of the air with sincere prayer.[80] In a sense, this looked like a Protestantisation process. Later, the Catholic writer Joseph Albert Alberdingk Thijm (1820-1889) went even further, maintaining that bell-ringing and prayer were no longer associated, separate partners but that the sound of the bells merely *symbolised* prayer,[81] a concept that in the nineteenth century corresponded to High Church Protestantism, in which prayer for God's protection against the storm was the key focus and the most that bell-ringing could provide was symbolic acoustic buttressing.[82] Inscriptions were no longer incantations but sonically-assisted pleas for divine help. A final example, and one that is more symbolic than magical, was the bell-ringing that in the late nineteenth century summoned the inhabitants of the area around Lake Constance in the summer months to the (daily!) prayers to ward off the storm, the so-called *Wetterzeichen* ("thunderstorm signs"). Men would then take off their hats and pray for the averting of *Hochgewitter* ("heavy thunderstorms").[83]

All in all, the fortunes of Baroque devotions such as bell consecration and bell-ringing show how much the Roman Catholic Church was in a dilemma, for on the one hand it wanted to formulate its truths clearly and therefore – to the joy of enlightened Catholics – allowed the prayer element in these rituals (without far-reaching allegorisation) to come into its own. Yet on the other hand the Church was also forced to leave enough room for all kinds of popular beliefs.[84]

▶ Cannon and bell founder, 1698, engraving by Christoph Weigel. From: *Abbildung und Beschreibung der gemeinnützlichen Hauptstände*, Regensburg 1698. The added poem calls for sincerity in prayers, as to ring the "prayer bells" when the heart is not engaged in the act leads only to a meaningless clamour

Der Stück und Glocken-Giesser.
Des Höchsten zorn reicht weit, erwägt ihn weil es zeit.
Die Glocke des Gebets anrühren,
und nicht das Herz zur Andacht führen,
macht ein Gebrumm: stellt dieses ein;
Soll sich ihr Schall im Himmel regen:
so muß der Glaube Hand anlegen,
Aufmercksamkeit der Schwengel seyn.

Enlightenment's Battle with Apotropaic Bell-Ringing

One of the most noteworthy features of the foregoing is the way in which an enlightened, more symbolic, spiritualised or internalised interpretation of bell-ringing was compatible with Baroque devotion itself – a development which, as we shall see, was typical of the Catholic Enlightenment and its offshoots in the nineteenth century.

Before that, however, a different approach – the rejection in principle of the whole notion that bell-ringing would avert a thunderstorm – was already in evidence. We have seen above that this view was largely a product of the Reformation.[85] For however much Martin Luther (1483-1546), for instance, may still have believed in devils he refused to accept that the demons in the air could be scared off simply by the clanging of bells.[86] Moreover, Lutheran theologians regarded the baptism of inanimate things as deeply reprehensible: as Luther himself had declared, it made a mockery (*Affenspiel*) of the sacrament proper.[87] Having said which, even Lutheran baptisms were often still accompanied by the pronouncing of exorcism formulas, though with more enlightened times they soon lost their conjurational character, evolving into a wish or a prayer.[88]

Thus, although there were demonological survivals in Lutheranism and Lutherans did sometimes retain the prayer when a bell was cast,[89] for example, they had no truck with bell-baptism and, by extension, the whole tradition of apotropaic bell-ringing. That standpoint was, of course, shared by all the other Protestants, including Anglicans.

▲ St Martin's Church in Bolsward. The bell bears a distinctly Protestant inscription from 1611, asserting that its sound has no power to drive away a thunderstorm and is only intended to call the congregation to church

To cite a single example, Hugh Latimer (ca. 1487-1555), bishop of Worcester and later chaplain to Edward VI (1537-1553), was a strong opposer of such popish practices.[90] Early in Edward VI's reign, anti-thunderstorm bell-ringing was banned in England,[91] as it was in other Protestant areas, or it disappeared on its own. In post-Reformation Friesland, for example, its demise was manifest in the wording on the bell, cast in 1611, of St Martin's Church in Bolsward. The bell could not drive away a thunderstorm, asserted the inscription boldly; its sound was intended to call the congregation to church, and the only voice that moved it was the power of iron.[92] And it will surprise no one to learn that in 1652 in Boxtel, in the south of the Dutch Republic, the bailiff was ordered by his Reformed superiors to "stamp out all superstitions," which included forbidding the beguines there to ring the bell "if it thunders and lightens." [93]

Traditional Protestant opposition to apotropaic bell-ringing stemmed from the more general denial of the efficacy of Catholic sacramentals, particularly consecration and exorcism.[94] Protestants readily perceived the ritual use of bells as magical. The only reason for ringing the church bell was to summon the faithful to worship. And while the bell could be tolled for a funeral, it was solely to announce the funeral service or convey the message of *memento mori* – it should never be a call to prayer for the deceased.[95] That bells were ritually baptised was bad enough, but that some made a point of using water from the River Jordan for the purpose made matters decidedly worse.[96] Nor were popish "insolences" limited to dealings with bells in church steeples. In Protestant eyes the Catholic habit of parading around the house tinkling tiny bells during thunderstorms to protect the building from lightning strikes was a particular source of vexation. The thimble-sized bells in question – "Loreto bells" as they were known – came from Italy and derived their name from the famous pilgrimage site of Loreto near the port town of Ancona. They were found over a very wide area and have been traced to such

◀ Donatus bell. During the Ancien Régime, when thunder rumbled Catholic believers would go through the house room by room ringing these little bells dedicated to Donatus, the "thunder saint", particularly if their home was beyond the range of the church bells or if weather-ringing had been forbidden by the authorities

diverse places as Tubbergen in the eastern Netherlands and Munich in Germany.[97] In the Southern Netherlands – present-day Belgium – where Loreto bells were often dedicated to St Donatus, protector against lightning strikes, the tradition persisted into the twentieth century.[98] The same kind of *Wetterglöckchen* ("little thunder bells") were also found in Austria, where they were probably used in lieu of church bells, which since the Enlightenment were often no longer rung during thunderstorms.[99]

As early modern Protestants saw it, to admit such tangible religious symbols – the results of Catholicism's material concept of grace – was to be more concerned with outward appearance than true repentance.[100] And although Catholics continued to stress that the bells were not baptised but consecrated, it was nonetheless the case that when confronted with a thunderstorm Protestants generally responded by seeking reconciliation with an apparently wrathful God while Catholics, with their less "rigid" concept of divine providence, initially focused on chasing off demons.[101] This generalisation is still valid despite the many penitential processions in Catholic milieus, while among Protestants it finds germane support in the *Wetterpredigte*[102] and traditional church hymns.

Taken all in all, in the early modern period bells were often a bone of contention between Protestants and Catholics. The Antwerp poet Anna Bijns (1493-1575) had not a good word to say about Protestants who referred to them as the "Antichrist's bells."[103] It was not only demon-expelling bell-ringing to which early Protestants took exception, but bell-ringing per se. The strictest among them simply considered it "popery". It is in that very Protestant intransigence that we can see a link to the Enlightenment, for there too we find a decided aversion to over-frequent bell-ringing.[104] In the Netherlands, that aversion was never more clearly manifested than in the years between 1795 and 1806, when the country – known during that period as the Batavian Republic – had become a vassal state of revolutionary France. On 18 August 1796 the Dutch National Assembly wrote to the Provincial authorities prohibiting every denomination from ringing bells to announce a service – an application of the principle of separation of church and state that since then has never been stricter.[105] The possibility of that enlightened equality being motivated by old-fashioned anti-papism cannot be ruled out, but it seems rather unlikely. At least none of the People's Representatives gives the impression of accepting a general ban on bell-ringing in order to prevent Catholic bells from ringing along with the rest. Times had certainly changed, for the fiercest opposition to bell ringing now also came from radical Catholic revolutionaries in the National Assembly.[106]

Yet it was not only enlightened egalitarianism that tempered the long-standing argument over bell-ringing; developments within the confessions themselves were equally responsible, making it clear once again that in this respect Protestantism cannot simply be regarded as a preliminary phase of the Enlightenment. For however vehemently Protestants lambasted the Catholic use of bell-ringing, there were many who happily held on to bell-related imagery. Take the Lutheran pastor Bonifacius Stöltzlin (1603-

1677), for instance: in his *Geistliches Donner- und Wetter-Büchlein* he urged Christians to respond with penitent "prayer bells" when God rang the heavenly thunder storm bell.[107] And however antithetical such internal bells were vis-à-vis Catholic church bells, it does not alter the fact that the intention and usage of Protestant prayer sometimes displayed magical traits, as if repentant supplication could hardly fail to turn aside the presumed manifestations of God's wrath and as if having a prayer book in the house was as the same as an amulet against thunder.[108] Enlightened Protestants made an emphatic point of setting right such misconceptions,[109] for in their view it was only replacing Catholicism's *Dingmagie* or magical treatment of things with Protestantism's *Wortmagie*, the magical treatment of words welling up from the heavy heart.[110] With the influence of the Enlightenment, therefore, Protestant prayer gradually changed in character, since for the enlightened citizen it was no longer a question of changing the direction of the lightning bolt so much as of easing the mind.[111] Hence traditional penance soon came to be interpreted as an expression of fear.[112]

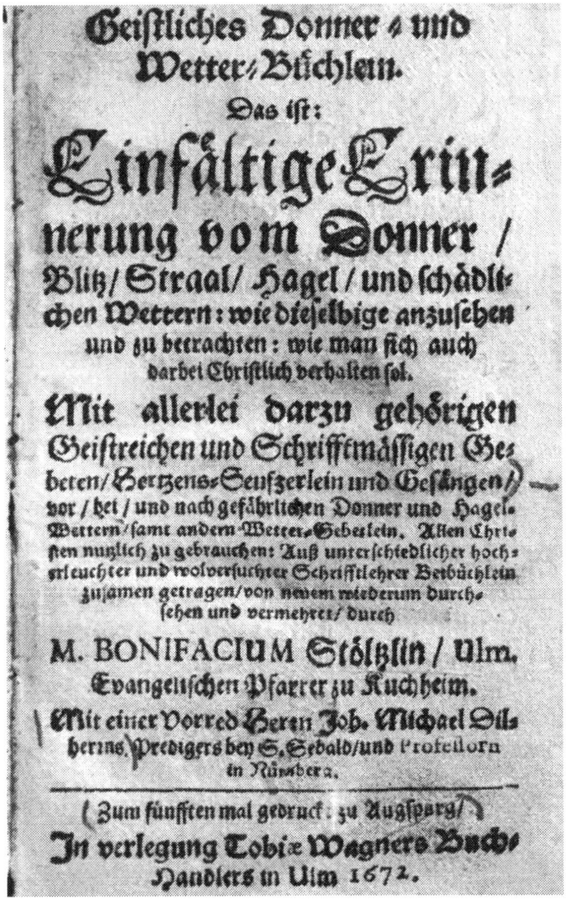
Geistliches Donner- und
Wetter-Büchlein.
Das ist:
Einfältige Erinnerung vom Donner/
Blitz/ Straal/ Hagel/ und schädlichen Wettern: wie dieselbige anzusehen
und zu betrachten: wie man sich auch
darbei Christlich verhalten sol.
Mit allerlei darzu gehörigen
Geistreichen und Schrifftmässigen Gebeten/ Hertzens-Seufzerlein und Gesängen/
vor/ bei/ und nach gefährlichen Donner und Hagel-Wettern/ samt andern Wetter-Gebetlein. Allen Christen nutzlich zu gebrauchen: Auß unterschiedlicher hocherleuchter und wolversuchter Schrifftlehrer Betbüchlein
zusamen getragen/ von neuem widerum durchsehen und vermehret/ durch
M. BONIFACIUM Stöltzlin/ Ulm.
Evangelischen Pfarrer zu Kuchheim.
Mit einer Vorred Herrn Joh. Michael Dilherins/ Predigers bey S. Sebald/ und Professoris in Nürnberg.
(Zum fünfften mal gedruckt. zu Augspurg/)
In verlegung Tobiæ Wagners Buchhandlers in Ulm 1672.

◀ Title page from Bonifacius Stöltzlin, *Geistliches Donner- und Wetter-Buchlein: Das ist: Einfältige Erinnerung vom Donner, Blitz, Straal, Hagel und schädlichen Wettern*, Ulm 1650 (reprinted many times)

The usual theological contrast between the Catholic concentration on the "external" and the Protestant emphasis on the "internal" thus came to seem rather pointless. Catholics had realised early on that presenting preventive bell-ringing in allegorical terms, explaining it away as a more or less symbolic support for faithful prayer, was an excellent way of neutralising Protestant criticism of its supposed magical character;[113] a way, moreover, that could be said to be in line with the Tridentine reforms referred to earlier.[114] Once the Age of Enlightenment arrived, an elite of Catholics and Protestants unanimously opposed practices such as bell-ringing, partly because they preferred to dissociate themselves from what they regarded as magical elements in their own traditions – in spiritualisation at least, Protestant Enlightenment and Catholic *Aufklärung* converged – but also largely for reasons both non-religious and entirely rational: bell-ringing was simply far too dangerous.

That elite interdenominational criticism, which was often expressed through the pages of journals and papers of learned societies, was not without consequences. Especially from the second half of 1783[115] – that year of frequent fearful thunderstorms – the rulers of numerous European states including Prussia, Austria, and Bavaria, a traditionally pious area, prohibited bell-ringing altogether.[116] France, where nearly a hundred bell-ringers had been killed or injured by lightning in 1783, followed the next year.[117] This remarkably quick action, even by such conservative governmental systems as France's Ancien Régime, makes it clear that the enlightened motivation of these administrative measures is open to some criticism. It is significant that such bans were not always directed against "superstition" but rather prompted by the high number of fatalities among bell-ringers.[118] One bell-ringing ban included a note that the metal of bells in motion actually attracted electricity – a view that (somewhat anachronistically) has been branded an expression of "technical superstition."[119] For the practical Franklin, that idea may have acted like a provocation. In any case, with a fine sense of irony he attached a set of little bells partly to the very first lightning rod he installed on his Philadelphia home, where they would tinkle out a warning of an approaching thunderstorm. Perhaps his device was a wry nod to the long-established custom of bell-ringing. In any case, he made the bells symbolically subordinate to – and indeed part of – his lightning rod.[120] For authors of Franklin's stamp – true advocates of the Enlightenment – were not content merely to point out the needless yet very real risk that bell-ringers ran[121] but also proposed an alternative: rather than maintain this outdated practice there should be a large-scale introduction of the lightning rod.

Nevertheless, not everyone was willing to see an end to storm-averting bell-ringing. In Passau in Bavaria, for instance, the rural population's persistent petitioning for the resumption of bell-ringing was already granted in 1798.[122] In that same year, in the commune of Châteaumeillant in central France, the *Garde nationale* even had to call in reinforcements to put down an uprising by farmers who had seized the bells to ensure that they could still be rung during a thunderstorm. The priest himself resisted the riot, only to be met with derision.[123] And, to mention one final example, the parishioners of Auzouer, near Tours, were of exactly the same mind, for since the parish priest had banned bell-ringing, their community had been completely flattened by hail.[124] It was not until well into the nineteenth century that the large-scale introduction of the lightning rod finally did away with the old custom.

Wetterschießen: Firing Cannon at Thunderclouds

Nevertheless, there were many Enlightenment thinkers who supported the lightning rod yet were still convinced of the efficacy of loud noise. Its supposed effect could be both direct and indirect: in the first case, such noise immediately "diluted" the muggy atmosphere by dispersing the clouds; in the second, the sound was thought to push the air upwards, thus keeping the storm clouds at a safe distance. Take the eminent natural philosopher and inventor of the Leyden jar Petrus van Musschenbroek, for example; open-minded enough to demonstrate a positive interest in the lightning rod with his experiments at a very early stage yet still a defender of bell-ringing as a means of dispersing the sulphurous vapours in the air – witness his entry on "Tonnerre" in Diderot and d'Alembert's *Encyclopédie*.[125] The loud noise that would "disrupt and divert" the thunder did not necessarily have to be made by bells, he pointed out; the same beneficial result could be achieved by firing a cannon at approaching thunder clouds.[126] *Wetterschießen*, as bombarding thunderstorms with cannon fire was known, was particularly favoured in the Alpine lands. The idea was to detonate the sulphurous vapours in controlled explosions or at least "dilute" them by dispersion.[127] One significant advantage of cannon fire over bell-ringing was the smaller risk to the gunner compared to the bell-ringer who had hold of a conductive rope; hence the preference for the artillery method.[128] Granted, *Wetterschießen* did increase the risk of avalanches in winter, but then so did bell-ringing.[129]

As with bell-ringing, the effect of *Wetterschießen* had long been the subject of scholarly debate, though it will be mentioned only briefly here. Already during the Renaissance it was believed that clouds heavy with thunder and hail could be dispersed by the noise of cannon fire.[130] Whether by drumming, bell-ringing, or firing artillery, the idea was that the cloud would be mechanically perforated,[131] as it were, so that the particles of "lightning matter" would fail to reach a critical phase and would be scattered in a timely fashion.[132] Francis Bacon in the early seventeenth century and Peter Ahlwardt in the eighteenth century had already noted that the putative beneficial effect was probably due not to the sound per se, but rather to the movement that the sound caused in the air.[133] But whichever opinion was held on the matter, the generally accepted idea remained: the louder the sound, the greater the effect.[134]

The logical outcome of that line of thought was that cannon fire was superior to campanology, even when the very largest bells were rung. In turn, and for the same reason, bell-ringing was much better than the ancient custom of banging "thunderstones" or "thunder chisels" (belemnites) on wooden boards.[135] In the eighteenth century some also saw a chronological progression reflected in this hierarchy. But they were not alone. According to contemporary Swiss historian Max Engammare, firing cannon at thunder clouds only began when gunpowder enabled man to make a sound that was louder than thunder.[136]

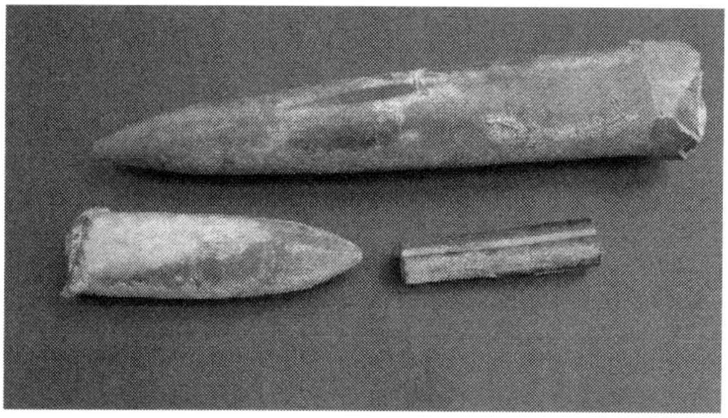

◀ According to an old and remarkably persistent folk belief, "thunder chisels" or "thunderstones" were the hammers and axes with which Donar wrought havoc and which fell from the sky during lightning strikes. They were often ascribed an apotropaic power. Science has identified these objects as belemnites

But is that response to gathering storm clouds really so modern? To be sure, it has been claimed that in Bavaria *Wetterschießen* only dates from the mid-eighteenth century and that unlike bell-ringing it was motivated not by a belief in demons but purely by physics.[137] That seems improbable, however, for everywhere else its magical roots are hard to deny. In neighbouring Austria in the early modern period, *Wetterschießen* was certainly part of a demonological context. So closely associated was it with belief in devils and witches that while the shot was in the process of being cast, all kinds of sacred objects were added and the gunpowder was blessed before firing began.[138] We have already seen how dairy farmers in Styria made as much noise as possible to chase away the thunderstorm demons.[139] On closer examination, it proves that producing counter-noise or counter-violence in the event of a thunderstorm is both an ancient and global human response. The ancient Greek historian Herodotus, writing in the fifth century BCE, tells us that when it thunders and lightens, the Thracian Getae aim their arrows at the sky, uttering threats against their god, Zalmoxis.[140] And in Uganda, during a thunderstorm, a villager with bells around his ankles would go around to ward off collective disaster with their counter-sound, while in Japan, during Shinto weather rituals, taiko drums were played both to generate rain and to drive away evil storm spirits.[141] All in all, it seems reasonable to regard not only bell-ringing but also *Wetterschießen* as a form of sympathetic or imitative magic.[142] In the late eighteenth century both customs were equally affected by specific prohibitions.[143]

▲ Five Thracian horsemen shoot arrows skywards at a thunderstorm and its instigator, the god Zalmoxis, while a sixth, killed by lightning, lies on the ground, watercolour by George Jones (1786-1869)

In defence of the modern character of *Wetterschießen*, some might point to similarities with today's hail cannon, which is also intended to limit the damage caused by storms – purportedly, the sonic shock waves generated by hail cannon break up storm clouds. Or with cloud seeding, which is meant to modify the amount and type of precipitation that falls from the cloud. Normally used with a view to producing rain, cloud seeding can also aim to prevent ice crystals from growing into dangerously large hailstones. To that end a projectile containing countless silver iodide crystals is fired into the atmosphere. Supposedly the icy water droplets in the cloud will be drawn to the hygroscopic silver iodide crystals, which will then almost endlessly divide themselves instead of clumping together. It should be said, however, that the effectiveness of both methods is hardly less controversial than the *Wetterschießen* of old.[144]

Conversely, another argument in support of the pre-modern character of *Wetterschießen* is that it is included in the traditional debate about sulphur just as much as bell-ringing is. By definition, of course, that debate is not magical – we have already seen how "superstitious" elements were sometimes unconsciously rationalised. Someone like Van Musschenbroek could therefore derive a "scientific" explanation for the effectiveness of bell-ringing and cannon-firing while at the same time giving a natural explanation for the rain of burning sulphur that destroyed Sodom. Even so, he opposed the demonological thinking from which the sulphur theory partly arose.[145] Talk of demons and devils was superseded by discussion of the light inflammable subterranean vapours (such as methane, discovered in 1776) that issued from their traditional habitat. Thus, this theory gave a physics-based footing to an age-old conception of the location of hell while simultaneously offering a "natural" scenario for the way in which the elements will be dissolved with fire on Judgement Day (2 Pet. 3:10).[146] Similarly it provided a basis in physics for St Paul's "spiritual forces of evil in the heavenly places" and the belief that many based on it that thunderstorms were caused by demons, while that kind of explosion theory also finally explained why lightning was so often a spectral blue colour and left a foul odour. However enlightened scholars such as Van Musschenbroek and Ahlwardt may have been and however much they manipulated the various opinions on sulphur to suit their rationalist thinking, the roots of this survival were still magical. But even if they were aware of it, their use of it in their scientific practice was not necessarily a hindrance.[147] Meteorological phenomena eventually originated in the subterranean world.

Beyond Demonology: the Cult of Saints

The arsenal of means – ecclesiastically condoned or not – by which thunderstorms could be averted was, of course, many times greater than just bell-ringing and cannon fire. What was to be done if neither had any effect? Of course, they had to be accompanied by parishioners' prayers. But even that was not always enough to stymie the thunderstorm demons, in which case (as we have already seen) the parishioners often blamed themselves rather than question the effectiveness of the custom, and assumed an even more humble and penitent demeanour. Occasionally, however, the response was anger or even aggression. There was an old belief that higher powers could be intimidated by being threatened with weapons.[148] In a surprising parallel with the aforementioned Getae, in certain areas of pre-Revolutionary France the parish priest would fling his shoe – that contemptible foot covering – at the heavens when neither prayers nor bells worked. Supplication was replaced by threat.[149] And one German pastor had no hesitation in shooting storm clouds with a silver pistol.[150] Meanwhile, this kind of reaction raised questions about who actually caused the storm: was it indeed a legion of more or less autonomous demons, instigators of the aforementioned *gemachte Wetter*, hotly followed by the Almighty? Or was it perhaps angels playing bowls, or St Peter or "our dear Lord" himself? – the variety of "explanations" is enormous.[151] But if thunder could be interpreted as the devil's yowling, lightning was the Lord's own forewarning to the people to arm themselves against evil with the sign of the cross.[152] And what role was reserved for the saints, with their relics and intercessions, in this spiritual dogfight?

As we have already seen, bells were usually dedicated to the patron saint of the church in which they hung, whose intercession was assuredly sought during thunderstorms and whose relics were venerated. In the Netherlands and parts of Germany and Austria no figure was more popular in that respect than the "thunder saint" and martyr Donatus. Many a bell was dedicated to him. One, in the seventeenth-century St Donatus Chapel in Dorne near Maaseik (a town between Roermond and Maastricht), was still credited by some with storm-averting power as recently as 1960 or thereabouts, in spite of all the marginalisation of St Donatus's veneration in the twentieth century.[153] It was the same in Rozenaken (Russeignies), in what was then East Flanders and is now Wallonia, where in 1769 a bell was recast and inscribed with a plea to Donatus that its ringing would dispel the lightning. So closely were bell and saint associated that whenever it stormed and the tocsin's deep toll was heard, the villagers would cry "Donatus rings!"[154] These are just two of an enormous number of examples.[155] In the eighteenth century the Rhineland was particularly well provided with bells dedicated to Donatus.[156] In some cases the bells also bore an image of the saint, in southern Germany, for example.[157]

It was not only church bells that were dedicated to Donatus; so were the aforementioned much smaller Loreto bells, intended for use at home. When a thunderstorm broke, the householder should not only sprinkle the home with holy water but, crucially, also process from room to room ringing the little bells – especially important if the church bells were too far away to hear – and reciting the St Donatus litany. Like St Donatus candles, which were often accompanied by the same sort of domestic prayer on being lit, the Loreto bells first had to be consecrated in church (a pilgrimage church for preference)[158] and if it was possible they should be touched to the saint's relics.[159] One might also acquire printed devotional material that had been brought in contact with Donatus's relics and would therefore act as a protection against lightning strikes.[160] That went some way towards getting around the problem of the scarcity of relics. Taken all together, prayer, bell-ringing, candle-lighting, prints, and images effectively formed a Catholic counterpart to the pietistic practice of the *Hausandacht* or domestic prayers.[161]

Despite all the devotion, nothing was (or is) known with certainty about the saint himself. According to a tradition that dates back no further than the seventeenth century, Donatus was a Christian, born around 150 CE, who rose through the ranks in the imperial Roman army, eventually commanding the Legio XII Fulminata or "Thunderbolt Twelfth Legion" (the thunderbolt had been the legion's emblem since the time of Augustus, 63 BCE – 14 CE). In that capacity he is said to have taken part in Emperor Marcus Aurelius's campaign against the Quadi (a Germanic tribe in the region of what is now Slovakia) around 177 CE. When the encamped army was ravaged by extreme heat, a severe shortage of drinking water, and contagious disease, and prayers to pa-

▲ Statue of Saint Donatus, maker unknown, 1905-1914. Donatus is represented as a thunder saint with lightning bolts in one hand and a martyr's palm branch in the other; the helmet at his feet alludes to his military profession

gan deities went unanswered, the emperor asked Donatus to pray to his God. Almost at once, clouds gathered, lightning flashed, and a mighty rain fell. The troops revived and the epidemic ended – clearly as the result of divine intervention. The thunder and lightning that accompanied the "rain miracle", as it was already being described in Antiquity,[162] also drove the Quadi to flight, giving Marcus Aurelius a resounding victory. Once back in Rome, however, and promoted to high position Donatus was condemned for his Christianity.[163] He was executed with a sword – an invariable part of a soldier's accoutrements – which is why he is often depicted with that attribute in his right hand (sometimes he also carries a shield emblazoned with lightning bolts), while in his left he usually bears a palm branch, symbol of a martyr's victory.[164]

In the second half of the seventeenth century, the emerging piety of the Baroque period evidently recruited Donatus into the early Christian version of Marcus Aurelius's "rain miracle". It was also around this time that devout Catholics were searching the catacombs of Rome – rediscovered after centuries of oblivion – for the remains of early Christian martyrs. Relics spread from Rome to southern Germany, Austria, and Switzerland, often through the agency of the Jesuits.[165] In 1649, for example, the tomb of a certain Donatus was discovered in the catacomb of St Agnes, just outside Rome. A year or two later, in 1652, a new Jesuit college church was founded in Münstereifel in Germany and several relics of this unknown saint were sent there. The last stop on the relics' journey was St Martin's Church in Euskirchen, some 15 kilometres from their final destination, and that is where they were on the morning of 30 June (the day on which the new church's foundation stone was to be laid) before completing their translation to Münstereifel accompanied by much religious display. The Jesuit Father Henricus Herde had arrived in Euskirchen the previous day to take charge of the relics and oversee their veneration. On the morning of 30 June, as he celebrated an early Mass, a violent thunderstorm broke overhead. Lightning struck the church; Father Herde was partially paralysed and badly burned. Whether it was because he had taken off the Caravaca cross[166] that he usually wore as protection against storms, the story does not tell us. In any case, he called out to Donatus to intercede and made such a rapid recovery that he was able to travel on (in company with a pious flock) to Münstereifel that very same day.[167] To many believers it was now perfectly obvious what the new saint was to be invoked for: had not God himself shown by that example of miraculous healing that Donatus was a "thunder saint"? Thus a new cult came into being. Soon there was a flood of stories describing deliverances due to Donatus's intercession, closely followed by tales of his exemplary life. As happens so often, the pious fiction, which was mainly passed on through devotional leaflets and prints, evinced a greater passion for edification than truth-seeking. Thanks to such religious propaganda and still more to the possession of those "miraculous" relics, Münstereifel became a popular place of pilgrimage in the second half of the seventeenth century.[168] Possibly the resemblance of the saint's name to that of the Germanic god of thunder, Donar, may have helped a little, too.

▲ Caravaca cross from the Oberhausmuseum in Passau, eighteenth century. Objects like this were worn as amulets against the harmful effects of thunderstorms; they were banned during the eighteenth century

Later events made the connection between Donatus and the escape from damage caused by thunderstorms all the more plausible for believers. When thunder rolled once again and lightning flashed dangerously, it was not uncommon for devotion to Donatus to blossom. For while its roots may have been in Münstereifel, the veneration of the saint gradually spread much further. It reached Arlon in the south of what is now Belgium, for example. On a hill in the centre of the town was a Capuchin monastery that had been struck by lightning so often in the early eighteenth century that its *custos* or guardian father vowed to erect a statue of Donatus and to petition Münstereifel for at least one relic. Although the former proved easier to achieve than the latter, the request was eventually granted. On the second Sunday in July in the year 1738 – the same day that Donatus's feast day was celebrated in Münstereifel – the holy relics were installed in Arlon and exposed for veneration in a silver shrine. A procession was also held. Soon Arlon too became a pilgrimage site, from which many pilgrims brought away sacred objects that they had touched to the relics.[169] In turn, the Capuchins became great promoters of the Donatus cult in the Netherlands. When the monastery was dissolved in 1796, the church (today's St Donatus's Church) remained in use and the relics in particular were spared. Which is why, to this day, Arlon is still regarded as the centre of Donatus's veneration in the Low Countries.[170]

In the Protestant northern Netherlands there was considerably less chance of St Donatus gaining a foothold in the eighteenth century. But when a fire started by lightning destroyed a large part of the village of Reek in 1779, loud appeals for a relic of Donatus were heard there too. At the time, Reek was not in the territory of the Dutch Republic but the contiguous Land of Ravenstein, then the possession of a Catholic German princely house and part of the diocese of Liège. So the Catholic population could practise their religion freely there. Reek's pleas for a Donatus relic were favourably answered, and before long the church received not one relic but two. They came through the mediation of the Capuchins, most likely friars from the nearby monastery in Velp, and were brought to the church in a ceremonial procession accompanied by music and hymns.[171] On 23 November 1779 the parish priest of Reek received permission from the *custos* of the Capuchin monastery in Arlon to establish a confraternity of St Donatus and thirty-five confreres were immediately enrolled. The confraternity brought together clergy and laity who sought protection from the damaging effects of

▲ St Donatus's Church, Arlon. Arlon is still the international centre of Donatus's veneration

thunderstorms by beseeching Donatus's intercession, both in private prayer and by celebrating Masses or having votive Masses said. Confreres who took part in the special veneration on the second Sunday of July were guaranteed a plenary indulgence. In 1785 Pope Pius VI extended that indulgence to those who were not members of the brotherhood but who did take part in that St Donatus Sunday pilgrimage.[172] The geopolitical upheavals that followed the French Revolution led to the Land of Ravenstein becoming part of the northern Netherlands. Hence Reek became the earliest centre of Donatus's veneration in the country.[173] It will come as no surprise to learn that the enrolment forms that the confraternity used from 1829 were touched to one of the relics.[174]

▲ Medallion containing a relic of Donatus, 1915-1924

▲ Donatus medallion or confraternity badge, brass, the Netherlands, 1890-1929. In Germany in the eighteenth century a Donatus medallion was already a popular pilgrimage souvenir. However, this is a membership badge inscribed with the petition for intercession: "St Donatus, pray for us"

Unfortunately, little is known about the way in which the early veneration of Donatus was practised in the Netherlands. What is clear, however, is the importance to the flourishing of the Donatus cult – which from 1779 developed many different forms – not only of the relics but also the images of the "thunder saint". Reek's was just the first of the many churches and chapels that soon had a statue for veneration. Sometimes (according to foreign data), the statues were also carried in processions. In a ceremony of that sort in Münstereifel in 1770, for instance, it seems that a silver reliquary bust was processed.[175] On the other hand, objects of the kind are not mentioned in the description of the so-called *Hagelfeiern* or "hail celebrations" that took place in the archdiocese of Trier and also involved processions: after the celebration of a solemn High Mass, a cross, the monstrance, and reliquaries were carried out of the church and through the fields in order to protect the farmland from meteorological misfortune. The Donatus processions were often the ecclesiastical successors to that centuries-old pagan tradition.[176] Whether the little statues of St Donatus that pilgrims could take home were already available in that early period is not clear; their presence is at least documented in the twentieth century.[177] At any rate it is obvious that since the end of the eighteenth century and certainly since the second half of the nineteenth, whether he was at home or on his travels, in his own country or abroad, the devout Catholic need never be at a loss for devotional material or a place in which Donatus could be venerated. The choice steadily grew in size and diversity: not only did the number of pilgrimage sites increase,[178] "hail crosses" incorporating a figure of Donatus were also erected at road junctions to ward off the evil effects of thunder and hailstorms.[179] By the eighteenth century a number of plays were already dedicated to the popular saint – certainly in countries neighbouring the Dutch Republic.[180] And, finally, there was the aforementioned wide array of printed matter, which from an early stage was used to promote and foster the cult of St Donatus, particularly by the Capuchins (in the Netherlands) and the Jesuits (in Germany). It included devotional pamphlets (containing prayers, litanies, a hymn to St Donatus,[181] and a hagiographic *vita*), as well as (pilgrimage) prints.

The way this devotional material was used sometimes went well beyond what was ecclesiastically proper. That the propaganda leaflets and Donatus prints that had touched the saint's relics were regarded as repellents or even amulets and were often pasted onto the doors of the house or stable, or worn around the neck, or placed in the fields,[182] was going far enough. But what was done with the traditional house blessings went even further. These were forms or prints that invoked a blessing on a house and those who dwelled in it and usually consisted of a sheet of paper with a woodcut depicting the Virgin and Child on the left and an image of the crucified Christ on the right, surrounded by a prayer text. Often a separate prayer to Donatus was added. Sometimes, however, the blessings were provided with all kinds of magical elements, which were pasted on in the middle in what was called the "sanctuary". The use to which these "little papers with a sacred part" were put was equally magical: folded up like packages they were often placed in the vulnerable parts of the house – often the easily-combustible roof – or in openings such as doors, cellar holes, and chimneys, to keep the home safe from witchcraft, lightning, fire, and disease.[183]

It was precisely because of the way in which devotional material that had been in contact with the relics was treated – prayers to Donatus were sometimes bricked into the foundations of houses[184] – that many believers saw little difference between "little papers with a sacred part" and regular house blessings, even though the Catholic Church disapproved of the use of such magical pendants. As far as Rome was concerned there was a fundamental difference, which lay in the scrupulous *consecration* of the devotional material. Be it a matter of house blessings or church bells, the rituals of consecration and exorcism – the two sacramentals that marked the differences from the Protestant approach[185] – were essential to the Church. Even so, in that respect the Age of Enlightenment did see a degree of denominational rapprochement. Some eighteenth-century Catholic clerics saw a very thin line between strict ritualism and superstition, and they therefore stressed that consecrated religious objects such as the leaves printed with supplications to St Donatus had no power in themselves but gained it through prayer. The German Jesuit Franz Hunolt (1691-1746), a renowned preacher, lashed out at this *Aberglaub in Heiligen Sachen* ("superstition in holy matters") and warned that carrying a picture of St Donatus was no guarantee of immunity from lightning strikes.[186] And while he did not reject the practice of touching relics as such, it is clear that to Hunolt and many other clerics influenced by the Catholic *Aufklärung* the "mechanical" interpretation of the principle of *ex opere operato* (in virtue of the ritual action itself, irrespective of the spiritual state of the priest or the recipient) had come under pressure. Gradually the emphasis shifted to faith and prayer, a process of change that in spite of all its connections with Catholic tradition (particularly the intentions of the Tridentine reforms[187]) seems almost to be verging on Protestantism. And what were house blessings anyway but so many stimuli to prayer? It was through that same faithful prayer (whether via the intercession of saints or not) that objects were truly consecrated. That belief united devout and enlightened Protestants and Catholics, and it applied to both devotional materials and secular instruments, not excluding the lightning rod itself. These objects were means in God's hands, to be used in prayerful dependence on the Lord; the "deification of things" was altogether wrong. No one put that more clearly than the nineteenth-century pastor Jan Jakob Lodewijk ten Kate, whose poem *De planeeten* "Christianised" Franklin by supposing that it was the inventor's prayers offered up during sleepless nights that had consecrated the lightning rod – as we already have seen.[188]

▲ Devotional print titled "Litany in honour of St Donatus, Martyr, Protector against Thunder, Lightning, Storm, and Tempests". This coloured woodcut was published by P.J. Brepols in Turnhout between 1817 and 1833. Remarkably enough, the bolts of lightning are still arrow-headed

▲ Agnus Dei, Italy, 1740–1759. Wax discs like this were made from the previous year's Paschal candles that were collected from the churches in Rome. Blessed by the Pope himself they were worn to ward off perils of all kinds, among them certainly accidents caused by thunderstorms

▲ Printed prayer to St Donatus beseeching his intercession, by Isabella Hertsens, Antwerp 1802, coloured engraving on parchment

The Protestant and Catholic elites may have moved closer to each other during the Enlightenment, but the battle of minds still raged among members of each denomination. According to their respective opponents, old-fashioned devout Catholics should cast aside their ritualism, which bordered on magic, while uber-orthodox Protestants should eschew their superstitious fatalism. In both cases traditional believers saw the lightning rod at the very least as a competitor, if not a downright foe.

It was typical of this element of rivalry that 1779, the very year in which the first Confraternity of St Donatus was founded in the Netherlands, also saw the establishment of the first lightning-rod installation firm, the Schaap company in Deventer.[189] We find the same suggestion of competition or conflict in cases where mistrust of the rod was still so great that relics of the thunder saint were placed in church towers, whether they were fitted with a lightning conductor or not.[190] In 1782, when St Martin's in Euskirchen – the same church in which Father Herde was said to have been so miraculously healed 130 years earlier – was struck by lightning again and a major fire broke out, the priest, Father Johann Wilhelm Carman (d. 1821), refused to have a lightning rod installed, maintaining that a relic of St Donatus was sufficient guarantee for the safety not only of the church but the entire town.[191] But the reverse also happened: sometimes folk were not so unconditionally trusting of the parrying power of a *Wetterkreuz* erected on a mountain top that they refrained from attaching a lightning rod to it.[192] Even the relics at the very heart of the Donatus cult in Arlon proved unable to provide adequate protection, for in 1774 lightning struck the Capuchin monastery.[193]

Although Andrew Dickson White's conflict thesis, with which we began this chapter, finds little support in today's historiography, the antagonism between trust in traditional devotional materials and reliance on modern technology only came to an end when both were equally consecrated by prayer or intercession and used accordingly in dependence upon the Almighty. Yet in daily life belief did not always follow such (theo) logical lines. In practice there often seems to have been a kind of compartmentalisation, the result of a certain innocence or thoughtlessness. In Germany (and elsewhere) belief in the power of relics and lightning rods to avert thunderstorms sometimes went effortlessly in tandem – even among Protestants, although in their case relics or other devotional items were generally replaced by the Bible and catechism.[194] Among Catholics, finally, there was the not unimportant consideration that acquiring a relic or a piece of paper that had been touched to a relic was many times cheaper than installing a lightning rod. That was an additional reason for the just-mentioned Father Carman to sideline the demands of the government in Düsseldorf to have a lightning rod installed and to do his utmost to obtain a relic of St Donatus for his poor parish.[195]

Donatus was certainly not the only saint whose intercession was solicited to avert the damaging effects of a thunderstorm – needless to say, the Virgin Mary was also frequently appealed to in all manner of afflictions and distress. The Jesuits promoted their own Francis Xavier as a specific patron of weather, too.[196] Often the veneration of saints had a local, regional, or national character. In the Dutch Republic and France, for example, St Barbara – the go-to saint for those in danger of sudden death, especially by lighting[197] – was a particular favourite. In Austria she was joined by Florian, invoked as patron saint against the dangers of fire and water, who was martyred there: his likeness is often found in Germany as well, on the fronts of houses.[198] In old Switzerland and elsewhere, Agatha was regarded as a storm-averting saint: the little loaves named after her that were baked at home on the eve of her feast day (5 February) then blessed in church, were supposed to ward off all kinds of fire hazards for a whole year, while blessed "St Agatha flour" was sometimes scattered from the window during thunderstorms for the same purpose, as if to feed the elements.[199] Clare, on the other hand, was widely invoked as the patron saint of good weather; a prayer to her was sometimes accompanied by an offering of eggs – especially propitious if laid on Ascension Day.[200] The list of saints could easily be lengthened to include Blaise, Fridolin, Lawrence, Mary Magdalene, or the Three Magi, for example.[201] But even without such a comprehensive catalogue it is clear that the cult of St Donatus tended to supplant the veneration of other saints. This is especially true of the Rhineland.[202]

▶ *Agatharingli*, a form of "Agatha bread". In old Switzerland and elsewhere, Agatha of Catania (d. ca. 250) was venerated as saint who could ward off a thunderstorm, among other things: the bread named after her and consecrated in church was believed to avert all kinds of fire hazards

On balance, the origin and flourishing of the cult of St Donatus seems to be a rather complex phenomenon. Based on a purported lightning miracle on the day of the translation of his relics, it is, in the first place, an expression of the intensifying popular piety that began in the time of Baroque Catholicism in the second half of the seventeenth century and which, remarkably enough, was still going strong in the period of increasing popular Enlightenment a century later. To what extent Father Herde's miracle was due to chance at the time or subsequent deliberate manipulation by the Jesuits is hard to say.[203] But in a more general sense, it is clear that the veneration of catacomb saints in Baroque Catholicism was intended as powerful polemical testimony vis-à-vis Protestantism to emphasise the unbroken connection of the Mother Church with early Christianity.[204] That it should be in that very respect that inadvertent Protestantising tendencies emerged in the second half of the eighteenth century – not only among leading representatives of the Catholic *Aufklärung*, but also, more subtly, among lower socio-cultural classes – is as ironic as it is remarkable. For also on a local level, many a village priest put diminishing emphasis on the outward features of sacred rituals and objects and focused increasingly on faith and the prayer that sanctified everything. Almost imperceptibly, in the century of Enlightenment, even at grass roots level, age-old ritualism not infrequently gave way to processes of allegorisation and spiritualisation. And that tendency, it was now increasingly and reassuringly realised, was unquestionably in line with the original intentions of the Council of Trent.

▲ A young mother with three frightened children seeks protection from the storm at an image of the Virgin and Child. Painting by Ferdinand Georg Waldmüller, 1832. On the left is a rake with its tines turned upwards to keep the thunderstorm demons at bay, reflecting an old peasant belief

Nature: Popular Beliefs and Agricultural Life

The lives of fishermen and farmers are inherently bound to nature. Their livelihoods often depended almost entirely on what the sea and land yielded. In the absence of an insurance industry, such vulnerability turned bad weather like gales, thunderstorms, and hail into immediate and formidable enemies.

In the hope of protecting vessels from accidents of any kind they were often named after a saint. St Erasmus of Formia (ca. 240-303), for example – St Elmo, as he is often known – was the patron of Mediterranean sailors from early days, and the electrical phenomenon named after him – a kind of ghostly dancing blue or bluish-white flame sometimes seen at mastheads – was interpreted as a sign of his protective presence. Elsewhere, however, St Elmo's fire was seen as a "demonic" spectacle, which could only be stopped by crossing oneself, at least according to the Breton sailors of around 1850. Even more marginal was a belief held by the mariners of Guernsey, who at around the same time confessed that they felt safer when a "thunder chisel" was hidden in the captain's cabin as an amulet.[205] Sometimes, moreover, seamen seem to have worn "lucky" (apotropaic) items of clothing to make themselves less vulnerable. A zigzag lightning pattern has traditionally been a popular motif on the jumpers worn by Dutch skippers, for example.[206] Might we see here the rudiments of sympathetic magic, used to avoid being struck by lightning?

▶ St Elmo's Fire, drawing, nineteenth century. This electrical phenomenon, named after the patron saint of sailors, St Erasmus of Formia (ca. 240-303), sometimes occurs when a thunderstorm approaches, though it is only visible in dark weather. Occasionally, if there are high-voltage differentials between the ship and the atmosphere, so-called spray discharges take place, which are particularly noticeable at prominent high points such as the mast

Whatever the case, while popular belief might have been prevalent among those who made their living at sea, it was omnipresent in the countryside. At sea, the threat was mostly to people and ships; on land the yield of a whole year's labour was at stake, which made farming a hugely fertile ground for popular belief. Naturally, those who lived so close to nature and the multiple threats it presented could first try to influence the weather somehow. As we have already seen, it was believed that the ringing of consecrated bells and the intercession of saints could help to redirect harmful precipitation, as could holding processions and sprinkling the fields with holy water. Moreover, that kind of ritual safeguarding of agricultural land was not even an exclusively Catholic tradition, for as late as the nineteenth century pietistic *Dienstgänge* ("religious rounds") were still being made, their purpose being to protect the fields against the effects of hail and thunderstorms. The Württemberg pastor Gustav Adolf Heyd (1835-1906), for example, still carried out such rounds.[207]

At the same time, that dependence on nature also underlined the need for a safe shelter. However primitive it was, the farmhouse, which was usually built largely of combustible material such as wood and reed, had to offer protection. To ensure that it would, methods and customs that were hallowed either by long tradition or the Church were indispensable. The central position, sacred in origin, was reserved for the hearth – not coincidentally also the point where in earlier centuries witches were supposed to enter and leave the house. Earthenware objects or glass bottles known as bellarmines or witch-bottles filled with urine or sharp metal objects were sometimes buried beneath the hearth to ward off witches.[208] When moving into a new farmhouse, hanging the chimney crook in the fireplace was a key moment: in the early modern and modern period, it became the occasion for a festive meal, for only from that moment did the new inhabitants feel as if they had really taken the farm into use. They also thought they were safer from then on, too. It was therefore at the hearth that they took refuge during thunderstorms.[209] Could that be connected with the old belief, rooted in Germanic mythology, that the hearth fire owed its origin to the lightning bolt of Donar, the protector of the household?[210] The frequent recommendations that were still being scrupulously made even as late as the eighteenth century to put out the fire as soon as a thunderstorm came up at least suggests that,[211] notwithstanding all the scientific argufying about the risk of rising sulphurous vapours. Moreover, in Ancien Régime France some of the hearth's sacred origins seem to have survived in the curious custom of taking the chimney crook from the fireplace during a thunderstorm and flinging it at the storm clouds, for that was the only way the thunder could be driven away.[212] There was sometimes little separating reverence and threat, as we have already seen in the case of the parish priests who would hurl their shoe at the heavens when prayer failed to work.[213] And the same ambivalence is also apparent in the fact that in other areas it was actually during thunderstorms that the fire was lit.[214] Another expression of sympathetic magic? In any case, a folkloric custom for whose effectiveness eighteenth-century science also believed it had an explanation, as put forward by Alessandro Volta.[215] But whichever character was attributed to the relationship between the hearth and the thunderstorm, the actual existence of such a relationship seems not to have been in doubt in any of the cases described.

At the same time, however, the flinging of the chimney crook is open to yet another interpretation, for it is not an autonomous tradition but a custom that can be seen as an expression of so-called *Metallzauber* ("metal magic"). Since time out of mind, iron has been considered a magical metal.[216] Its anti-necromantic action was particularly effective if the iron had been sharpened.[217] Which is why, during thunderstorms, a peasant family would sometimes put a knife – preferably a large one[218] – on the table and would often place all sorts of objects and tools in and around the farm, such as horseshoes, axes, scythes, sickles, dung forks, even harrows, ideally with the teeth or tines upwards to unnerve the demons and keep them at a safe distance,[219] though in this extraordinary back-and-forth play of meanings intimidation could just as easily turn into timidity.[220] The threatening aspect also appears in the way the chimney crook was deployed against the storm clouds, for like the harrow, it too has a serrated edge. The same thinking holds true for the scythes that were placed edge upwards on the farm threshold, although when these tools glinted in the sunlight metal magic and sympathetic magic might seem to have been in harness. Was this a case of placing oneself under the protection of powers whose wrath one ought to fear?[221]

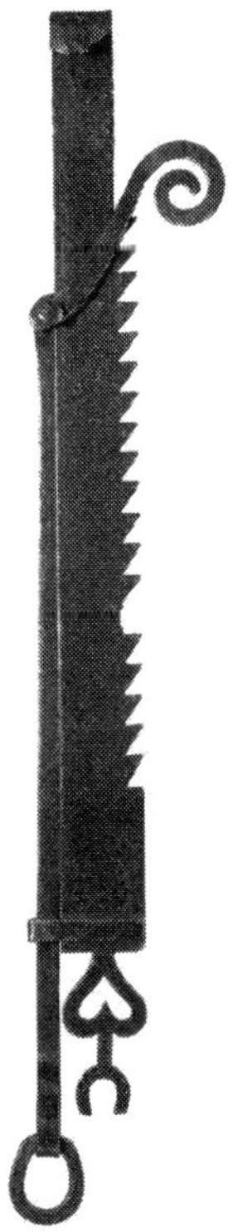

▲ Wrought iron chimney crook decorated with a volute, a heart, and a horseshoe, Southern Netherlands, 1750-1850

Be that as it may, in the event of a thunderstorm the doorway, being a vulnerable part of the house, merited magical reinforcement. That might be positive – hanging up a horseshoe as a kind of talisman – or negative, positioning scythes or axes to ward off evil. For the axe, too, the traditional attribute of Donar, god of thunder, was ideally placed at the threshold, albeit not always sharp edge up – sometimes the farmer drove it into the doorstep.[222] In the same connection, it should be recalled that the aforementioned "little papers with a sacred part" were also placed in doorways.[223] Interestingly, well after the end of the Ancien Régime there was a similar popular belief, which while still occurring within the rural community was not met with in the farmhouse or in Catholicism. For however exceptional it may be it can hardly be coincidence that as recently as 1940 to 1945, the years of the Second World War, in the firmly Protestant village of Krabbendijke in the Province of Zeeland, fathers, as heads of households, still took up their post on the step of their open front door during thun-

▲ A farmer prays before the door opening of his hut during a thunderstorm, etching by engraver F. Ramberg, after a design by D.N. Chodowiecki, ca. 1795. Illustration from J.G. Jacobi, *Taschenbuch von J.G. Jacobi und seinen Freunden*, Königsberg-Leipzig: F. Nicolovius, 1795

derstorms to fend off disaster, as if it were a question of warding off the destroyer that passed over the land to strike down the firstborn of Egypt.[224] However much the local headmaster deplored such behaviour as superstition,[225] that survival illustrates that fathers followed in the footsteps of the English Puritans as a sort of minister in the microcosmic church of the family[226] – a "spiritualisation of the household",[227] as an expression of which in Germany, for example, the thunderstorm prayers on the aforementioned *Gewitterbänke* may also be considered.[228] German *Andachtsbücher* ("devotional books") also addressed patresfamilias in particular about their religious responsibility for the family.[229]

Even more vulnerable than doorways were farmhouse roofs. They were also reinforced with a variety of magical means. The original function of the tree-shaped "thunder brooms" decorating the gables of farmhouses in Twente, in the east of the Netherlands, for instance, was to avert lightning strikes.[230] And the use of that kind of ornament was by no means limited to Twente. Its earlier history becomes clear in Brunswick in Germany, for example, where deformities in pine and spruce trees consisting of a dense mass of shoots resembling a broom or bird's nest and known as *Donnerbesen* or *Hexenbesen* – thunder brooms or witch's brooms – were cut down and fixed to farmhouse gables before roof-builders copied them and made them permanent parts of the top of the house.[231]

▲ A "thunder broom", a tree-shaped decoration found mostly on the gables of farmhouses around Twente, originally intended to avert lightning. Thunder brooms were also found in the masonry of farmhouses

Like the gable, the farmhouse roof ridge also had its apotropaic appendage. This would often incorporate a thunderstone.[232] There was an old and remarkably persistent belief[233] that thunderstones were the hammers and axes with which an angry Donar wrought havoc and which fell from the sky when lightning struck. As a medium of sympathetic contagious magic they were subsequently used as a defence against all kinds of misfortune,[234] especially storm damage. Thunderstones are in fact belemnites – fossils that were sometimes fashioned into prehistoric tools in the Neolithic and have nothing whatever to do with thunderstorms. Used as amulets, they were placed in all the pregnable parts of the farm,[235] especially the roof. Sometimes, in Denmark for example, they were hidden just under the roof,[236] elsewhere they were put in a jug or clay vase and fixed or immured in the most vulnerable part, the ridge, or attached to the gutter.[237] Of course, there was also no harm in carrying a thunderstone in a pocket as a personal amulet during a thunderstorm, as some French peasants were still doing in 1880, while reciting "Pierre, pierre, garde-moi de la tonnerre" ("Stone, stone, guard me from the thunder").[238] And even in those areas where thunderstones were no longer used as magical repellents they were often still the favourite type of whetstone for sharpening the scythe.[239] The church viewed such "superstitious" practices with dismay and sought to Christianise them in some way. Until the late the nineteenth century in the province of Béarn in south-west France, for example, an earthen pitcher containing holy water was sometimes incorporated into the farmhouse roof ridge to ward off lightning strikes.[240] And around 1800 in the vicinity of 's-Hertogenbosch there were no thunderstones either but small wooden crosses instead, little copies of the pre-eminent Christian symbol that were installed on house roofs to safeguard them from demonic powers.[241]

In addition to thunderstones, certain plants were also thought able to protect the roof. In that context by far the oldest and best-known plant in all of Europe was the "thunder plant" or "thunder beard" – the common houseleek, *Sempervivum tectorum* – also known as *Hauswurz, joubarbe* or Jupiter's beard. It covered many farmhouse roofs, particularly those that were thatched.[242] With pointed reddish-brown flame-like leaves reminiscent of lightning, it was originally dedicated to Donar, and would sometimes be placed in a bowl attached to the roof, as was the case in the region just south of Hamburg, for instance.[243] Some scholars rationalised the ancient magical roots of the houseleek's prophylactic efficacy, for they thought that the little plant, especially in combination with sunlight, was particularly good for purifying foul air by removing phlogiston.[244] In the Mediterranean area laurel leaves and olive branches were placed on the roof to avert lightning.[245] Like the houseleek, they too were credited with lightning-resistant power from as early as Antiquity. The laurel was dedicated to Apollo, whose lyre could calm Zeus when he was disposed to hurl his thunderbolts. And the Roman emperor Tiberius quickly donned a laurel wreath when a thunderstorm approached.[246]

▶ *Sempervivum tectorum*, the common houseleek, also known as the "thunder plant" or "thunder beard", was a succulent that many peasants and farmers encouraged to grow on their roofs as the pointed, reddish-brown, flame-like leaves were reminiscent of lightning and said to have an apotropaic effect

In short, plants, shrubs, and trees were important, and certainly not just for making the roof lightning-proof. Palm branches – since the Revelation to John the Christian symbol of victory of the blessed (Rev. 7:9) – also bore particular significance. We have already seen that the palm branch and apotropaic power over lightning were both part of the St Donatus imagery.[247] But besides being an attribute of martyrs, the evergreen palm branch also had (and has) a special liturgical function in regular Roman Catholic worship, as does the twig of boxwood.[248] On Palm Sunday they would be consecrated during the church service and taken home afterwards as a defence against all kinds of disaster. Often, the head of the family would perform a short ritual,[249] carrying the consecrated palm branch through and around the house and touching the vulnerable places – the chimney, roof beams, fields, and so on. Throughout the year, in extreme weather such as thunderstorms the palm branch was dipped in holy water and carried around the house while prayers were said.[250] Later on, in some homes in North Brabant, the palm branches were even placed behind the electric wiring.[251] Clearly the lived faith that was manifest in the Palm Sunday rituals not infrequently strayed off the ecclesiastically-approved paths. Particularly so when the palm branch acquired a comparable double function as the bunch of herbs that was specially picked at the height of midsummer on John the

▶ A bunch of herbs specially picked on John the Baptist's feast day (24 June) and consecrated on Assumption Day (15 August), then hung above the door or chimney as an amulet. The posy of herbs was believed to offer protection against lightning strikes and other disasters

Baptist's feast day (24 June) and consecrated (in an otherwise official church blessing) on Assumption Day (15 August):[252] on the one hand, a general defence against thunderstorms and other disasters – the bunch of herbs was often hung above the door or chimney to act as an amulet[253] – and on the other, a second protective role when the palm branch, like part of the bunch of St John's wort and other herbs, was thrown into the (apparently unextinguished) fire during thunderstorms.[254] Not coincidentally, in addition to St John's wort (*Hypericum perforatum*[255]) and houseleek, that bunch of herbs often contained Aaron's rod (mullein; *Verbascum thapsus*) and nettles.[256] The church frowned on both practices, decrying them as superstitions.[257]

The idea of sympathetic magic is not far to seek here. With houseleek, Aaron's rod, and flame-coloured poppies the parallel is undoubtedly in the colour – it is even present in the blue of ground ivy (*Glechoma hederacea*)[258] – but with the nettle something else is going on. The plant was already sacred to Donar among the Germanic tribes,[259] being known as *Donnernessel* in Mecklenburg and Silesia, for example.[260] It should therefore be classed with the hawthorn branches – often consecrated and set alight on Holy Saturday[261] – whose thorns (and presumably the smell as well) were supposed to keep demons at bay – an expression of the aforementioned aggressive variety of sympathetic magic. And that it was a crown of thorns with which Christ was martyred during his Passion only increased the hawthorn's putative prophylactic value.[262] Meanwhile, prevention was of course better than cure: by the same sympathetic imitative line of reasoning, for example, cooking peas during a thunderstorm was regarded as a bad idea as it was bound to provoke hail showers.[263]

The point of that line of thought was to collect twigs or branches from those plants, shrubs, and trees that were thought to be more or less immune to lightning strikes. Hazel twigs were popular inclusions in the bunches of herbs, for according to a (partly German) legend the Virgin Mary had found shelter from a thunderstorm under a hazel tree's branches during the flight into Egypt.[264] And as we have seen, olive branches and laurel leaves were also favoured (in the bunch and elsewhere). Beech twigs too were sometimes considered a repellent,[265] because lightning would no more strike a beech tree than a laurel. In the agrarian world of the Ancien Régime, laurel trees would be planted close to the home if possible.[266] Evidently it was considered such a potent defence that the risk of lightning hitting the house itself was deemed to be very small.

While beech and laurel stopped an approaching storm in its tracks the oak did just the reverse and actually attracted lightning. Since time immemorial the oak tree had been sacred to the god of thunder, be that the Greek Zeus, Roman Jupiter, or Germanic Donar.[267] Grown at a little distance from the farmhouse, a tall oak could act as a natural lightning rod, as is apparent from a poem by Willem de Perponcher.[268] It is no coinci-

◀ Thor or Donar with his hammer *mjölnir*, battling giants, who symbolise evil. The word *mjölnir* can mean both sledgehammer and lightning. Painting by Mårten Eskil Winge, 1872

dence that pieces of oak were popular as amulets. They were believed to be particularly protective if they came from a tree that had been struck by lightning.[269] Sometimes that was because it was thought unlikely that lightning would strike the same tree or even part of it twice.[270] But more often it was due to a kind of sympathetic reasoning: whether they came from an oak or some other species, pieces of wood from a tree struck by lightning derived their repellent action from that direct contact with an almost supernatural power, just like thunderstones, and therefore contained as much as *mana*. This "contagious" energy could manifest in many different ways: it could turn a chip of oak into a thaumaturgic toothpick to help prevent toothache,[271] or give the wood magical herbicidal properties,[272] or ensure that when shooting one's aim was as accurate as the lightning bolt itself.[273] It will surprise no one to learn that in Silesia, when thunderstorms raged, splinters of "lightning wood" were thrown into the fire along with part of the bunch of herbs.[274] On the other hand, the back-and-forth play of meanings that has already been mentioned was sometimes in evidence too, because timid or fearful people thought such wood could also attract lightning. To advocates of that argument it was sheer folly – taboo, in fact – to bring it into the home or incorporate it into new buildings.[275]

Whichever side of the argument was taken, sheltering under the oak tree itself was regarded as exceedingly dangerous. Find a beech, not an oak, counselled old Dutch and German folk wisdom. Of course, enlightened eighteenth-century scholars had an explanation for the different frequency with which the diverse species of tree were struck by lightning: the sap of the oak was "watery" whereas in other trees, such as beech, olive, and laurel, the sap was "resinous". Thus argued the eager installer of lightning rods, Cornelis Kraijenhoff,[276] and both the Reverend Johannes Martinet in the Netherlands and the Abbé Giuseppe Toaldo (another keen advocate of the lightning rod) in Italy agreed with him.[277] That resin should have been associated with electricity at that time is not really surprising. It was already known to be non-conductive; moreover, when studying the properties of electricity the French chemist Charles Dufay (1698-1739) had discovered that there were two kinds of charge, positive and negative, which he called vitreous and resinous respectively.[278] Furthermore it had been known since Antiquity that amber – which is the fossilised form of resin – could be made static by being rubbed against an animal's fur. Not by chance is the Greek word for amber *elektron*.

How far eighteenth-century scholars were aware of the magical roots of the "resinous" and "watery" tree debate is an open question. We can no longer look into their souls. The Protestant Martinet was probably not, for otherwise, given his uncompromising rigour in pursuit of truth, he would have expressed a strong aversion to the very idea. And if, on the other hand, the Catholic Abbé Toaldo suspected that said debate embodied the rationalised remnants of an ancient magical belief – as seems indicated – he would more likely have incorporated and reformulated such ideas.[279] However that may be, modern science dismisses the notion that safety from lightning can be found under a beech but not an oak or willow as mere fable. Perhaps the misunderstanding stems from the fact that the electricity in lightning penetrates the bark of some trees more easily than others: the smooth strong bark of a beech tree is comparatively immune, for example, while other species suffer considerable damage.[280]

It was not only plants, shrubs, and trees that were associated somehow with thunderstorms but farm produce as well, sometimes again in a sympathetic and subsequently rationalised way. For example, according to an old belief, a thunderstorm would "turn" the milk. Unsurprisingly, in earlier centuries it was not muggy weather that was blamed for the growth of bacteria but the witches or demons responsible for the thunder and lightning.[281] And because it was not only during storms but also the hours of darkness that witches and demons were abroad it would be foolish indeed to carry milk or yeast through the streets or leave the stable door open once time had been called at the inn. For then the devil's minions had free rein; food would spoil and the cows would stop giving milk.[282] The second edition of the Dutch version of Noël Chomel's well-known dictionary, published in 1778, gave a rational explanation for why various foodstuffs rot as the result of a thunderstorm. It clearly shows a relationship to the discussions about sulphur that have already been mentioned several times in these pages: "Thunder is detrimental to everything that involves fermentation in its preparation, such as bread, beer, wine, etc., because electrical air prevents fermentation; one should therefore close the cellars [compare the open stable doors and the witches in the painting by Roelant Savery, JWB]; the same lest the butter and especially the cream or the milk, should spoil, because the electrical vapours [which were deemed to be sulphurous, JWB] quickly turn all these things sour."[283]

Earlier centuries, however, felt happy about the possession of a large arsenal of magical, partly sympathetic resources. If, "thunder" had unfortunately turned the milk sour, then ipso facto it was believed to be the perfect medium for putting out a fire caused by lightning.[284] In addition to this "contagious" sympathy, our ancestors also had the imitative or homeopathic variant. If there was a thunderstorm while beer was brewing it was an easy matter to prevent souring: a bunch of nettles placed on the edge of the beer vat was all that was required.[285] And, to cite another example, a particularly effective form of *Gegenzauber* ("counter-magic") was to use the moisture that seeped from the dunghill, the equally "sour" liquid manure, which can be interpreted as an expression of the aggressive variant of homeopathic sympathy, although a magic-expelling effect was attributed to the faeces of humans and animals anyway.[286] In any case, it is clear that according to popular belief a fire caused by lightning was very hard to extinguish with water;[287] milk or liquid manure were much better. Examples can be found in Germany and – as late as 1783 – in Domburg in Zeeland.[288] Indeed, such a view was even more widely held: paragraph 48 of the official *Allgemeine Feuer-Ordnung für Altbayern* ("General Fire Regulations for Old Bavaria") of 1791 states that in the case of fires caused by lightning

▲ Cows in a byre, with a witch depicted in each corner of the painting, by Roelant Savery, 1615. The cows are being milked. Witches or demons were traditionally deemed responsible not only for the occurrence of thunderstorms but also the resultant souring of the milk, especially when the barn door was open, as it is here. In Scandinavia it was also thought that witches stole the milk

either caustic alkaline from soap boilers or fresh milk should be the first extinguishing agent to be used. In Lüneburg, finally, in 1767 the preference was for vinegar, a barrel of which was kept on top of every church tower to fight fires resulting from a lightning strike.[289]

That the magical background of such ideas was not always evident also appears from an incident that occurred in Edam in the Netherlands. On the evening of 2 February 1643, lightning struck the tower of the Grote Kerk. Milking his cows nearby was a Mennonite farmer who saw the event through an opening (!) in his barn. Taking a bucket of the fresh milk he rushed to the church and he and the sexton climbed the tower to put out the incipient fire.[290] Although the eighteenth-century author and eighteenth century editor of this story both stated that they were not entirely convinced of its historicity, it should be noted that their doubts were not focused on the credibility or effectiveness of the age-old belief in milk as a fire retardant but only on the somewhat shaky consistency of their sources. Mutatis mutandis, the eighteenth-century natural philosophers who associated the souring of milk with sulphur and therefore with thunderstorms also seem unaware of its ancient magical context. The aforementioned warning against sweating during thunderstorms because such a "sulphurous" exudation would attract lightning therefore seems to be a rationalisation of the overly cautious fear evident in this form of imitative sympathy.[291]

Other dairy products were similarly ascribed an important role in warding off damage caused by thunderstorms. Reputedly, cheese made on Ascension Day[292] and the eggs laid (and blessed) on that day and especially on Maundy Thursday would never go bad; more importantly they were also believed to be particularly effective against disease, witchcraft, fire, and thunderstorms.[293] Which is why such auspicious eggs[294] were often placed at vulnerable spots around the farmhouse: sometimes (in a ceramic bowl) on the roof,[295] sometimes by the hearth or in the chimney,[296] or on the windowsill,[297] in the corner of a recessed door,[298] under the doorstep or, if necessary, bricked up in a wall.[299] In a certain parallel with the sacramental vessels used during Mass, they should never be held in the bare hand, for that could deprive them of their sacred power.[300]

In an agrarian environment eggs were often seen as a symbol of life and fertility, even of the universe itself.[301] The vital power they contained could be transferred to man, animal, and field, it was believed, so they also became a demon-repelling force.[302] If saints were involved, so much the better: as late as the twentieth century in Ghent, eggs were still offered to St Clare, the patroness of good weather.[303] But an egg's vitality was fragile, for it was widely believed that in some mysterious way thunder made eggs go bad and in fertilised eggs it would kill the embryo. To prevent that, an iron object[304] was sometimes placed underneath the nest of sitting hens – a specific form of *Metallzauber* whose usage had already been recommended by Pliny the Elder.[305] Yet although it may seem like an obvious comparison could be made with milk turned sour by thunder, as far as I am aware there are no examples of "lightning-rotted" eggs being used as a means of contagious sympathy to drive away demons. Rather the contrary: the dominant feeling seems to have been timorous fear, for folk would try to placate the storm demons by preparing food and leaving it outdoors or in an open attic window, as if to feed voracious beasts with large licking tongues and so prevent them from allaying their hunger by setting fire to the house with lightning bolts.[306]

On the other hand, the way in which rural communities regarded and treated animals sometimes seems to bear traces of sympathetic magic. The custom of protecting hearth and home from terrifying (thunderstorm) demons by hanging equally monstrous animal skulls on barns or under the roof ridge, as occurred in the southern part of the Black Forest,[307] is probably a case in point. Elsewhere the carcasses of vultures, hawks, owls, magpies, crows, or bats might be nailed to the door, ideally with wings outspread.[308] And from the sixteenth to the nineteenth century antlers were hung in the bell tower of St Stephen's Cathedral in Vienna, no doubt for a similar purpose.[309] When this type

of *Gegenzauber* was used in the northern half of Germany, however, it was sometimes thought enough to hang up a horse's head carved from wood.[310]

Birds were not used merely as a deterrent, though. Sympathetic magic could also have a more or less positive character. For example, a red cockerel, the traditional herald of fire, was simultaneously known to give protection from lightning and fire.[311] Hence an earthenware cockerel was sometimes mounted on the roof ridge.[312] It was also thought that lightning would never strike where the stork – with its red legs and red beak – built its nest.[313] It was not mere coincidence that caused Germanic cultures to dedicate that bird to Donar. Birds' nests in general, and of red birds in particular, were considered valuable preventives against lightning strikes.[314] But the nest built by the swallow, already regarded as a bringer of good luck, also had an apotropaic effect, according to popular belief.[315] The cuckoo, finally, was valued as a harbinger of thunder.[316]

Pets also deserved careful weather-related consideration. Cats with three colours in their fur were believed to bring good luck,[317] but otherwise it was best to banish pets from the house during thunderstorms. The dog was sometimes seen as a lightning devil incarnate,[318] and together with the cat (and the horse) would be chased away from the farmstead when there was a thunderstorm.[319] There were several reasons for it. In the dark, for instance, dogs' and cats' eyes have an eerie glow that was once associated with lightning. And secondly, dogs, cats, and horses were said to be "magnetic" because the hairs in their coats could easily be charged with static electricity: was that not the very reason why those creatures might attract lightning?[320] For eighteenth-century people electricity once again proved to be a mysterious force, indiscernible in itself and only becoming manifest at the moment that discharges took place. What natural or even supernatural energies were hidden in animals and in the world as a whole? As we will see below, there were some – not infrequently from the higher socio-cultural milieus – who went so far as to suggest that in the static electricity thus evoked a deep cosmic or perhaps even occult power had been unearthed.

◀ Landscape during a thunderstorm with stork's nest, overdoor by Derk Jacobs Postma, 1831. In the imitative magic of popular belief, storks were often regarded as protectors against lightning strikes on account of their red legs and beaks. The stork was already dedicated to Donar among Germanic tribes

▲ *Magnetism Unveiled*, anonymous cartoon, France 1784–1785. On the left, Franklin flourishes a copy of the report on mesmerism that he, Lavoisier, and other members of the royal commission of inquiry had prepared, while at the top right Mesmer and a colleague flee on broomsticks, accompanied by the devil. On the ground a mesmeric tub has toppled over, around which two ladies, one of them blindfolded and naked, almost fell victim to the mesmeric "therapy" of two men

6

Intermezzo: Electrical Nature? The Animated Nature of Theosophy

To the ordinary man of the eighteenth-century, electricity was a mysterious force. But to others it was a natural elemental energy of universal cosmic significance, a force, moreover, with excellent potential for therapeutic use. The Austrian physician Anton Mesmer claimed to have discovered a vital principle, which he called "animal magnetism" or "animal electricity", with which he professed to effect spectacular cures. Theosophists, meanwhile, maintained that a "spiritual incarnation" had occurred on the fourth day of Creation when the sun was formed and the light that came into existence on the first day of Creation "sank" into matter itself to become an embedded principle, and that consequently natural science and theology were in perfect harmony. Some saw the same harmony between nature and grace; likewise between electricity and spiritual rebirth. This "strange enlightenment" or "natural magic" contradicted both reason and wonder, and therefore theosophy operated in a kind of twilight zone, mistrusted by the Church and secular science alike: not pious enough for the devout, too magical for the enlightened.

Part of the wondrous energy of nature mentioned in the previous chapter was revealed, according to some, when a person injured by lightning was believed to have acquired healing powers as a result. We find instances of such contagious sympathy occurring well into the nineteenth century if not later. On Maundy Thursday in 1865, for example, a widow from North Brabant who had been hurt by lightning while in the Old St Willibrord's Church in Middelbeers claimed to have received the power to cure rheumatic pains by stroking the affected parts with her left hand – and she was not alone in making such assertions.[1] *Mana*, the healing power that was said to be present in lightning, had turned the Brabant widow into a female magnetiser.

Some people, scholars included, went beyond that popular belief, however. They had long been certain that electricity in general – so not just lightning strikes and the *mana* they allegedly bestowed on the unfortunates they injured – was not merely a curious peripheral phenomenon but a natural elemental force of universal, cosmic significance. A force, moreover, with excellent potential for therapeutic use in medicine and which could be transferred from person to person by stroking and therefore in a certain sense by "sympathy".

◀ *Mesmeric Therapy*, painter unknown, possibly Claude Louis Desrais, between 1778 and 1784. A group of patients undergoing "magnetic" or "electrical" therapy, gathered around a large *baquet*, a tub-like apparatus devised by the Viennese physician Franz Anton Mesmer. Into the *baquet*, Mesmer maintained, he had transferred "magnetic" fluid from his body. His patients hoped that the contact between the "magnetic" fluid in the *baquet* and the volatile "nerve juices" in their own bodies would cure all manner of ailments, such as afflictions of the eye. As a therapist, Mesmer asserted that he could also transfer the "fluid" of the "animal magnetism" believed to be present in the human body to patients individually, especially by stroking the affected parts or even by his eye movements

In late eighteenth-century Europe no one was more influential in this respect than the Austrian doctor Franz Anton Mesmer (1734-1815), renowned for his ideas on so-called animal magnetism. Rooted in older, Neoplatonist-inspired cosmological views that suggested that magnetism and sympathetic action were rather interchangeable concepts[2] while simultaneously foreshadowing the cosmic-religious notion of sympathy in Romantic natural philosophy, he and his followers had done much to remove stroking

◀ A smaller variant of Mesmer's *baquet*. Here too, the combination of ropes and iron rods that connected the patients to each other and to the vessel is striking

from the realm of popular belief and transfer it to the academic sphere – a remarkable turn-about at a time of *Volksaufklärung*, even though there were many who would soon call Mesmer a charlatan. By making "magnetic passes" or sweeping movements of the hands over certain parts of his patient's body, Mesmer claimed, he could transfer the animal magnetism cached in his own body to the patient and thus remove whatever was obstructing the smooth flow of vital fluid. What in later mesmerism was referred to as *Nervenvermählung* or neurogamy (Greek for "marriage of nervous systems"), the outcome of a "harmony" between the nervous systems of the therapist and patient, should, according to Mesmer, lead to a "crisis". It was true that the crisis could be accompanied by violent convulsions, but, so claimed this father of all magnetisers, it was precisely the reaching of such an extreme state that had enabled him to effect a spectacular cure in some of his patients.[3]

Before the Leyden jar was invented it was not always easy to distinguish magnetism from (static) electricity due to the minute energy of the observable electrical discharges. And even when it did become possible to tell the difference it was still many decades before the exact relationship between the two was understood. So it is not hard to see why Mesmer himself took such a long time to decide whether the vital principle he had "discovered" should be called animal magnetism or animal electricity.[4] It was clear, however, that neither phenomenon could be explained purely mechanically.

▶ Portrait of Franz Anton Mesmer, engraving by Louis Legrand (1723-1807), after a design by André Pujos, last quarter of the eighteenth century

▲ *Patients in Paris Receiving Mesmer's Animal Magnetism Therapy*, print after Claude Louis Desrais, ca. 1780. To the left of the *baquet* a lady receives individual treatment

Meanwhile, the discovery that fish like the electric eel and electric ray stunned their prey not by some mysterious method but by delivering an electric shock (from a short distance), had given the notion of animal magnetism a certain plausibility. That being the case, it was wondered, why should the human nervous system not accommodate electricity just as well as the animal's did.[5] While there would be little point in making sharp distinctions between "real" science and pseudoscientific magnetism at a time when scholarly minds were still blithely preoccupied with imponderables such as phlogiston, the weak point in mesmerism was, nonetheless, precisely its lack of empirical evidence. That many advocates claimed the contrary, maintaining that their own experiences provided proof, changed few minds on that score. The royal commission appointed by Louis XVI of France in 1784 to examine the matter and which included such luminaries as Benjamin Franklin and the disprover of the phlogiston theory Antoine Lavoisier (1743-1794), concluded that the existence of an electric or magnetic fluid in humans and animals had not been demonstrated. Indeed, the commission dismissed the effects of mesmerism as imaginary, although the worldly-wise Franklin was willing to admit that illusions were sometimes useful.[6] The commission's unequivocal judgement, which in a broader sense was an expression of the progressive professionalisation of the various scientific disciplines, led to mesmerism being discredited in France, just as it had been in Austria, which Mesmer had been obliged to leave in 1778 after a failed attempt at a cure. Moreover, eminent scholars working in Vienna had spoken out against Mesmer's theories – perhaps all the more emphatically in the case of Jan Ingenhousz since his own pioneering work in electricity had been branded as necromancy by the people of Breda two decades earlier.[7]

There is no room for even a brief history of animal magnetism in these pages, nor is this the place for it. It is clear, however, that the same scholarly electrical research that led to the invention of the lightning rod also gave a powerful boost to the revival of ancient theosophical ideas about the animate nature of the cosmos, with particularly remarkable consequences for the exegesis of scriptural passages on creation and regeneration. And it is equally clear that as a typical eighteenth-century variant of the theosophical tradition, Mesmer's ideas found the most response in Germany and little in England.[8]

Friedrich Christoph Oetinger (1702-1782), for example, a Lutheran theologian and founder of the theosophical school of thought in Württemberg Pietism, published on what he called "the electric fire", a superfine fluid that animated man, animal, and indeed all life. It is not surprising that his description of it bears a close resemblance to Mesmer's explanation of animal magnetism.[9] Oetinger was convinced that with the discovery of electricity, science had unearthed nothing less than the light that came into existence on the first day of Creation. On the fourth day, when the sun was formed, that

▶ Portrait of Friedrich Christoph Oetinger, founder of the theosophical school of thought in Württemberg Pietism, print by Jacob Andreas Fridrich the Younger, 1724-1779

first light was "sunk" into matter itself, as an embedded principle, so Oetinger believed.[10] Thus this intangible animating principle had been built into God's creation right from the start. That there should be a very direct connection between spirit and nature and thus between theology and the natural sciences was therefore inevitable.

In addition to his own prolific writings, Oetinger also expounded his views when he published a manuscript by his Moravian friend, the Premonstratensian priest, theosophist, and electricity pioneer Procopius Diviš, whom we encountered in earlier chapters. Towards the end of his life Diviš had produced a manuscript with the revealing title *Magia naturalis*. But its publication had been prevented by Gerard van Swieten (1700-1772), Maria Theresa's personal physician, prefect of the Imperial Library, and president of the book censorship commission. Objections had been made by the ecclesiastical censor as well.[11] After Diviš's death, however, Oetinger and a number of like-minded *theologi electrici* translated, edited, and published it.[12]

Diviš, who had also had a medical practice decades before Mesmer, subscribed to the theosophist principle that a "spiritual incarnation" had taken place in all creation. "Electric fire" was already present in the clay from which Adam was made and therefore, as an "electriser", Diviš endeavoured to activate that fluid in an attempt to stimulate blood circulation and particularly to heal the paralysis caused by a stroke.[13] At the

◀ Portrait of Augustin von Balthasar, professor at Greifswald University, print by Johann David Schleuen the Elder, 1740-1774

same time, given the purported harmony of theology and natural science, all kinds of things fell into place from an exegetical point of view as well, for Diviš and every other theosophist.[14]

That was not only in the case of the biblical passages relating directly to the Creation but also those about regeneration. In 1745 Augustin von Balthasar (1701-1786), a professor at the university of Greifswald in north-eastern Germany suggested that the "tongues, as of fire" which according to the Bible rested on each of the apostles on the day of Pentecost (Acts 2:3) had been expressions of burning electric fluid[15] – an opinion shared five years later by the Dublin-trained Church of Ireland clergyman Richard Barton (1706-1759), who was presumably also influenced by theosophical ideas.[16] For most theosophists, however, a straightforward identification of nature and grace was going too far if only because the Gospel according to Luke, who is presumed to have authored the Acts of the Apostles as well, states that Christ would baptise people with the Holy Spirit *and* fire (Lk. 3:16).[17] On the other hand, the respective realms of nature and the Spirit were at least very analogous and mellifluously harmonious. In this connection, therefore, electricity should be seen as an accompanying and possibly even an intermediate phenomenon – thus argued the theosophists. Those who adhered to the latter view were no longer surprised that according to the New Testament the reception of the Holy Spirit was effected by the laying on of hands. And certainly in later times, when a powerful religious experience or conversion was accompanied by convulsions, there were some theosophists for whom the analogy with the "crisis" induced in mesmerism was well nigh complete. Retrospectively they interpreted the violent convulsions experienced by Jansenist pilgrims when visiting the tomb of the ascetic François de Pâris (1690-1727) in the churchyard of Saint-Médard in Paris in the 1730s as mesmeric crises.[18] And, mutatis mutandis, in the eyes of some American evangelicals there was if not a direct relationship then at least a metaphorical one between experiencing electric shocks and undergoing the pains of a spiritual rebirth.[19] Did they not both emanate from the same ethereal fire which, as a natural and spiritual medium, permeated and nourished all living beings and subjected them to God's laws? For some, such as the late eighteenth-century itinerant physician, electrical scholar, and millenarian T. Gale of New York State, rebirth thus became an integral part of a natural process despite its supposed transcendent causation.[20]

▶ In the 1730s Jansenist pilgrims were seized with convulsions at the tomb of the ascetic François de Pâris (1690-1727) in the Saint-Médard churchyard in Paris. Anonymous French print, eighteenth century, from Bernard Picart et al., *Cérémonies et coutumes religieuses de tous les peuples du monde* (9 vols.; Amsterdam: J.F. Bernard, 1723-1743)

Even enlightened critics of "enthusiastic" Christians of that type were sometimes unable to unplug themselves from "electrical" assumptions. For while one party interpreted a sudden conversion metaphorically or theosophically as the lightning strike or the electrical shock issuing from the Holy Spirit,[21] the other saw only the natural operation of electrical phenomena or what was worse, a bigoted zealotry with a dubious resemblance to the artificially generated current of an electrical machine.[22] In short, while both parties acknowledged God's omnigovernance, the choice seemed to be either to claim command of electricity in true Enlightenment fashion or to meekly accept that this animate cosmic force had been man's master since the Creation. In the first case, religious enthusiasm might be explained away as, at best, an abuse of natural electrical irritability if not utter self-delusion and vain fanaticism. In the second, opting for a theosophical interpretation of the natural phenomenon was obvious.[23] In practice, however, such a dichotomy was too simplistic. For on one hand, even a rationalist like Franklin was humble enough to admit that electricity had the power to amaze mind and body time and again,[24] while on the other the cures effected by the Austrian-born priest Johann Joseph Gassner (1727-1779) using the ancient sacerdotal practice of exorcism, for example, were appropriated by Mesmer, his rival, and consequently secularised as expressions of animal magnetism.[25]

Even more important than the breadth of the spectrum of views, however, is the capricious course of the lines that divided them. Something strange seems to be going on, particularly with the theosophical, often rather concrete interpretation of the Bible. For if the Holy Spirit literally used electricity in man's conversion, not only did the new natural science take on the old function of the handmaid of religion but that dependent relationship also led to an increase in the importance of science![26] Whereas Enlightenment and enthusiasm were antagonistic just about everywhere in the eigh-

▶ Illustration of the *Beatifikation*, in which a subject was insulated and electrified. As a result, when seen in the dark a halo appeared around his head. Possibly the theosophist Augustin von Balthasar was inspired in his interpretation of the Pentecostal "tongues, as of fire" described in Acts 2:3 by this experiment by his compatriot Georg Matthias Bose

teenth century, curiously enough they were compatible in this re-sacralisation.[27] That "strange Enlightenment",[28] or, if preferred, "natural magic"[29] turned against both reason and wonder. For while on one hand, theosophy's religious concretism also needed visible signs of God's power in everyday life, on the other there were ecclesiastical protests against biblical miracles being naturalised as the simple consequences of the action of animal magnetism or electric fluid. To regard Christ as a magnetiser as a result of such an instrumental, intermediate point of view was therefore mistaken, argued the young Dutch minister Guillaume Henri Marie Delprat (1791-1871) in 1817, for example.[30] And while enlightened minds could, in turn, appreciate such a "disenchantment" in its own right, at the same time they had a sneaking suspicion that enthusiasm was being spruced up with a touch of the scholarly. In other words, theosophy operated in a kind of twilight zone, mistrusted by the Church and secular science alike: not pious enough for the devout, too magical for the enlightened.

All in all, eighteenth- and early nineteenth-century theosophy ultimately remained a rather marginal phenomenon, despite its transnational and trans-denominational character. It is certainly true that it had stemmed partially from science, for Mesmer was to some extent justified in referring to Newton and his ether theories[31] and it is obvious from Diviš, for example, that theosophy, the Church, and science could sometimes see eye to eye. Initially the Italian anatomy professor Luigi Galvani (1737-1798) even seemed to provide sound arguments for the existence of what he called animal electricity, for in 1786, when he caused a dead frog to twitch and jerk without an external electrical source but only by touching it with curved rods of various kinds of metal, he concluded that the small cadaver functioned as a kind of Leyden jar.[32] Yet there was an important difference, for unlike speculative mesmerism, which had come to be seen by numerous people as quackery, Galvani relied on scientific experiments with correctable conclusions.[33] And from the religious point of view, furthermore, the direct animate connection that theosophical philosophy postulated between nature and supernature ultimately satisfied only a relatively few. If nature (as in physico-theology) was not only the indirect but even the direct revelation of God, very many saw both realms as belittled. Such total or partial identification, somehow implying the equating of creation and re-creation, or of electricity and rebirth, was unacceptable to both the pious and the enlightened. For divergent reasons, both groups preferred a looser relationship. Even those who were more cautious in that respect and posited a halfway, instrumental relationship could sometimes count on opposition. For much as the more pietistic or evangelical Christians were willing to compare conversion with the thunderbolt of the Holy Spirit, and much as the convictions of a Christian theosophist like Willem Bilderdijk (1756-1831) were drenched in metaphorical analogical thinking in this respect,[34] when it came to the Pentecostal miracle itself, orthodoxy gladly left the strictly figurative interpretation of the fiery tongues to more enlightened minds. Likewise the naturalistic interpretation of other biblical miracles.[35]

▶ Portrait of the enlightened minister and professor Jacobus Albertus Uilkens (1772-1825) of Groningen, painting attributed to Wessel Lubbers, 1820-1825. Uilkens more or less combined theosophical and physico-theological concepts, and consequently he sometimes leaned towards a metaphorical interpretation of biblical miracles. As a typical enlightened minister, Uilkens also published a *Verhandeling over de vooroordeelen omtrent het onweder* [Treatise on the Prejudices about Thunderstorms] (Groningen, 1808), sadly now untraceable

By now it might be clear that no other eighteenth-century phenomenon harboured such a wealth of metaphorical possibilities as the newly discovered, mysterious phenomenon of electricity, of whose central significance the invention of the lightning rod was both an effect and a cause. Attempts to describe its action in all its intangibility soon necessarily devolved into metaphoricity. The interaction between knowing and not-knowing, or, if preferred, between mastery and mystery in the study of this natural phenomenon, thus formed the juncture for its poeticisation.[36] For conversely, the electrical metaphor also lent itself perfectly to denoting the most diverse things. That almost inexhaustible metaphorism, which manifested not only on the theological level but extended into many other areas and finally still animated nature – now in a literary-aesthetic sense – merits a chapter all to itself.

▶ The Italian scholar Luigi Galvani initially seemed to provide sound arguments for the existence of what he called animal electricity, for in 1786, when he caused a dead frog to convulse without the aid of an external electrical source but simply by touching it with curved rods of various kinds of metal, he concluded that the small cadaver acted as a kind of Leyden jar. In this regard, however, Galvani's ideas were corrected by Alessandro Volta: the dead frog's convulsions were caused not by animal electricity but by the electromotive force of various kinds of metals or other heterogeneous conductors. This experiment captured the literary imagination. Soon Mary Shelley would use this mysterious galvanic electricity in her gothic novel *Frankenstein* (1818), turning its eponymous protagonist into a modern Prometheus as he creates life itself

Fig: 1.

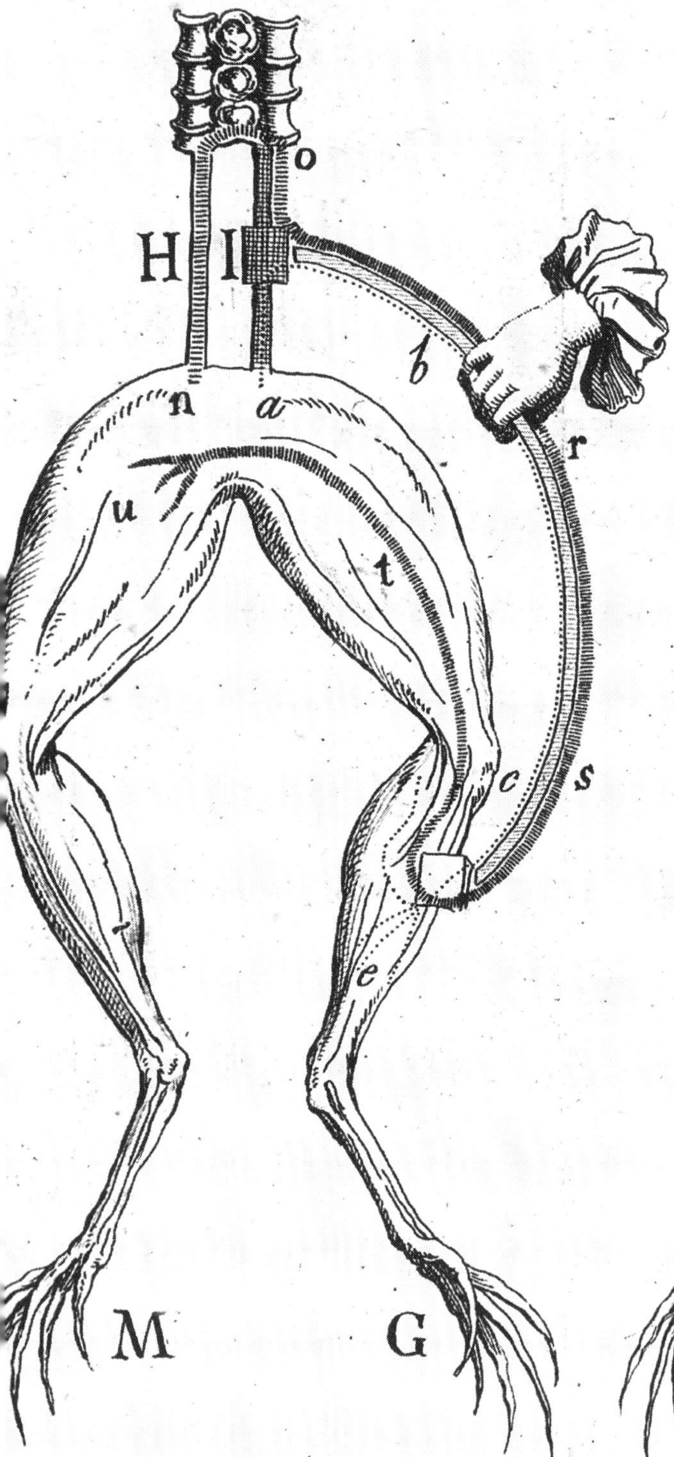

Fig: 2.

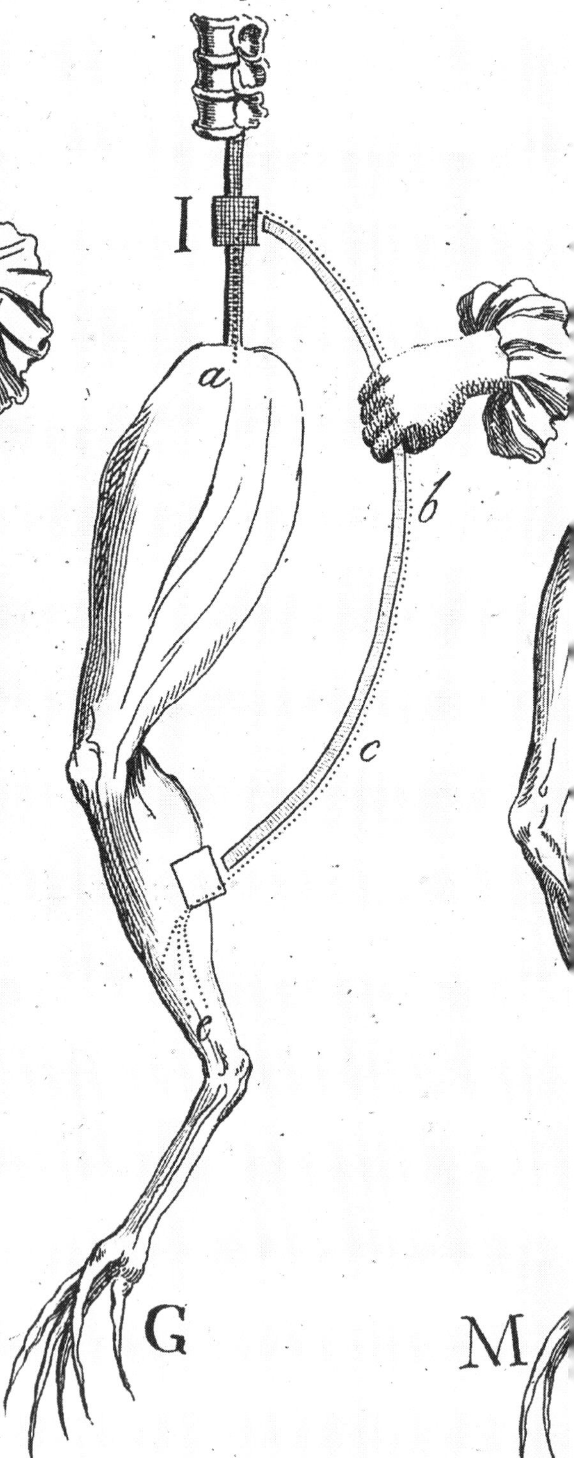

Fig: 3.

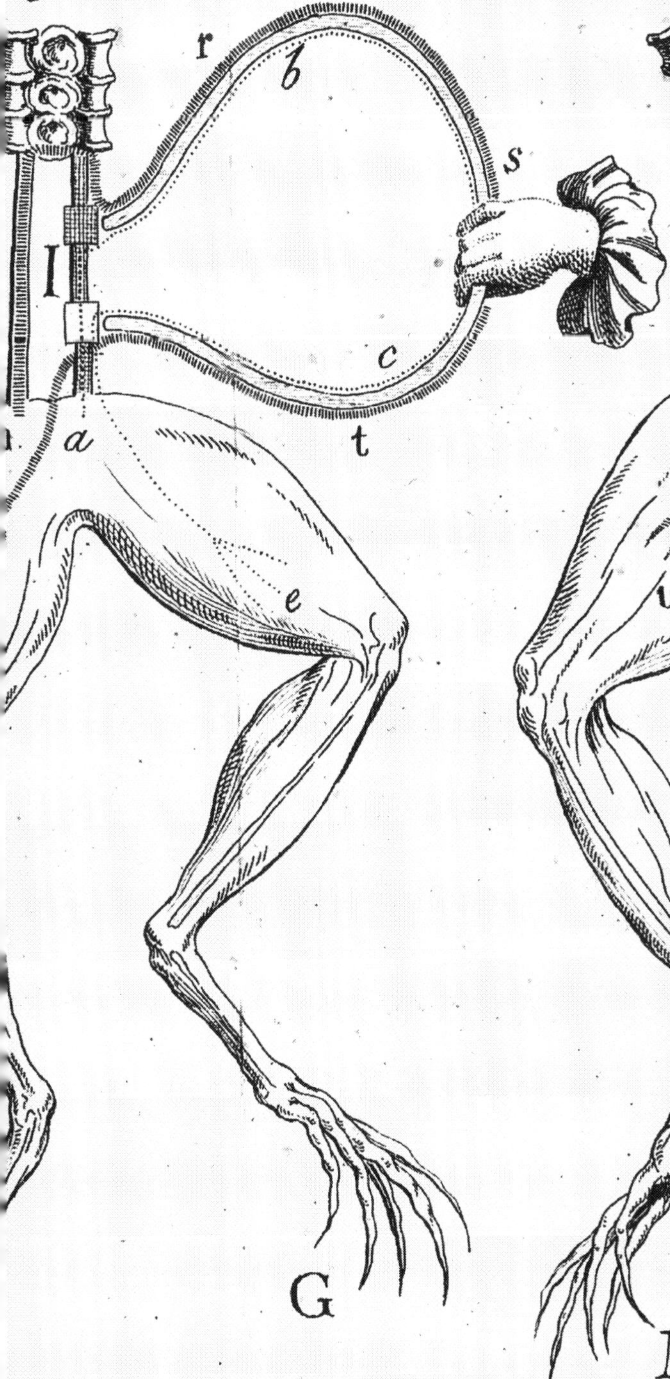

Fig: 4.

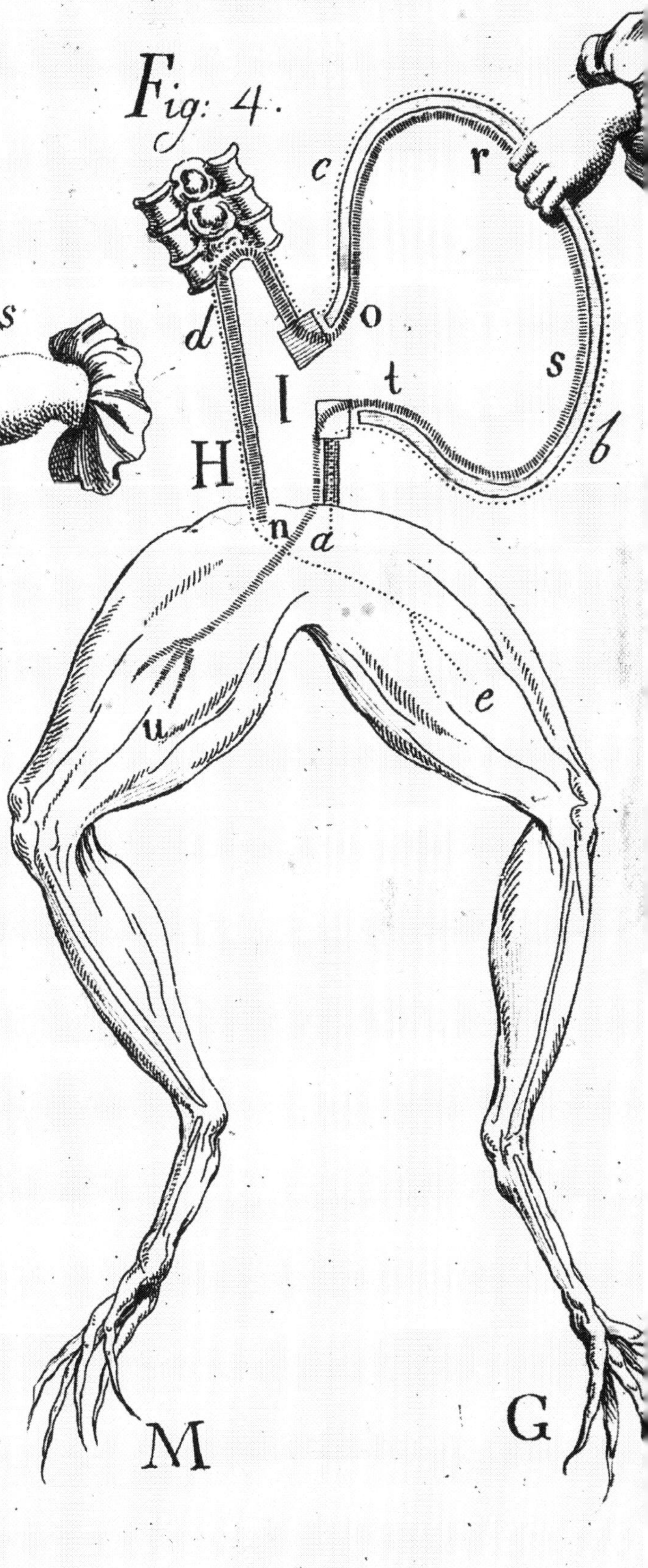

HOMERI OPERA

Art

◀ Portrait of Willem Bilderdijk, painting by Charles Howard Hodges, 1810

▲ The "Venus electrificata" or "electric kiss", French print, eighteenth century. This experiment, devised by the German poet and physicist Georg Matthias Bose, involved having a lady stand or sit on an insulated base and charged with electricity, then inviting gentlemen to kiss her, which of course resulted in a rather painful shock as electricity discharged between their bodies

7

Thunderstorms and Electricity in Poetry, Music, and Painting

In the eighteenth century electricity, especially lightning, was often employed to symbolise power – in both a political and a sexual sense. Revolutionaries used thunderstorm symbolism to denote tyrannical rule, while more conservative propagandists deployed it against the movements of revolutionary masses. Symbolically the thunderstorm also lent itself to all kinds of war, violence, and quarrel. With the dawnings of Romanticism in the last decades of the eighteenth century the electrical metaphor came into its own. With its unpredictability and spontaneity, atmospheric electricity became a trope for artistic inspiration. Sometimes it worked the other way round – the thunderstorm itself took centre stage and was represented through poetry, music, and painting. Edmund Burke introduced a new version of an age-old aesthetic category, the sublime, which was now used primarily in reference to the grandeur of nature. Beholding a sublime scene could simultaneously evoke both terror and fascinated delight when viewed from a place of relative safety. Immanuel Kant, on the other hand, focused on the observer's autonomy, the moral superiority that enabled him to master nature's violence by his own energies. To many enlightened minds the invention of the lightning rod was one of the clearest proofs of that moral triumph.

The Sexual Metaphor

The dividing lines between theosophy and metaphorism were not always equally clear. Even those who did not immediately associate a vigorous conversion experience with the effects of real electricity could often appreciate the imagery of a lightning strike, as if such listeners were "electrified" by a "thunder preacher" in religious gatherings.[1] But the border between theosophy and metaphorism was even thinner in the sexual realm, for hardly anywhere did the animate character of nature, or to put it more earthily, the close bond between mind and body, seem more tangible than there.

▲ *The Electric Spark*, painting by Louis Léopold Boilly, ca. 1791. In this allegory of courtship, Boilly depicts an original combination of the ancient alchemical search for the elixir of life with the operation of the modern electrostatic generator. The latter is connected with a statue of Cupid, whose arrow the girl somewhat anxiously touches with her fingertip. The spark that is about to jump will give her an erotic charge. Or does the human body contain "voltage" anyway? According to Mesmer, the fingertips and eyes are the main points of transference of "electric fluid". Nonetheless, it hardly takes a mesmerist to make a connection between electricity and sexuality. Long before Mesmer, Albrecht von Haller had already written about the amazing significance of the finger in the transmission of "electricity". "Could one believe that a lady's finger (...) should send forth flashes of true lightening?" he enquired in the April 1745 edition of *The Gentlemen's Magazine*

According to theosophists, the human body contained a great deal of electricity and sometimes there would be a surge in the current, as it were.[2] Women were deemed to be particularly susceptible to the phenomenon, although the London physician John Shebbeare (1709-1788) did believe that the male erection also resulted from the action of a "vital fire", while the French scholar Charles Rabiqueau (active around 1750-1785) saw the entire process of human reproduction as electrical. The controversial Scottish therapist James Graham, who opened a "Temple of Health" in London in 1780 and famously touted his "celestial bed" with its range of stimulating aural and olfactory embellishments, advertised electricity – natural and artificially-generated – as a major feature in his sexual therapy, though arousing his clients' imagination was also a factor in the "treatment".[3]

Making a connection between electricity and sexuality was not the exclusive province of mesmerists and theosophists, nor of artists or scholars. Anyone who regarded electricity in all its volatility as no more than one of the symbols of male or female passion, ipso facto ascribed great importance to such energy in the human body. Nevertheless, we shall see that atmospheric electricity sometimes had an equally important part in that discourse. In the first case it was often a direct metaphor, in the second usually an indirect one.

In the Dutch Republic, undisguised sexual symbolism could be found very early on, as in the burlesque entitled *De electriciteit; of Pefroen, met het schaeps-hoofd. Ge-electriseerd* ("Electricity; or Pefroen with the Muttonhead. Electrified") from 1746, for example.[4] It is still there half a century later, in the pages of a *Vrolijke reis van een Engelschman, door Holland* ("Merry Journey of an Englishman, through Holland") penned by the versatile writer Gerrit Paape (no theosophist himself) but published anonymously in 1796. At one point in the story the protagonist visits a fancy haberdashery to buy gloves. As the shopkeeper presents her wares for closer inspection she touches the traveller's fingers in such a sensual way that he repeatedly receives "a beatific electric shock (...) which had its effect directly on the heart". As the shopkeeper entices him into trying on her goods, one of the glove's fingers "accidentally" splits and the disconcerted Englishman is surprised to find her tenderly squeezing his exposed digit[5] as she gazes at him "with an eye that pierced my whole being like lightning." It soon becomes clear that the haberdashery is nothing but a genteel knocking shop.[6] It seems likely that in the passage quoted here Paape is making an implicit reference to mesmerism, for it was well known that the eyes and fingertips were the places where the transfer of electrical fluid supposedly occurred.[7]

▶ Portrait of Gerrit Paape, radical author and journalist, print by Mathias de Sallieth, 1788

Whatever the case, it is clear that electricity – particularly static electricity, which could be generated by friction – easily leads to direct "physical" metaphorism. For how suddenly and readily sparks of love leap from one person to another (a favourite theme in elite Italian epithalamia, for example[8]), and how much of human relationships consist of irresistible attractions and repulsions, not to mention the building up and release of tension, let alone convulsive spasms. The English author John Cleland (1710-1789), who was far more unrestrained than Paape, makes abundant use of that sort of straightforward erotic imagery in his notorious novel *Fanny Hill, or the Memoirs of a Woman of Pleasure* (London: G. Fenton, 1749), in which electricity becomes a metaphor for the vital force behind sexual libertinism.[9]

The Marquis de Sade, in turn, used this kind of imagery to such an extent that even atmospheric electricity takes on a direct function in his sexual symbolism. So direct, in fact, that lightning becomes a phallic substitute[10] and the woman is primarily a passive object of pleasure, which arouses a libidinous "thunderstorm" in the man that is then discharged in her womb.[11] Although De Sade usually left a little more to the imagination in his novels, in one way or another virile sexuality is invariably associated with violence in nature. In other words, he employed lightning – now made manageable by man – as an expression of licentious and often perverse desire, whose victims were inevitably virtuous women.[12] In many other eighteenth-century French novels, too, it is notably women who are the sufferers of tempests and thunderstorms,[13] although the sexual metaphor (if any) is almost always indirect.

It is quite a step from De Sade to the Dutch poet Willem Bilderdijk (1756-1831). Yet in his works, too, atmospheric electricity sometimes seems to be gravid with sexual suggestiveness. For not only did that Christian theosophist consider ethereal electricity so materially present in man's body that he shudderingly repudiated mesmerism as an immoral and even satanic monstrosity,[14] he was also writer enough to turn the thunderstorm to account as an indirect sexual metaphor. The impetuous Bilderdijk knew well what he was talking about, for though a married man he had fallen in love with his much younger student Katharina Schweickhardt (1776-1830). When his mistress kept her distance in the year 1796, the anguished Bilderdijk sought solace in poetry. In that tormented state of mind, and borrowing freely from an epitaph composed by Lady Mary Wortley Montagu (1689-1762),[15] he penned a macabre little poem entitled "On the marriage bed of a bride and groom, struck by lightning on their first wedding night."[16] Evidently that was how close passionate love and gruesome death seemed to him. But – and that is the point in this context – not coincidentally (and entirely in keeping with classical tradition[17]), atmospheric electricity and sexuality are still closely linked here, albeit via an indirect metaphor. The harmonic cosmic parallel that the theosophical poet draws between the "flame" which united the young couple and the thunderbolt with which the all-good God sealed that love is typical. Nor does it take much imagination to conclude that Bilderdijk, like many other writers of his time – certainly the authors of popular novels – projected his own sense of disharmony onto storms or other devastating scenes of nature.[18] Later on, Bilderdijk continued to be preoccupied with the theme of death in the bridal bed or the prematurely terminated marriage.[19] But in contrast to comparable sixteenth-century stories, such as the tale of a wealthy farmer's riotous son whose wedding night was abruptly ended when he was fatally struck by a divinely directed bolt of lightning,[20] his image of God is invariably a mild one.

Metaphor is not always involved, therefore. For while in the early modern period the thunderstorm was still regarded as the punitive instrument of the Almighty, without literary detours, in the literature of the late eighteenth century it is sometimes little more than an accompanying stage effect that amplifies the drama of love or sexuality. Unlike Feith's *Julia* (1783), in which thunder still symbolises the temptations of looming lust, in *Het onweeder*, a poem by Jacobus Bellamy (1757-1786) in which an amorous couple in a rose arbour are surprised by an oncoming storm, the mighty electrical force of nature is not even an indirect metaphor.[21] The most it does, in that poem and elsewhere, is serve to underline fateful events or as a "symphonic accompaniment to the emotional heights."[22]

▶ As a storm rages, Dido and Aeneas take shelter in a cave, painting by Pierre-Henri de Valenciennes, 1792. Here the painter takes up a theme which shows that even in Antiquity atmospheric electricity and sexuality were already linked through an indirect metaphor. In the fourth book of Virgil's *Aeneid* the lovers hide from the storm in a cave and in that safe place they plight their troth

Even without a metaphorical context, electricity and sexuality were sometimes connected, as was divertingly demonstrated by the famous "Venus electrificata" or "electric kiss" experiment, invented by the German poet and physicist Georg Matthias Bose (1710-1761). It involved seating a lady on an insulated stool, charging her with a moderate dose of static electricity, and then inviting gallant "suitors" to kiss her – which naturally resulted in a rather painful shock as electricity discharged between the two bodies. "Scientific" demonstrations of the "electric kiss" were extremely popular, particularly in aristocratic circles.[23] For if there was no natural electricity present in the female body, there was certainly a willingness to introduce it artificially – the stereotypical notion that female nature was driven by passion (rather than reason) was virtually emblematic of the contemporary zeitgeist.[24] It is typical, then, that here, with the invention of the electrostatic generator, static atmospheric electricity was artificially converted into its "physical" equivalent – in fact a disillusioning reversal in which the metaphor of unfulfilled desire was reduced to a painful, prosaic reality.

◀ Portrait of Georg Matthias Bose, the German pioneer of electricity, anonymous print, eighteenth century

The Political Metaphor

In the eighteenth century atmospheric electricity, particularly lightning, was often used to symbolise power – in a sexual sense, certainly, but equally if not more so in the political arena. We need only recall Turgot's epigram on Franklin[25] to realise that notions of wresting lightning from heaven and the sceptre from tyrants not only licensed brazen libertinism in the likes of De Sade, but also emboldened many would-be republicans.[26] Occasionally, the two were combined, as in a salacious French political cartoon in which the virile charmer Franklin as proxy violator forces himself on a female representing the American republic, with an electrostatic generator in the background.[27]

▲ Franklin as a political and sexual plenipotentiary, anonymous French cartoon, last quarter of the eighteenth century. As proxy violator the virile charmer Franklin coerces a woman who represents the American republic, with an electrostatic generator in the background

He may have snatched the flickering sceptre from tyrants, but even in the case of the sexual and political plenipotentiary Franklin, the metaphor of such "individual" electricity remained linked to power. Moreover, the presence of the electrostatic generator did not bode well, since to the loyalists who opposed the American Revolution, Franklinian republicanism was the equivalent of artificially generated electricity that violated "the spontaneous work of nature and of God."[28]

For their part, revolutionaries in both America and Europe not only recalled with horror the thunderbolt-hurling, despotic Jupiter – the "archetype and patron of monarchs"[29] – not to mention what they saw as the misuse that Christian princes made of such images of a punitive deity, but were also quick to contrast that capricious and intimidating regal figuration with a diametrically different and often collective metaphor. For did not the reverberant thunder of the revolutionary masses rise at least as rapidly, and did it not evoke at least as much terror? Indeed, was it not as uncontrollable and devastating as the lightning bolt of every tyrant known?[30] But since even the destructive power of nature could be seen as something positive,[31] the storm of revolution was also simply a part of the natural order of political things. In the Age of Enlightenment, that evocation of their supposedly "natural" character gave an inherent legitimacy to revolutionary forces.[32] A justification so self-evident that such a more or less affective appeal was sometimes deemed to make extensive substantive argumentation superfluous[33] – in their implicit suggestiveness metaphors not only reveal something but also conceal it. An additional advantage of thunderstorm imagery was the implication that it would be futile to resist such autonomously destructive natural phenomena. It was better to bow to the inevitability of revolutionary power, for the status quo was doomed.

That confident message was as reassuring to revolutionaries as it was alarming to representatives of the Ancien Régime. At the same time, however, contemporary concepts of makeability and free will were at odds with the thinly-veiled fatalism that such metaphors concealed. While it might indeed be the case that very little could be done about the cause or course of a thunderstorm (any more than earthquakes and volcanic eruptions[34]), the consequences of atmospheric electricity were now another matter – which is why the lightning rod sometimes became a counter-revolutionary symbol. On the other hand, the amount of static electricity could also be influenced, and so, as we have seen, the electrostatic generator became a revolutionary emblem, because in order to drive out tyrants revolutionary potential could be artificially increased considerably. And then, of course, all kinds of intermediate gradations were possible. For even though, to cite just one example, the German thinker Johann Gottfried Herder (1744-1803) was no enemy of revolution, he was anti-fatalistic enough to loathe the widespread interpretation of the French Revolution as a thunderstorm or an earthquake. He preferred to see it as man's carefully conceived effort to achieve the best form of government.[35]

Finally, there was also disagreement about another implication of the thunderstorm image, that being the belief that once all the revolutionary showers had blown over a period of beneficent, idyllic tranquillity would naturally ensue because the political air had cleared and all kinds of annoying old-fashioned "clutter" had been tidied away.[36] That idea can be found among initial admirers of the French Revolution such as Friedrich Gottlieb Klopstock and Friedrich Hölderlin, for instance.[37] In the Netherlands the revolutionary-minded Cornelius Rogge (1761-1806), a member of the so-called "Patriots", was of the same opinion – political revolutions, however exceptional they may be, were as necessary a means of clearing the air as thunderstorms and hurricanes.[38] There were, however, many for whom such an interpretation, in which the verbal evocation of natural phenomena was used to legitimise social and political changes, was definitely going too far. Yet even a Romantic-conservative author like François René de Chateaubriand (1768-1848) was willing to admit afterwards that there was some truth in it. For all his criticism of the French Revolution he was enough of a realist to observe, with a certain resignation, that even in the (thunder)storm of the revolutionary period a kind of order was still discernible after the event – an idea to which the parallel with the ancient cyclical meaning of the word *revolutio* in astronomy had already led him.[39] Order, but also a new fertility, up to a certain point. For whether one joined the revolutionaries in emphasising the emancipatory consequences of the political storm, or the conservatives in decrying its devastating results, there was one thing on which both sides often agreed: the atmosphere was unquestionably clarified by it, and that beneficial refreshment offered opportunities for new initiatives. In that sense, during the Restoration period Napoleon could well be appreciated as the lightning rod that safely carried the high voltage of revolutionary blows for freedom to the earthiness of practical politics.[40]

▲ The electrical generator of the rights of man produces a "spark of liberty", a "republican electricity" that will topple all the thrones of the "crowned scoundrels", print by François Marie Isidore Quéverdo, after a design by Dupuis, 1794

Meanwhile, the idea that the thunderstorm provided an analogy for collective political unrest and lightning for the individual exercise of power is too simple. For not only did the lightning bolt – the classic attribute of the tyrant – sometimes become the power symbol of the sovereign revolutionary people,[41] but conversely, among members of the anti-Orangist "Patriot" movement in the Dutch Republic the "thunderstorm" of Stadholder William V of Orange functioned as the emblem of his purported arbitrary autocracy.

The latter is apparent in a poem by Gerrit Paape in the *Hollandsche historische courant* of 8 March 1785, for example. Here it is the "power of the pen," the freedom of the press, that disperses the tyrannical storm clouds:

> *a glowing bolt of lightning went forth from the Tyrant's throne*
> *(...) What Deity, they cried, will halt the destructive storm?*
> *Then came Freedom forth; "My Citizens," she said, "stand"*
> *And turn that thunderstorm with your swan-feather pens.*[42]

At the same time, however, it becomes clear here (and elsewhere[43]) how often the lightning, thunderstorm, and thunder metaphors were sometimes used interchangeably in the political discourse.[44] If a generalisation can be made, it is mainly a chronological one. While lightning was initially used primarily as a symbol of tyrannical princely power (which consequently made the lightning rod a subversive symbol), later, bourgeois atmospheric violence turned with equal force (and often with the suggestion of greater political enthusiasm[45]) against those presumed despots.[46] Typically, in the Netherlands, the many political journals that used thunderstorm symbolism in their titles were not only published during the most radical years of the Batavian Republic (1795-1806), the French-dominated successor state to the Dutch Republic, but virtually all of them were also revolutionary in character. In propagandist periodicals such as *De politieke blixem* ("The Political Lightning") (1797-1798), *De politicque donder* ("The Political Thunder") (1798) and *De weerlicht* ("The Sheet Lightning") (1798) the articulate citizen took a robust stand for his freedom and often criticised the government of the time for not being "Patriotic" enough, for being insufficiently revolutionary.[47]

In short, therefore, despite all the differences in interpretation, atmospheric electricity in general and lightning in particular invariably symbolised power, be it in the hands of God, tyrants, or the people. Not infrequently it also signified hidden[48] or sudden power[49] – an energy that could accumulate in people's minds as it did in a Leyden jar, to be unexpectedly discharged at some later date.[50] For our ancestors, well-versed in the Scriptures as they were, the transition from the religious to the political sphere or vice versa was quickly made. In their political discourse Dutch Patriots blithely deployed the traditional idea that God's omnipotence was manifested in the storm, with thunder and lightning his awesome tools.[51] And a comparable, more concrete transition occurred when the German revolutionary Georg Forster, a Protestant pastor's son, drew a parallel between the legislative activities of the French National Assembly and the handing down of the Ten Commandments to the people of Israel, accompanied by thunder and lightning, on Mount Sinai (Ex. 19:16).[52] Yet for every time that lightning was used as a means of augmenting authority, the lightning rod was, by definition, just as often a subversive device. For if lightning itself could be bridled, why not princely or revolutionary tyranny?

▲ *La Constitution Républicaine*, here compared with the tablets of the law given to Moses on Mount Sinai, in the midst of a lightning storm (cf. Ex. 19:16). The mountain has been turned into a volcano. Print by Louis-Jean Allais commemorating the French Republic's Festival of Union which officially celebrated the *Montagnard*-backed constitution of 1793 on 10 August of that year. The tablets bear the simple words "rights of man" and "constitution of 1793"

Metaphor for War, Violence, and Quarrel

How easily political strife could degenerate into violent struggle, at home or abroad. The thunderstorm was often used to symbolise all kinds of war, violence, and quarrel. As a metaphor it boasted an impressive history: Homer had used it, likening the dense throngs of an approaching army to a thunderstorm from the west.[53] Parallels between the warrior and the thunderbolt had also been made in the ancient East, derived from the flash of the sun's reflection on the metal parts of armour (such as helmets, lances, and shields) and the speed with which chariot-borne warriors moved.[54]

But there was more. War and thunderstorms resembled each other in their often devastating consequences. Indeed, those who took the cannon as *pars pro toto* for war per se even saw a causal similarity in the two phenomena. For according to the eighteenth-century discourse, both involved sulphurous vapours that exploded – as the result of human action or otherwise. Moreover, the parallel was often a double one, both visual and auditory in swift succession. The poet Joost van den Vondel (1587-1679) had already compared the initial bright flash of heavy artillery fire and the deafening sound that followed with lightning and thunder.[55] It was an obvious and much-favoured metaphor, as well as a long-lived one. We find it, for instance, in the traditional poem on the four seasons composed by the regent of Zierikzee Jan Daniel Macquet (1731-1798) and published in 1761, in which he mentions in passing the Battle of the Downs, a naval encounter that took place 21 October 1639 (NS) when the Dutch trounced the Spanish: in this instance, lightning and thunder are not only an expression of God's majesty but also a metaphor for the flashing and crashing of the guns.[56] In the late eighteenth and early nineteenth century, by which time the thunderstorm theme was virtually ubiquitous in Western culture, its imagery was used even more frequently by artists of all kinds, in the Netherlands and elsewhere. In 1808 the Rotterdam-born Hendrik Tollens (1780-1856) produced a poem (kitschy by today's standards) on the Four Days' Battle, another Dutch naval victory, this time against the English, fought between 11 and 14 June 1666 (NS), in which the thunderstorm predictably symbolised the roar of the cannon of Cornelis Tromp's (1629-1691) ships:

The thunder growls with fierce clatter
And rolls to and fro along deck and gunnels;
To mountains swells the flying water
And bursts asunder down to the bottom.[57]

▶ Portrait of Heinrich von Kleist, coloured miniature on ivory by Peter Friedel. Kleist had it made in 1801 for his fiancée Wilhelmine von Zenge (1780–1852)

In the same year the figurative language used by the German poet and dramatist Heinrich von Kleist in his tragedy *Penthesilea*, to cite another example, was certainly inflated and stereotyped, too. He took his material not from more or less recent history but turned instead to the epic tradition of classical Antiquity. In that instance the storm symbolises the thunderous tumult of battle itself. In the midst of the Trojan War, Penthesilea, the Amazon queen, loves, loses, and ultimately kills Achilles (reversing Homer's narrative). In the opening scene, which is set on the battlefield before the gates of Troy, the princely Greek warrior Diomedes speaks of the fierce battles that have already been fought:

Since that day
The battle has been rumbling on this plain
Without a pause, a constant rage, as of
A storm hemmed in between steep wooded cliffs.
When I came yesterday with reinforcements
Recruited in Etolia, I arrived
Just as, with thunderous crashing, she let loose
An elemental blow, designed, it seemed,
To split all Greece down to its very roots.[58]

Thus the thunderstorm was used not only to symbolise the roar of modern artillery but also the clamour of classical war. Kleist was not the only writer to exploit the metaphor in this sense: in 1804 the learned Dutch Remonstrant Paulus van Hemert (1756-1825) described the tumult of war as sounding like "the last blow of pregnant thunderclouds," a nocturnal thunderstorm that reverberated amongst the mountains.[59]

◀ Portrait of Paulus van Hemert, the learned Dutch Remonstrant, print by J.E. Marcus, after a design by H.W. Caspari, 1816

Meanwhile, the thunderstorm metaphor was also employed to denote more subtle forms of violence and din. Sometimes those who dug deeper into psychological factors and seized on the old ingrained partisanship as the cause of disagreement also found it a useful symbol. The dissension that had brought the Netherlands to the brink of civil war in the 1780s was a traumatic experience for many. And it was undoubtedly with that painful memory in mind that Hendrik Daniel van Hoorn (1731-1802), a moderate member of the Batavian Republic's National Assembly writing in 1796, described that deep discord as the rattling of thunder.[60] Later, when the old antagonisms seemed to have been overcome and the pathos ebbed away, the oh-so-domestic nineteenth century would also apply the metaphor to "crackling" discord in marriage and family: the term "domestic storm" coined by the writer Jan Jacob Antonie Goeverneur (1809-1889) may serve as an example here.[61]

Many of those of an earlier generation would have found such a commonplace use of the thunderstorm metaphor terribly tame. Hölderlin, for instance, who, as we have seen, associated the thunderstorm with the monumental shock of the French Revolution and the concomitant clamour of war, also saw in the idyllic conditions that followed its termination a metaphor for the renewal of poetry in the coming age, a phase of renewed fertility, not just in a military and political sense but also and primarily in an artistic one. So it was hardly a great step from there to making (atmospheric) electricity itself a metaphor for poetic creativity or spontaneous poetic inspiration.[62] That "creative storm", however, is a subject in itself.

Metaphor for a Sparkling Mind or Tormented Inner Life

When the study of electricity was still in an early phase, the aforementioned German physicist and poet Georg Matthias Bose postulated an analogy between the respective roles of the imagination in poetry and electricity in physics. Poetry and electrical theory both sought to point to something that was hidden. In his didactic poem, *Die Electricität*, written in 1744, uncertain availability and difficult evocability were points of comparison between the electric and poetic fires. Artificial means could only help to a limited extent.[63]

It was not until the last decades of the eighteenth century, however, that the electrical metaphor really came into its own. With its unpredictability and spontaneity, electricity, particularly the atmospheric or cosmic kind, became a trope for artistic inspiration, whether as earthly energy which "shocked" the artist or as a bolt from the blue. In the former instance, references might be made to mesmerist influences, although parallels were also drawn with the creative tension evoked by two poles in the artist's innermost being. In the latter case, the grace of divine inspiration could sometimes strike without warning; alternatively, however, it might be called the result of a single-handed conquest after a titanic battle.

More than ever before, the late eighteenth century saw connections being made between creativity and electricity, for that mysterious phenomenon activated material nature just as much as it did the poet's mind.[64] Sometimes those connections were even represented as "scientifically" physiological. For those who attributed spiritual qualities or even animation to nature it was a relatively simple matter to follow the opposite path and interpret mental processes as the result of electrical or chemical stimuli.[65] We have already seen how theosophists of every stripe led the way along this route. To anyone even slightly influenced by them, therefore, employing electrical metaphors and metonymies was virtually self-evident. That was true of the German poets Friedrich Hölderlin, Heinrich von Kleist, Novalis (see below), and Ernst Hoffmann (1776-1822), not to mention Samuel Taylor Coleridge (1772-1834) and – mutatis mutandis – Lord Byron (1788-1824).[66]

Take Georg Friedrich Philipp Freiherr von Hardenberg – better known as Novalis – for example. He used electricity as an metaphor for creative sparkle. In 1798, in one of his most famous aphorisms, he described "Witz" ("wit"),[67] – a concept entirely typical of him – as "spiritual electricity," adding somewhat cryptically, "solid bodies are necessary for that."[68] That last comment had a dual meaning. On one hand, Novalis sought to indicate that a good meal stimulated "Witz und Laune" ("wit and cheerfulness") and it was precisely for that reason that portly people were often quick-witted and good-humoured. The dinner table was where friendships tended to develop and that infectious human interaction could be seen as the result of "contact electricity": "who does not sense the magnetism of souls here?"[69] On the other hand, however, with the addition of those "solid bodies", Novalis (who had a background in science as well) referenced what were then highly topical experiments that would soon lead to the discovery of the "voltaic pile", an early type of battery that for the first time made it possible to generate a continuous electric current. For just as a current was the result of the chemical reaction between metals (the electric potential difference between two poles, which would come to be called anode and cathode), Novalis declared, the sparkle of original minds should be regarded as the fruit of creative tension between the powers of judgement and imagination. It was precisely the tingling preliminary phase that was important, for a full and facile fusion of the two things soon petered out into mediocrity.[70] In 1784 Novalis's older contemporary Johann Paul Friedrich Richter (1763-1825), better known by his pseudonym Jean Paul, had already opined that "Witz" was much like a volatile electric spark, which in the lively mind of a wit jumped from one idea to another and so made original connections visible.[71]

◀ Portrait of Novalis (Georg Friedrich Philipp Freiherr von Hardenberg), print by F.E. Eichens, 1845

▶ The "voltaic pile", shown here in various stages of development, was the precursor of the modern battery. It made electricity available not as a short-lived spark but a continuous current. Illustration from the *Philosophical Transactions* 1800, Pl. XVII, p. 430. Unlike Galvani, Volta believed that electricity was inorganic in nature. To prove it, he conducted an experiment in which a number of bowls were filled with a saline solution and connected with wires, one end of which was copper and the other zinc. When the ends touched, current began to flow through the wires. The last three figures show a simplified version of the previous experiment, the actual "voltaic pile"

Spark analogies of that kind often underlined the sudden, incidental nature of discoveries in the field of art and science. But other authors believed that there were also sparks of a more sustained kind. According to the American periodical *The Nightingale*, which in 1796 published a retrospective survey of the life of Benjamin Franklin (who had died in 1790), there had been an electric spark at the genius inventor's birth which "sympathised" with the "fine aether of his intellect."[72] If this all sounded rather theosophical and "horizontal", the fact remained that Franklin was a self-made and self-taught man who had begun to wrest lightning from heaven entirely on his own initiative. In this characterisation of the role of the genius inventor, therefore, the Romantic pen-

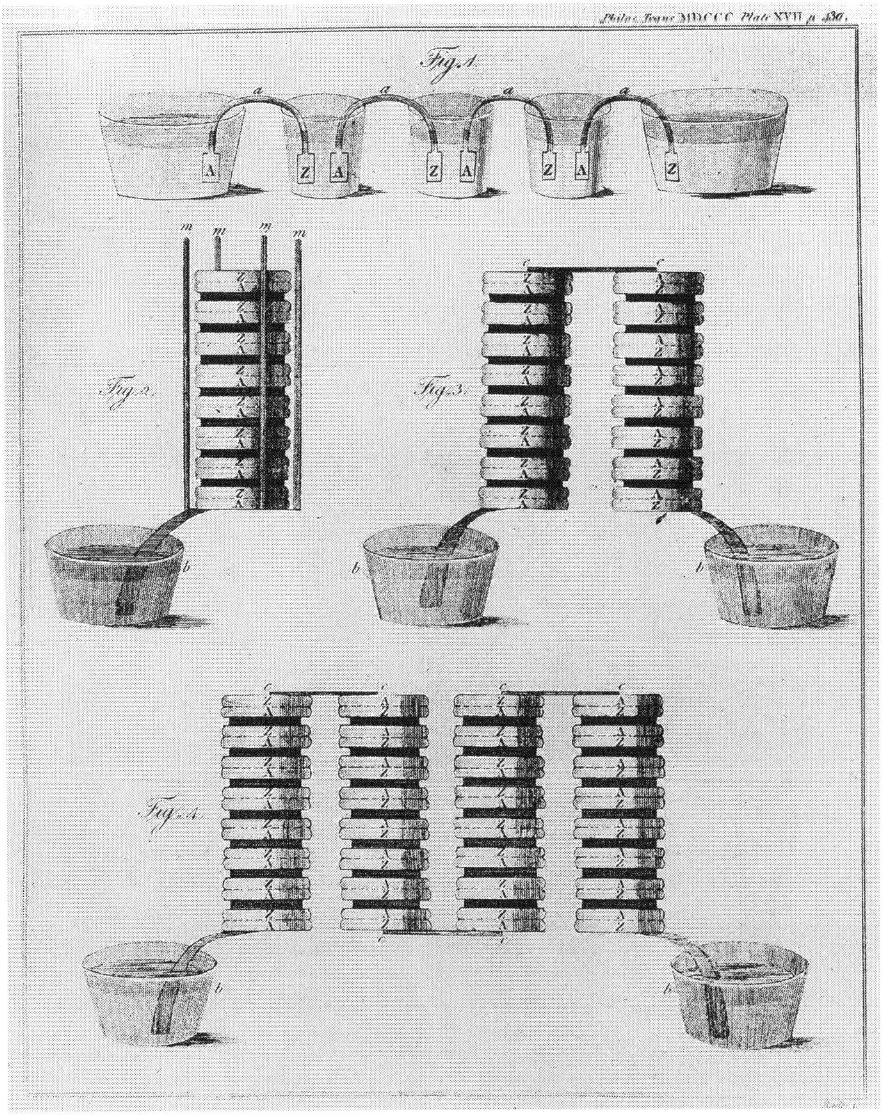

chant for the "wild", artless, and dauntless talent was also manifest: here the emphasis was on the independence of the creative genius. For just as electricity – that mysterious phenomenon without mass or weight – seemed to scoff at what were assumed to be the natural laws of mechanical motion and was apparently untethered to time and space, so too did a truly great and original writer freely flout the established rules of literary art. Therefore, argued Johann Gottfried Herder, the role of the genius in art ought to be the same as it was in science, for a truly great mind should, like a second Prometheus, "fetch the electric spark from heaven" and inspire his work of art with it – a mysterious process of creation whose causes were as hidden as those of the inexplicable natural phenomenon of electricity itself.[73]

Many contemporaries were so convinced of the autonomy of genius that it mattered little to them whether artists and scholars were deemed to have come up with their brilliant ideas themselves or whether inspiration had simply struck, bestowed by some otherworldly agency. Divine inspiration and human autonomy were not always seen as antithetical, if only because both could involve an inner struggle. Despite all the stressing of sublime inspiration, man's independence – now vis-a-vis the higher powers – sometimes remained sturdily upright.

Meanwhile, given the emphasis on the "vertical" aspect of poetic inspiration, the thunderstorm metaphor in general and the symbol of lightning in particular were obviously the natural choice, even if they did not always need to have religious connotations. The French philosopher Diderot, for example, had already made secular use of this metaphor in 1757 to indicate the absolute inevitability with which an artist must unburden himself of his creations.[74] And more than half a century later, Percy Shelley characterised poetry as a "sword of lightning, ever unsheathed, which consumes the scabbard that would contain it," certainly if said scabbard was held by representatives of the past enlightened generation who sought to impose their orderly stamp on it.[75] More often, however, the lightning motif did have a more or less religious hue. Hölderlin, for instance, invoked classical mythology. For he tended to associate the poet's divine inspiration with the lightning of Zeus, who, though he had most unwillingly incinerated the Theban princess Semele, pregnant with his son, did save the unborn Dionysus, deity not only of wine but also passion in poetry and music. That divine inspiration, wrote Hölderlin, echoing the Greek lyric poet Pindar (522-443 BCE), brought with it a special obligation for the poet, namely to pass on the salutary lightning:

And it is our duty, poets, to stand
Bare-headed under the storms of God,
Grasping with our own hand
The Father's beam itself,
And to offer the gift of heaven,
Wrapped in song, to the people.[76]

▲ Portrait of Friedrich Hölderlin, nineteenth-century German steel engraving, after a pastel from 1792 by Franz Karl Hiemer

▲ Portrait of Percy Bysshe Shelley, print after a painting from 1819 by Amelia Curran

Typically, in this context, it was from the 1750s onwards that the word *Gedankenblitz*, denoting an unexpected poetic intuition or a sudden violent burst of insight,[77] began a spectacular and steady rise in use in Germany.[78]

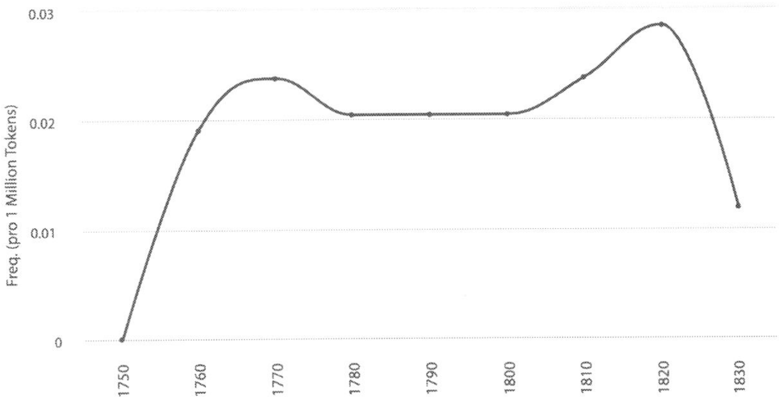

▲ Graph of the frequency of the use of the word "Gedankenblitz" in the German language area 1750-1830

But whether that more or less sudden creative irruption was the result of a lightning bolt from above or of polar tension in the bosom of the artist himself, in both cases there was some form of electrically-launched dialectic, whether in a dazzling play of heavenly light and earthly darkness or a process, often represented as galvanic, in the "battery" of that same artistic mind. On balance, vertically or horizontally, the electrical metaphor more often than not formed an image of the Romantic genius' inner torment[79] as he was tossed to and fro between divine inspiration and earthly coarseness, between the poetic fire stored as if in a Leyden jar and the mundane life that is so often its direct opposite. Folk such as Novalis and Johann Wilhelm Ritter (1776-1810), echoing the Hegelian zeitgeist, blithely declared that it was precisely this dynamic dualism that sparked poetic afflatus and led to an ever higher creative synthesis.[80] But there were also the suffering souls like Kleist, who emphasised that the artist must draw his inspiration from his own inner self, without hoping that the aforementioned antitheses would ever be "aufgehoben" ("sublated") in a higher synthesis; at most the poet could make them manifest in all their rawness.[81]

But the scope of electrical metaphorism extended even further. For not only did it serve to indicate the arduous creative process in the poet's mind but was also used in a broader sense to give expression to a tormented interpretation of reality per se. The dialectical play of heavenly light and earthly darkness opened a new philosophical perspective, as it were, by which the Romantic sought to rise above an all too sober and arid "normality" by downgrading it to the veil of an unknown higher reality.[82] And even when the electrical metaphor was not used to symbolise inspiration from above or in allusion to a metaphysical reality, it was still invariably a tellurian trope for the inner rending of Romantic souls. This dialectic, this choking constriction, found its favourite

◀ Portrait of the author Joseph von Eichendorff, woodcut from 1880 by H. Geday (later tinted), after a miniature by Karel Joseph Raabe from 1809

expression in an image that ultimately united "horizontal" and "vertical" electricity – not only among literati but Romantics in general: *Gewitterschwüle*, the threatening oppressive atmosphere that presaged the eruption of a thunderstorm. For it was exactly at that point that the ambivalence that was so rich in contrast was manifested in all its intensity: on one hand in the threat of a fierce thunderstorm, which appealed to the Romantic anyway, like any other unusual phenomenon,[83] and on the other the unreal empty silence that often preceded it and whose description presented receptive souls with a chance to infuse even the dullest aspects of "normality" with Romantic implications. In the end, no one summed up this distressful feeling with more striking aptness than the high-Romantic poet Joseph von Eichendorff (1788-1857), when he wrote that "the world is caught in a gloomy, oppressive atmosphere."[84]

From Metaphorising to Aestheticising

Given the many metaphors to which electricity – not least atmospheric electricity – gave rise in late eighteenth- and early nineteenth-century literature it is clear that the thunderstorm was aestheticised in all manner of ways. As we have seen, authors could effortlessly associate a whole range of subjects with lightning. But on occasion this observed or suggested similarity led to the reverse route being followed, in which case – at least initially – the thunderstorm itself took centre stage and was represented just as aesthetically by means of poetry, music, and painting, sometimes with the inclusion of metaphor or symbolism. Drama, on the other hand, offered very little scope, for the theatre was the pre-eminent locus for artifice and rigorous stylisation. Those looking for the theatrical preferred to make nature itself a stage by conducting experiments – with copies of Franklin's "electric kite", for example – thus abandoning the realm of art for the domain of science. In poetry, music, and painting, on the other hand, the path was often trodden in both directions, for naturally those arts offered not a scientific but an artistic description of the thunderstorm and so represented the feelings that thunder and lightning evoked in the observer. It goes without saying that there were very great variations in the level of art produced.

The aforementioned change of perspective can be seen in, for example, the curious practice of thunderstorm-imitation on what is possibly the most impressive musical instrument in existence, namely the organ. *Orgelgewitter* became all the rage in Europe, especially from the late eighteenth century. The German virtuoso organist, composer, music teacher, and cleric Georg Joseph Vogler (1749-1814) made a particularly significant contribution to the nascent tradition, so much so that in 1859 he was still being described as the "Vater aller Orgeldonnerwetterexecutoren" – the "father of all organ thunderstorm players".[85] His improvisations – *Spazierfahrt auf dem Rhein, von Donnerwetter unterbrochen* ("Excursion on the Rhine, Interrupted by a Thunderstorm"), for instance – enjoyed immense popularity.

▲ Portrait of the musician Georg Joseph Vogler, who imitated thunderstorms on the organ all over Europe, painting by August Friedrich Oelenhainz, 1790

Vogler performed all over Europe, the Dutch Republic included. And although at the time the country lacked the sort of well developed infrastructure for the public enjoyment of music that was found elsewhere, there was certainly room for such a commercial and partly charitable initiative. Vogler was not one to hide his light under a bushel. Not only did he perform his thunderstorm fantasies on the Schnitger organ in the Grote Kerk in Zwolle in 1786[86] but also demonstrated his improvisational skills a couple of years later on the famous Müller organ in St Bavo's Church in Haarlem – a shrewd choice, as on its completion in 1738 it was the largest organ in the world. On that same tour he also stopped off in Leiden to play the Van Hagerbeer organ in the Pieterskerk and that performance too was a great success. Indeed, poet, painter, and polymath Johannes le Francq van Berkhey was stirred so profoundly by it that he promptly dashed off a *Dicht-verrukking* (...) in which he expressed his admiration for the faithfulness of Vogler's representation:

The hiss of the lightning seemed to sizzle;
The crackle of the electric fire;
Seemed through the shock of the organ pipes,
To spark on the temple wall.[87]

▲ The Müller organ in St Bavo's Church in Haarlem, the largest organ in the world when it was built in 1738. It was on this instrument that Vogler performed his improvisations in 1789

Vogler's performances were both an expression of and a stimulus to a development in which the organ was increasingly seen as a solo instrument on which unaccompanied recitals could be given. The status of the instrument was improving, especially as old, mainly Calvinist objections wore out and the organ gradually became a fixture in church, used to accompany congregational singing. It is no coincidence that during the last decades of the eighteenth century many (Dutch) churches took possession of a new organ.[88] Indeed, so closely were church and organ associated that the instrument came to be seen as inherently sacred, which sometimes led to protests by ecclesiastical authorities when profane thunderstorm numbers were played in such sanctified surroundings.[89] In other words, from being reviled during the (Calvinist) Reformation the organ underwent an unmistakable process of re-sacralisation in the late eighteenth century.

Vogler's performances were obviously in an imitative tradition. Yet something had changed, and it was precisely in those changes that his renderings reflected the contemporary zeitgeist, for "sterile" imitations were no longer in vogue. Although earlier musical imitations had naturally never been so perfect that they left little if anything to the listener's imagination, the celebrated organist aimed to achieve more than the largely timeless descriptive and imitative tone painting that had been common in Baroque music, and in the depiction of thunder and lightning, for instance, had worked with jangling tremolos and melodic lines that lunged violently from high to low.[90] And although such sound paintings could sometimes still be heard in later music, in the case of Vogler and many other (performing) artists imitation gradually segued into expressivity as they sought to depict the impressions that the external world made on the human soul.[91] In order to create a varied palette of moods Vogler, apostle of late eighteenth-century *Sturm und Drang*, focused on the programmatic portrayal of the storm's successive phases.[92]

In both respects Vogler was admired and imitated, in the Netherlands and elsewhere, although in their musical depictions of thunderstorms his followers sometimes employed not only the organ or other instruments but also the human voice. With that qualification in mind, the Zwolle city organist and musician Johann Carl Röhner (1774-1837) proved himself a true disciple on the several occasions on which he performed *Het onweder* – a cantata of his own composing (1807) with a libretto by Rhijnvis Feith – often using twice the normal number of instruments, so as to render the sound of the unleashed violence of nature as faithfully as possible.[93] He also introduced a clear phasing in his cantata.

Nowadays Röhner is all but forgotten. Nevertheless, in many respects his thunderstorm cantata was typical of the contemporary zeitgeist. Its composition and performance more of less coincided with the creation of one of the best-known instrumental pieces of weather-depicting music, Beethoven's (1770-1827) Symphony No. 6, "Pastoral".[94] What is more, its arrangement not only displays a remarkable parallel with Beethoven's symphony but also moves along the then customary line: an idyllic gathering of countryfolk on a beautiful summer's day is abruptly interrupted by a thunderstorm, after which refreshing rain and the calm of the setting sun evoke feelings of gratitude and relief.

That storyline – in which the musical representation of the thunderstorm (likewise the gale or tempest) follows a clear progression – had been evolving since the late eighteenth century. At first the phasing involved nothing more than a straightforward bipolar development from *fureur* to *apaisement* – the same scheme that is encountered at the end of the second act of Mozart's (1756-1791) opera *Idomeneo,* which premiered in 1781, for example, and in Beethoven's ballet music *Die Geschöpfe des Prometheus* ("The Creatures of Prometheus"), composed in 1800-1801. And to descend for a moment from this high classicist art to a completely different level of works, those in the domestic, pedagogic, and edifying category, that same jointed progression is also found in the setting produced around 1823 by the Leiden composer and university *Kapellmeister* Christian Friedrich Ruppe (1753-1826) for Hieronymus van Alphen's poem for children, *Het onweder.*[95] What is always missing in these and similar cases is the opening idyll. Only when Joseph Haydn (1732-1809) included a thunderstorm scene in his oratorio *Die Jahreszeiten* ("The Seasons") (1801) did the three-part structure become the norm.[96]

▲ Jan Pieterszoon Sweelinck walking in the dunes. In 1879 the Romantic representation of the thunderstorm written for the organ was retrospectively and anachronistically credited to the Dutch composer Jan Pieterszoon Sweelinck (1562-1621). According to a biography written more than 250 years after his death, Sweelinck was walking in the dunes near Haarlem with his children and pupils when a big storm came up. Greatly impressed by its beauty and majesty he vowed that on the following Sunday, as parishioners left the church, he would play Psalm 29 as the voluntary, adding that "I have never yet played that psalm properly, now I know how to do it, yes, now I know how to do it!"
See Pieter Louwerse, *Jan Pietersz. Sweelinck of de toonkunst in Nederland in de 17e eeuw* (Schiedam: Roelants, 1879), 172-74. Cited by Jeroen van Gessel, *Muziek beleven in het negentiende-eeuwse Nederland* (Utrecht: IJzer, 2020), 32-34. Van Gessel has evidently failed to realize that Louwerse's misinterpretation related to an entire organ tradition in the footsteps of Vogler

◀ Joseph Haydn in the period between 1790 and 1795, aboard ship en route to England, during a storm, nineteenth-century mezzotint engraved by P. Allais, after a design by Édouard Hamman

In orchestral practice the changing moods, which thanks to the introduction of such phasing could now be evoked, should be largely the result of a delicate succession of soft tones and hard sounds. In the period of *Sturm und Drang*, however, great emphasis was placed on the powerful effect of the latter, be it produced by the organ, timpani, or trumpets. That obvious partiality for heavily accentuated crescendos that built up to an explosive climax[97] often led to somewhat coarse "thunder music". Only in this overheated way could the listener's heart be touched, or so it was thought – a view that later generations would often dismiss as bombast or striving for cheap effect. Yet even in the classical period, that kind of musical inflation was still alive and well. Take Beethoven's famous sound painting of *Wellington's Victory* (Op. 91; 1813) with its thundering 188 cannon firings, for example. First performed in Vienna in December 1813 and conducted by the composer himself, it was Beethoven's musical monument to the English routing of the French at the Battle of Vitoria in northern Spain earlier that year, portrayed with the help of a battery of percussion instruments and numerous blasts from the brass section.[98] Or, to cite a more direct example, Haydn's oratorio *The Creation* (1798), which depicts and celebrates the creation of the world as described in Genesis. Though in itself quite a balanced composition, in the beginning violins and timpani produce a good many flashes of lightning and rolls of thunder, of which, curiously enough, there is no mention in the biblical narrative.[99] But whatever the artistic appreciation of these compositions may be and however that period is labelled, what had come into being here was nothing less than a new version of an age-old aesthetic category, namely the sublime, which was now used primarily in reference to the grandeur of nature.[100] According to contemporary evidence, loud noise should be considered an essential component of it: "in some measure, elevated feelings are inseparable from unusually loud or tremendous noise. There is something sublime in the beating of drums, the roaring of cannon, the crashing of thunder, the mighty eruption of a fire-breathing mountain: (...) loneliness greatly magnifies that emotion."[101]

Clearly musical instruments and the (natural) phenomena they imitated (in phases or otherwise) not only produced sound but also awakened feelings. That most versatile thinker Edmund Burke (1729-1797) was one of the first to interpret these sentiments in *A Philosophical Enquiry into the Origin of Our Ideas of the Sublime and Beautiful*, a treatise on aesthetics published in London in 1757, in which he states: "Excessive loudness alone is sufficient to overpower the soul, to suspend its action, and to fill it with terror. The noise of vast cataracts, raging storms, thunder, or artillery, awakes a great and awful sensation in the mind, though we can observe no nicety or artifice in those sorts of music."[102] From Burke's investigation of the beautiful and the sublime it appears that

▶ Portrait of Edmund Burke, the versatile founder of the modern aesthetic category of the sublime, print by Wilhelm Arndt, after Joshua Reynolds, 1792

the two are significantly different things; they are, in fact, each other's opposite. The delicacy of the beautiful produces a pleasant sensation, while the excessiveness of the sublime initially produces feelings of terror and disorientation, especially in the case of an unexpected confrontation or the sudden beginning or cessation of the cause. Yet besides overwhelming pain, the sublime can simultaneously evoke fascinated delight (which he distinguishes from "pleasure") when the observer is able to hear or see the sublime scene from a relatively safe place. Therefore, posits Burke, the experience of the sublime is inextricably linked to the striving for self-preservation. His philosophical and psychological analysis of aesthetic experience uniquely recognised that in such cases pain and delight are two sides of the same coin,[103] and for that very reason they intensify each other. Burke's ideas, conceived shortly after the devastating Lisbon earthquake of 1755, would come to form the theoretical basis of (pre-)Romantic art.

Consequently, thanks in part to Burke and his followers, "ugly", inharmonious subjects such as the thunderstorm also entered the list of acceptable artistic themes,[104] especially when it broke loose on a vast and stormy ocean cresting with tumultuous waves or a rugged Alpine landscape with inaccessible mountain peaks and fathomless abysses. Even so, in that context the lightning rod was the very thing that provided puny human beings with the necessary safe place in which to undergo the well-nigh transcendent[105] borderline experience of the sublime amidst the overwhelming violence of nature. Hence, when seen in that light, terror amounted to risk-free delight.[106]

On the other hand, those who reckoned without the lightning rod's protection were soon assailed by the indefinable dread inherent to the *Unheimliche* – the eerie or uncanny – the aesthetic background of the *romans noirs*, the gothic novels, and the *Schauerromane*.

The ideal settings for the ghastly tales of gothic literature were rustling vaults, subterranean labyrinths, or sinister chambers at dead of night. Should a storm occur it was the sound – the cracks and booms of thunder – that was emphasised, for in the darkness of those chambers, labyrinths, and vaults, the senses of touch and smell were retained but more importantly, so was the sense of hearing – the sense to which Romanticism had a special receptivity.[107] In this case, however, unlike the "safe" experience of the sublime, the fear induced by the sudden appearance from the darkness of thunder and lightning was not actually overcome, because the beholder had failed to find a safe place from which to survey nature's violence with a sovereign gaze. In this sense the sublime is essentially the domain of the eye.[108]

Yet that double generalisation requires some qualification. To begin with, it goes without saying that the entire difference between the eerie and the sublime is relative, for the simple reason that the reader of the *unheimliche* novel or the viewer of the *unheimliche* painting is usually in a safe position – culture is a game. Consequently aesthetic pleasure is also possible in the case of the eerie, albeit not for the figures in the work of art. But there is more. We have seen above what a major role the sublime also played in music,[109] while on the other hand the role of the visual was not entirely lacking in the *Unheimliche*, neither in literature nor in painting. For although Pliny the Elder may have declared that depicting the dazzling and deafening thunderstorm was simply not possible,[110] the artists of the late eighteenth and early nineteenth centuries were just as convinced that it was, precisely because of the dramatic contrasts inherent in the sublime – for in both the sublime and the *unheimliche* thunderstorm light and dark polarised each other, the disharmonious flashes of lightning creating momentary shadows and in that very way making the darkness clearly visible.[111] Because the landscapes that Romantic literature and painting so fondly depicted were often laden with meteorological symbolism,[112] something hidden was illuminated in a flash – in other words, the negative character of the darkness came to light: darkness as deprivation.[113] And because a similar dialectic operated between silence and sound in the storm, both the literature and music of the late eighteenth and early nineteenth centuries likewise left much to the reader's or listener's imagination. Here, nature taught art an age-old lesson – the power of suggestion by omission.

In the physico-theological epistolatory novel *Het land*, by the Dutch author Elisabeth Maria Post, to name but one, somewhat atypical, literary example, the dialectic of thunder and silence has a religious purpose. One night the protagonist, Emilia, is startled awake by a violent thunderstorm and is petrified with fear. But as the rain stops and the storm moves off a little, she goes outside and experiences a very different feeling as the terror of the phenomenon recedes and she becomes aware of "the beauty, the imposing dignity, the magnificence of the scene."[114] The whole earth is "wrapped in the darkness of the night," now and then brilliantly illuminated by streaking flashes of pale lightning. At the same time, Emilia notes a similar interaction between the rattling thunderclaps and the ensuing silence. "The whole air seemed to be an echo that repeated the tremendous noises, until the sound was lost in a deathly silence! A silence that seemed to call out to me: 'The God of glory thunders!' and gave an indescribable majesty to every swift successive bolt."[115]

Although, judging by this citation, Post as yet makes no distinction between the beautiful and the sublime, it is in the notion of a calling silence that she seems to have been subtly influenced – perhaps without realising it herself – by Edmund Burke.[116] For even though that oxymoron is a relative detail in the novel as a whole, it nonetheless suggests that in Post's vision there is a simultaneity rather than a succession in the sharp contrast the thunder creates between pain and (religious) delight[117] – evidently the ability to synchronise contradictions is not the painter's exclusive province.[118] In any case, in (pre-)Romanticism, when a sudden turn-about took place between contrasts a field of tension was created in which imagination and emotion (to a varying degree of intensity) were stimulated, sparking a growing awareness of the mystery.[119]

On a different aesthetic level the same simultaneous effect of trepidation and fascination emanates from *Thunderstorm and Lightning Strike at the Lower Grindelwald Glacier*, a well-known work by the Swiss artist Caspar Wolf (1735-1783), painted in 1774-1775. Thunderstorms in the high mountains is in itself a doubly sublime subject, which would only become popular in later Romantic painting,[120] but given the spectral light, the desolation, the inimical nature of this virtuoso piece of dramatised landscape painting, the typifying *unheimlich* is perhaps better in place. Here, in any case, potent nature almost transcends human perception, for even in the flat, rigid, and restricted plane Wolf touches the dynamics of the sublime. We see lightning streak from a gaping ochreous opening in a menacing mass of dark clouds to strike a second core of light in the vividly lit mass of ice, while at the edge of the forbidding and dizzying spectacle the tiny figures of animals, perhaps chamois, bound away. The Danish artist Jens Juel (1745-1802)[121] was living in Geneva when he painted *Thunderstorm in the Alps* (1777-1779): though similar to the work by Wolf, here we see only the silhouettes of a couple of mountain goats, while in a much later and more colourful painting entitled *The Storm* (1843), by the odd man out of Dutch Romanticism, Johannes Tavenraat (1809-1881), merely a fleeing deer and one or two birds are depicted.[122] If human figures do appear in this Romantic "untamed nature" pictorial type, their role is purely marginal,[123] as can be seen in the if possible even more intense and exalted work by the Russian painter Maksim Vorobyov (1787-1855). His *Oak Shattered by Lightning*, which dates from 1842, still has the vague shape of a man making a desperate attempt to save himself from the demonic natural violence to which the mighty oak has already fallen prey. In the chiaroscuro of this melancholy canvas the icy light of the thin, glittering streak of lightning rips through the work to leave a mighty gash, not only in the tree but seemingly in the painting itself – and indeed, in the life of the artist too, who had lost his beloved wife not long before.

▲ *Thunderstorm in the Alps*, 1777-1779, painting by the Danish artist Jens Juel, who lived in Geneva for several years. Like the painting by Caspar Wolf, this work expresses the doubly sublime subject of thunderstorms in the high mountains

◀ *Thunderstorm and Lightning Strike at the Lower Grindelwald Glacier*, 1774, painting by the Swiss artist Caspar Wolf, a combination of two sublime themes, inclining towards the *Unheimliche*

▲ *The Thunderstorm*, painting by Johannes Tavenraat, 1843. This Rotterdam painter's highly dramatic use of lurid colours and chiaroscuro effects made him something of an outsider in Dutch Romantic painting and brought his art more closely akin to developments in France

▶ *Oak Shattered by Lightning*, painting by Maksim Vorobyov, 1842. This painting has a literary precursor in the first chapter of Mary Shelley's *Frankenstein* (1818): Victor witnesses a terrible thunderstorm in the course of which a beautiful oak is struck by lightning, leaving nothing but "a blasted stump" – a true "catastrophe of the tree"

Paintings depicting a storm at sea complete with thunder and lightning – one of the few lofty subjects in the flat Netherlands – are also doubly sublime.[124] In seascapes of that sort it is not always clear which is the main subject – the mighty wind, the tempestuous sea, the pitch-black clouds, or the lightning – but sometimes human figures are depicted. Many of them seem to be the playthings of the immense forces of nature, a notion augmented by a restless handling of line. In 1772 the French artist Claude-Joseph Vernet (1714-1789) painted a ship wrecked on a rocky coast: some of the unfortunate souls the vessel carried have reached the safety of solid land while others, dangling from the ropes of the capsized ship and aided by their surviving companions, are still in the process of extricating themselves from their perilous predicament. Meanwhile, in the comparatively safe distance, lightning forks from a rent in the clouds.[125] In *The Shipwreck*, painted in 1775 by the Dutch artist Hendrik Kobell (1751-1779), the handful of men who have managed to get ashore watch from a rocky promontory as an English warship is about to founder and a Moroccan xebec is struck by lightning.

▲ *The Shipwreck*, painting by Claude-Joseph Vernet, 1772. In this doubly sublime scene, in which a turbulent sea and a violent storm have led to shipwreck, fear and pleasure are combined. The simultaneity of the contrast is underlined by the lovely moonlight that illumes parts of the coast and the darkness of the storm clouds. At the same time, moreover, relief at the rescue of some of those who might have drowned is coupled with fear for others still fighting for their lives

▲ *The Shipwreck*, painting by Hendrik Kobell, 1775. Off the North African coast English warships and a Moroccan xebec, a privateering vessel, are in distress. The six men who have managed to gain the shore watch from a promontory as an English warship seems doomed to be dashed onto the rocks and the Moroccan xebec is struck by lightning

But whether on land or at sea, in the paintings just mentioned the focus is always on the thunderstorm at its height. While the safe place that is needed to experience the sublime is provided in these paintings by the viewer's identification with the survivors on their comparatively safe rock, the same effect can be produced by depicting a thunderstorm advancing or receding.[126] Somewhat calmer scenes were often the obvious result. A couple of examples will suffice – a work by the aforementioned Danish artist Jens Juel and a picture by the Austrian painter Joseph Anton Koch (1768-1839). Although the people and animals that Juel depicted around 1793 are not yet in acute danger, dark storm clouds are already gathering behind the Zealand farmstead that forms the work's setting. Juel has adroitly caught the threatening atmosphere by combining the advancing shadow that skims across the ground with the last of the sunlight. Thus he gives the farmer in the foreground time to hurry home with his horse and cart; meanwhile, the farmer's wife is already awaiting her husband.[127] Alternatively, besides the depiction of such an ominous *Gewitterschwüle*, the aesthetic portrayal of the first easing of anxiety once a thunderstorm has passed can also provide the safe place that is a requisite of the sublime experience. This is exemplified in Koch's 1830 painting, *Landscape after a*

Storm, from which we see that he has largely departed from the classicist tradition – he had been producing scenes like this since the start of the century – and was now ready for some Romantic drama: man and beast have escaped a great danger. In paintings of this kind the rainbow often provides both the meteorological and religious sign of the relative safety required.[128]

In this context there is no longer any question of "delicious shudders" or a delightful intertwining of fear and euphoria. The more that those changes in the weather were represented and experienced in a more gradual, more phased manner, the more the artist and his viewers moved away from the Burkean blend of simultaneous delight and pain. It is, of course, both perilous and occasionally even purely speculative to try to detect the influence of an underlying aesthetic theory in actual works of art. But where man – in spite of all his insignificance and notwithstanding his sometimes fierce struggle[129] – is finally represented as the master of nature, we might sooner suspect the influence of Kant's aesthetics than Burke's.

▲ *A Storm Rising behind a Farmstead in Zealand*, painting by Jens Juel, 1793. Although the picture exudes a very threatening atmosphere, the farmer will still be able to get home in time. Thus, Juel allows him to reach the relatively safe place that is requisite for the sublime experience

▲ *Landscape after a Storm*, painting by Joseph Anton Koch, ca. 1830. In this late work, Koch shows that he has largely left the classicist tradition behind and is open to some Romantic drama: man and animal have escaped a great danger. In paintings of this kind the rainbow is often both the physical and religious sign of the relative safety required for experiencing the sublime

There was, in any case, a tradition that in the analysis of the sublime experience a narrative pattern was preferable to simultaneity of any sort. Many maintained that not only the storm itself but also the aesthetic experience of it proceeded in phases – we have already seen how Beethoven used both propositions in the "Pastoral", for instance. Aestheticians usually split the sublime experience into three successive stages: rest, crisis, and (climactic) resolution. It was the line followed by the philosopher Immanuel Kant himself in his *Critique of Judgement* (1790).[130] Unlike Burke, Kant and his adherents focused not on the relatively safe position of the viewer of the sublime natural scene but rather on his autonomy or nobility of soul – as much as the "self-maintenance of reason."[131] For no matter the extent to which the force of nature (electrical or otherwise) apprised man of his physical weakness and dependence and maybe knocked him off balance for a bit, that physical inferiority ultimately activated the reason-rooted moral superiority that enabled him to master nature's violence by his own energies – to many enlightened minds the invention of the lightning rod was one of the most obvious proofs of that moral triumph. Did not history teach "what man *can* do, if he earnestly *wants* to?"[132] In sublime subjects there was something which (in the words of the dedicated Kantian, Paulus van Hemert) made us discover "the immeasurable treasure of our own powers of mind," something that "*elevates* us, as far as we are *sensible*, to the *Infinite*."[133] While our imagination was bound to our senses, reason, on the other hand, was deemed able to contemplate even the infinite.[134] Like Kant and other Kantians, Van Hemert argued that on balance the sublime resided not in the phenomena of nature but in our own minds[135] – a liberating self-insight which caused him to exclaim that "we joyfully rejoice in the nobility of human nature."[136]

◀ Portrait of Immanuel Kant, print by Lambertus Antonius Claessens, after Antoine Charles Horace Vernet, ca. 1792-1808. In his analysis of aesthetic perception, Kant attributed a crucial role to man's reason, which could ultimately rise above the violence of nature

Strikingly, of all natural phenomena it was the thunderstorm that Van Hemert saw as an obvious example by which these aesthetic reflections could be demonstrated, as is apparent from his choice of words: even a dull mind is sensitised when struck by "the lightning bolt of the Sublime," is it not? To Van Hemert, in other words, just as to his philosophical kindred spirit, the theologian, mathematician, and physicist Johannes Friedrich Ludwig Schröder (1774-1845),[137] lightning became a metaphor for the sublime experience itself![138] The reassuring invention of the lightning rod would not have been alien to the use of that imagery.

Whether late eighteenth-century man adhered to the aesthetic ideas of Burke or Kant, the fact nonetheless remains that a relatively safe (starting) point was an essential condition for experiencing the thunderstorm aesthetically – a conclusion that applies to literature and music as well as painting. In every case the emergence of the sublime as an aesthetic category – with all the contrasts that went with it – meant an intensification of both sensibility and expressiveness.

▶ *Mountain Landscape with Rainbow*, painting by Caspar David Friedrich, 1809-1810. Whether the dark clouds represent a night sky or a thunderstorm is unclear. But the rainbow (also) indicates the latter possibility. The lone wayfarer – perhaps Friedrich himself – finds himself in a threatening, almost uncanny environment, but at the same time in a safe place. Friedrich may have been inspired by a passage from the novel *Franz Sternbalds Wanderungen*, by the Romantic writer Ludwig Tieck (1773-1853), which was published in Berlin in 1798: "At last the storm stopped and a lovely rainbow appeared in the sky, the forest was fresh and green and all the leaves sparkled with raindrops, the sultriness of the day was past, a cool breeze blew through all of nature, all the trees, all the flowers were glad"

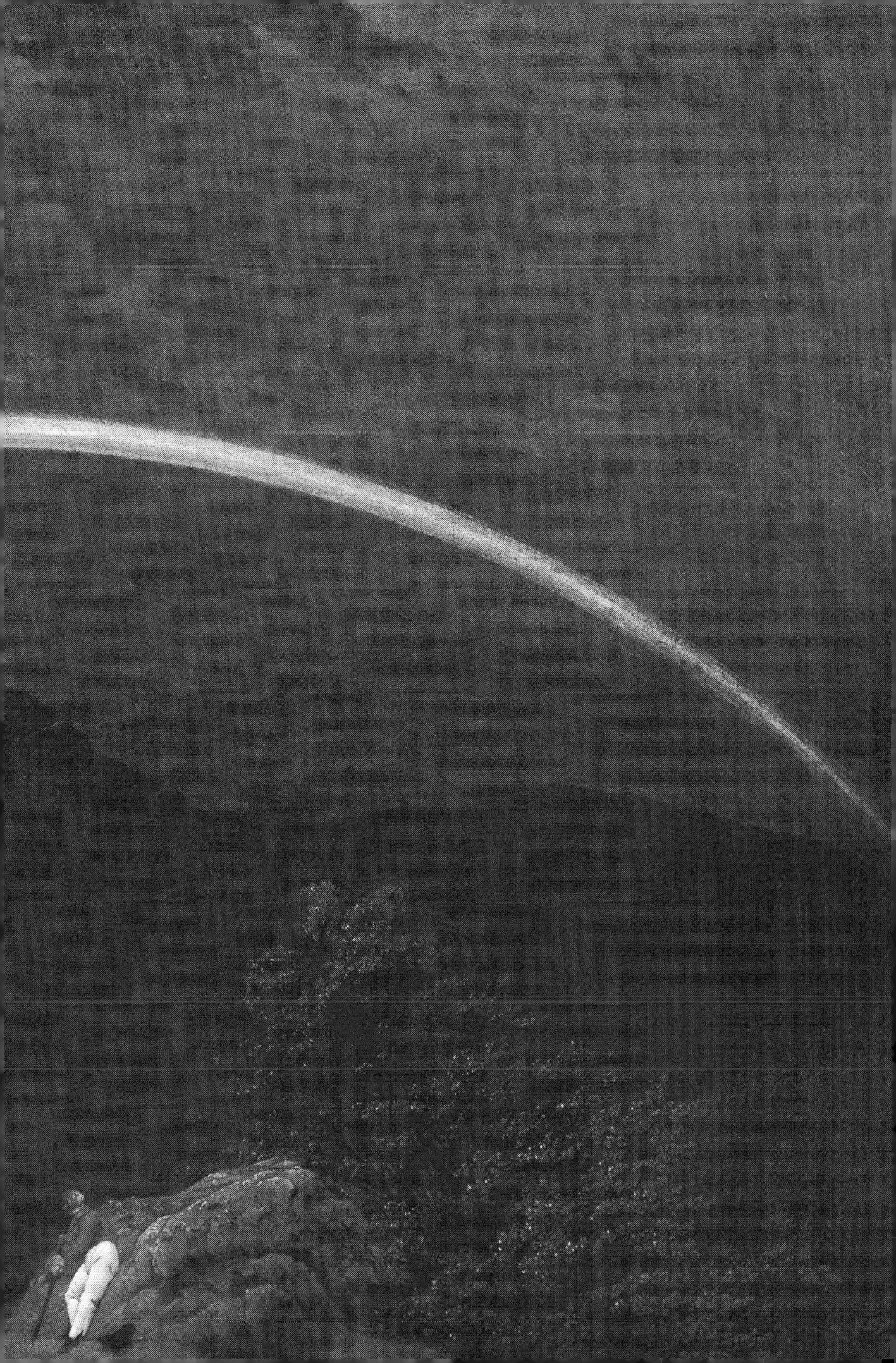

▲ *The Widow of an Indian Chief Watching the Arms of Her Deceased Husband*, painting by Joseph Wright of Derby, 1785. This work incorporates at least three sublime themes: a turbulent sea, a lightning storm, and a smoking volcano as the most inhospitable example of high mountains. The widow, on the other hand, is safely ensconced on a plateau

By Way of Conclusion

The systematic study of electricity steadily evolved into a discrete branch of scientific and technological research, particularly from the 1740s. Artistic interest, expressed in diverse media, increased likewise: the wholesale aestheticisation of the thunderstorm that took place in the late eighteenth and early nineteenth century was unprecedented in Western culture. While it seems highly likely that those two developments were connected there are considerable differences in interpretation, though these are not mutually exclusive. Some think artists' fascination with the thunderstorm should be understood as a reaction to the over-technical approach to nature. Others describe it as a way of allaying fear or as a means of conjuring inner uncertainty. The most plausible interpretation, however, sees that aestheticisation as (implicit) evidence that a wild danger had been domesticated, explaining the emergence of the sublime as a category by man's increasing mastery of nature – epitomised by the invention of the lightning rod. Once lightning was, in principle, rendered harmless, it could be played with, in poetry, music, and painting. Images of nature, man, and God underwent a major parallel change, in a more optimistic direction.

The Relationship between Fear, the Control of Nature, and Pleasure

By now it will be abundantly clear that electricity – both static and atmospheric – held a great fascination for the inquiring minds of the eighteenth century. The systematic study of the natural phenomenon steadily evolved into a separate branch of scientific and technological research, particularly from the 1740s, while artistic interest, expressed in diverse media, also increased – so much so that such a wholesale aestheticisation of the thunderstorm as took place in the late eighteenth and early nineteenth century was entirely unprecedented in Western culture.

A connection between those two developments seems very likely. The question we now have to ask is precisely what role the invention of the lightning rod played in them. As that new invention was clearly part of the general flourishing of the nascent science of electricity there is bound to be more than one answer. Even without the lightning rod there would undoubtedly have been an artistic preoccupation with the mysterious phenomenon that was sometimes so spectacular in its consequences, certainly in the broader context of the growing interest in the sublime. Yet there does seem to be some sort of relationship between Franklin's invention and artists' preoccupation with wild weather. From a historiographical point of view it remains to be seen whether that increased artistic interest came about despite or because of the new discovery. The first case, notwithstanding the optimism that hallmarked the Enlightenment, suggests the perpetuation of the old fear or even the introduction of a new one; the second implies mainly a matter of cultural game-playing.

Even after 1752, the year of the lightning rod's invention, there was still good reason to dread the consequences of a thunderstorm. The new device was, after all, controversial in scholarly circles, let alone less learned milieus. Moreover, it was decades before even the most at-risk buildings had one installed. And even in those cases, the apparatus was sometimes technically defective. For another thing, as long as the thunderstorm was interpreted as a displeased Deity's ineluctable punishment, it was pointless to try to avoid it. Nevertheless, since the means of doing so seemed to be at hand it was up to man himself to make effective use of it (with or without a serious prayerful nod to God). Sometimes the oppressive weight of that great human responsibility at a time when awareness of risk was increasing, simply sparked new fears.[1]

Accordingly, the many metaphors to which the storm gave rise should be interpreted as just so many means of exorcising fear.[2] Thus aesthetics operated as an outlet for the (presumed) new fear. The human psyche performed a *tour de force*, domesticating the terrifying into an aesthetic experience.[3] In this often Freudian-hued view, the sublime is little less than the externalisation of heightened inner fears (of one's own conscience and one's own urges) that were projected onto savage nature, even if that nature was less menacing than before.[4] For while factual fear may have diminished thanks to the triumph of enlightened reason, existential objectless angst had grown.[5]

Although there is more to this view than what Jürgen Habermas has called, in the horror literature context, the "ritualised repetition of historically endured fears"[6] it nonetheless fails to satisfy, precisely because of its complexity. Granted, there are certain advantages to it, for even without interpreting art's great preoccupation with thunderstorms in too Freudian a way, as an exorcism of intangible angst and inner uncertainty, it can be acknowledged, for example, that the Enlightenment's increased mastery of nature was not only a boon – in any case a linear perspective in which reason simply fills the gaps that fear has left is too reductionistic. Yet it seems to me that there is a different narrative with somewhat better credentials. The more science and technology unravelled "real" nature, turning it thus into a "dead" object, the more emphatically artists seem to have sublimated it[7] or even transmuted it into the sublime. It has rightly been noted that "where in principle everything becomes observable, testable, and explicable, (...) the Other [disappears] from the experience." Seen from that point of view, therefore, the rise of the sublime and the *Unheimliche* – the eerie or uncanny – may be interpreted as the settlement of one of the Enlightenment's outstanding debts.[8] A settlement not only in an artistic but also a broadly cultural sense, for the more that scholars and technicians gained control even over lightning and thus created the conditions for greater freedom and distance, the more man lost his familiarity with his natural environment and spontaneously longed to live the simple life, in close proximity to nature.[9]

Nonetheless, it appears to me that however plausible they are in their relative simplicity, such generalising compensatory ideas are also missing a dimension. There seems to be more going on, chiefly because – at least as I see it – the thunderstorm could only be deeply aestheticised once its effects could be controlled to some extent by means of the lightning rod. That man was becoming less helpless in the face of the forces of nature was the essential condition for the (re)birth of the sublime as a category, for as both Burke and Kant, for instance, maintained, being in a safe and safely-distanced position was the absolute precondition for an aesthetic experience.[10] Fear of a phenomenon eliminates the possibility of experiencing it as sublime,[11] but with a lightning rod on the roof the "beauty" of a thunderstorm could be more easily appreciated.[12] Only when that natural phenomenon could in principle no longer do any harm did man feel free to play with it, sometimes literally, with scale models such as "thunder houses" (which therefore not only had a function in experimental science but, as the ultimate and sometimes even mascot-like token of mastery over nature, also represented a typical cultural expression), or in paintings or notes of music – a carefree pastime for *homo ludens*, "man the player", freed from the need to exorcise fear. Surely one who was really afraid of the thunderstorm could not compose a symphony or even a movement that emulates one, still less find an audience able to appreciate it.[13]

Regarded from that angle, therefore, artists' fascination with the thunderstorm can be understood not as a reaction to a too technical approach to nature or the projected allaying of fear or inner uncertainty, but rather as (implicit) evidence that the danger has been tamed: the emergence of the sublime as a category is explained by the increasing mastery of nature. The German literary historian Richard Alewyn (1902-1979) put this hypothesis most eloquently[14] and – interestingly enough – in a way wholly consistent with Burke's "delightful horror". Seen objectively, he contended, there was less and less reason for fear, yet the need for fear continued to exist – a profound need that found its way into literature. Thus, in "high" culture at least, the *Lust an der Angst* became a surrogate for the fear that had been lost. This was not merely a matter of compensation, for according to Alewyn, fear was transformed into pleasure. As a result of that conversion one could have the best of both worlds: safety in real life, which in turn allowed one to enjoy a kind of pleasurable terror in literature.[15] Viewed thus, the lightning rod made it possible to flirt with fear yet still enjoy what enlightened culture regarded as its forbidden fruits – in gothic tales, for example. For if man really was to be called the lord of nature, such *angenehmes Grauen*, such delightful horror, was only a cultural game.[16]

Meanwhile, it is not clear whether and, if so, to what extent there is such a thing as a human need for fear that would have survived the capacity for fear.[17] Moreover, it is

quite a leap to conclude that less fear of nature means less fear per se,[18] the more so, because it is, after all, questionable whether in this connection technological factors alone are involved[19] – and even in that respect we have already seen that the introduction of the lightning rod was not without its teething troubles. For that and similar reasons it would be wrong to suppose that with Franklin's invention, eighteenth-century man had subdued (that aspect of) nature entirely,[20] however much its enlightened advocates might have wished that to be true. That rightly or wrongly the thunderstorm could still evoke a degree of unease was a good thing, according to Burke's rationale, for experiencing the sublime does not require every danger to be removed: a degree of trepidation only intensifies the sublime pleasure. Even volcanoes – utterly uncontrollable in their effects, extremely frightening, and therefore much less likely contenders than thunderstorms for contemplation of the sublime[21] – were evidently not so terrifying that in exceptional cases they could not lead to playing with scale models. In the eighteenth-century Wörlitzer Park at Dessau, a seventeen-metre-high activable model of Mount Vesuvius was built between 1788 and 1794 by the Prince of Anhalt-Dessau, who as part of his grand tour had climbed its smoking original in 1766, guided by his friend Sir William Hamilton (1730-1803), then British ambassador to the Kingdom of Naples.[22] In other words, eighteenth-century man was quite used to some degree of fear, and even without a lightning conductor it was not all that he felt. It is worth noting in this context that Joseph-Ignace Guillotin (1738-1814) extolled the humane nature of his eponymous device by comparing it with lightning: apparently a lightning-fast death was seen as a merciful method of despatch.[23]

So it would seem that in practice, lightning was less of a challenge for the Enlightenment than has been thought – an impression enhanced by the open-mindedness with which a handful of seventeenth-century artists portrayed stormy scenes, among them Jacob van Ruisdael (1628/29-1682), admired by many (pre-)Romantic painters. Evidently those Baroque artists were not so afraid of thunderstorms that they were put off painting that violent natural phenomenon, although it seems to have been chiefly a case of competition in craftsmanship and individual sensitivity to the sublime. On balance, the context was not yet that of a conscious culture of the sublime.[24]

◀ The eruption of the artificial volcano in the Wörlitzer Park near Dessau in August 2010, photo by Heinz Fräßdorf. The park, laid out in English landscape style between 1769 and 1773 for Leopold III, prince of Anhalt-Dessau, was embellished between 1788 and 1794 with a seventeen-metre-high activable model of a volcano built next to the Villa Hamilton on Stone Island. By igniting sulphur and employing other light, sound, and water effects, the prince sought to simulate a volcanic eruption and thus evoke the experience of the theatrically sublime in a relatively "safe" and controllable fashion. As a young man he had made his grand tour, including an ascent of Vesuvius in 1766, guided by the English ambassador to the Neapolitan court Sir William Hamilton, whose study of volcanoes had been recognized by the Royal Society

Yet neither a Saxon prince's bravura, nor a Guillotin's humanitarian argument, nor the genuine interest of seventeenth-century painters who took the storm as their subject despite the harm it might do to them, can obscure the fact that Franklin's invention marked a fundamental turning point. For however justified the relativisation of fear (of thunderstorms) may be and however incomplete the control of nature, what really matters is not the safety actually achieved but the heartening realisation, not to say the almost intoxicating rushes of perspective, that safety was really within reach, not only in the matter of thunderstorms but in other fields as well. The lightning rod's paradigmatic significance has rightly been pointed out: it emitted a very strong, if not always specific, suggestion.[25] I have previously alluded to that state of mind (in distant emulation of Peter Gay) as the "recovery of nerve."[26]

The picture that emerges from all of the above, therefore, is one of complex ambivalence. On the one hand we have the more or less ideological nature of the lightning rod debate, the invention being celebrated by enlightened minds as the visible triumph over prejudice.[27] On the other, however, that euphoric emphasis on its symbolic value does not alter the fact that in society in general a certain amount of anxiety still remained. And I am not referring here to the fear of thunderstorms that found such various expression in popular culture, and which caused the enlightened part of the nation such head-shaking despair – often with good reason, given the many manifestations of it described in Chapter 5. No, even in the higher socio-cultural echelons a certain unease in anticipation of a storm can sometimes be detected, despite the distance created by science and technology. Yet in that group, ironically enough, the lightning rod occasionally seems to have acquired the traits of an infallible talisman.[28] The grim dogmatism with which the need for strict adherence to safety regulations was stressed points in the same direction.[29] Was it not the case that from time to time the enlightened elite of scholars, artists, and educated citizens loudly tried to deny its own insecurity? For even if, in principle, the thunderstorm had been robbed of its threatening character, it seems to me that certain artistic expressions could rather be interpreted as reflections of inner unease than as gratuitous games made possible by a greater control of nature. That could be said of many late eighteenth- and early nineteenth-century novels, for example. Their authors exploited the attentive and meticulous observation of natural phenomena, particularly those that were terrifying, to excite emotions and prompt introspection.[30] The result, conversely, was that those individual emotions often came to be reflected in the description of natural phenomena, especially when a sense of unease was involved, for it was that very feeling that was frequently associated with the thunderstorm.[31] In a more general sense, Romantic art – certainly the *unheimliche* – eventually raises the question of whether, at that time, man could really be called lord and master of nature – a kind of doubt that, in my opinion, should not be dismissed as mere play.

That uncertainty merits further analysis. In my view, it is only by relatively overestimating the amount of fear remaining that the widespread aestheticisation of the thunderstorm in the period we have been looking at could be explained. Seductive though that theory may be, it does seem somewhat contrived, firstly because it appears to (necessarily) derive its arguments mainly from rather speculative Freudian-based theories about old fears and new fears and secondly because in consequence it often magnifies cultural reflections of fear into attempts to conjure fear itself. After all, there is far more thunderstorm art and metaphor than seems to justify the remaining fear – on this point Alewyn's surrogate thesis offers a more plausible explanation. But insofar as real fear or at least uncertainty remained in society – not just fear of the storm, but of nature in its broadest sense – another theory may be more convincing. It posits a fundamental change in mindset that occurred fairly early in the eighteenth century and which the French writer Annie Le Brun has characterised as a *revolution copernicienne*. Gradually, man became less inclined to see himself as the nexus of the landscape, which he had always thought of as surrounding him like a backdrop, and to realise that he himself was a participatory part of what he had initially seen merely as staffage: it was no coincidence that around this time many painters left the confines of their workshops and studios and ventured out into nature. Between 1720 and 1760 that altered perspective was already reflected in the change from the geometrically laid out, enclosed French garden devised by the human brain to the open, variable, "natural" English landscape garden. The result of that *orage végétal* was a diminution of anthropocentrism, characterised by Le Brun as nothing less than a *catastrophe intérieure*, the loss of a sense of ontological security, evinced in feelings of menace and undefined danger – a new sensibility, which in a sense ultimately discovered nature as nature.[32] Or, as the politician and man of letters Horace Walpole (1717-1797) wrote of one of the "inventors" of the English landscape garden William Kent (ca. 1685-1748), "He leaped the fence and saw that all nature was a garden."[33]

Even without amplifying on the new sense of nature – which we shall come to in a moment – it should already be clear that Le Brun's interpretation implicitly puts into perspective the importance of the lightning rod's invention and the control of nature and diminished fear of thunderstorms it brought about. The consequences of Franklin's invention for the spiritual climate were certainly embedded in a much broader development that was already underway before 1752 and would eventually lead to the emergence of the new, Romantic concept of nature. Admittedly, as regards the Netherlands, a metabletician could point to the remarkable concurrence of the lightning rod's breakthrough and the triumph of nature over art in Dutch literature in the years between 1775 and 1780,[34] but such a direct connection is easily shot down when we view the matter from a European perspective. For while the classicist theory of art may long have assumed that it was art that turned nature into something beautiful, Elisabeth of Orléans (1676-1744) was already describing her delight in watching a thunderstorm as early as 1722.[35] And even before that, in the years after 1709, Anthony Ashley Cooper, 3rd earl of Shaftesbury (1671-1713), had also argued in a more general sense that "the wildness pleases."[36]

◀ Portrait of William Kent, one of the founders of the English landscape garden, painting by William Aikman, ca. 1723-1725

▶ Portrait of Anthony Ashley Cooper, 3rd earl of Shaftesbury, politician, philosopher, and writer on aesthetic subjects, print by James Hopwood the Elder, ca. 1800

On the other hand, however, Le Brun's interpretation should itself be put in perspective, and in two respects. The gradual growth of a new concept of nature is evident, but it is worth questioning whether the uncertainty so often referred to would not be better construed as disorientation rather than deeply felt doubt or outright fear. And on a more fundamental level, it is also unclear whether such feelings can be regarded as the direct causes (or consequences!) of the new sense of nature. Even if that assumption was correct, it would not really matter much whether that unease was genuinely felt or simulated: in both cases, after all, the delicious shudder of the sublime is the same!

Nevertheless, it cannot be denied that in the second half of the eighteenth century pleasure overcame fear. In the discovery of nature as nature, there is not, as Le Brun seems to suppose, the deep uncertainty or unease of a *catastrophe intérieure* but rather an intense joy associated with or arising from generously yielding best-friend status to Mother Nature. In the course of the eighteenth century, when enlightened man "leaped the fence" with shining confidence, a supposedly blissful identification with "wild" nature occurs, the removal of a boundary that may lead to a passing dizziness – certainly with regard to phenomena experienced as sublime – but soon to a new enthusiasm for nature. Apparently, the (suggestion of) safety was now great enough not only to allow a cultural game with the thunderstorm and even with a working model volcano, but also to enable the new sense of nature itself to emerge in all its complexity.

By now it will be clear that this new cultural matrix came about in several phases. And if we are to summarise that development in a nutshell, the German situation can serve as a first guideline, for in many ways it can be said to be exemplary. Sharp boundaries cannot be drawn, of course, but what is clear is that in the early eighteenth century, artists were already deliberately going outside, albeit not yet to seek the sublime but to look for the beauties of nature. In that sense, these were already discovered by the time of the Enlightenment.[37] Full of self-confidence, scholars and artists penetrated the natural world to unlock its scientific and aesthetic secrets, not yet to escape a society that was perceived as too civilised.[38] In this early phase, particularly since the publication of the poem *Irdisches Vergnügen in Gott* ("Earthly Pleasure in God") in 1721 by the aforementioned German poet and physico-theologian Barthold Heinrich Brockes, the search for beauty and the contemplation of nature had already supplanted the shallow concept of utility which is often associated with the Enlightenment. At all events, this is not the case with Brockes.[39] And although he himself announced a new era[40] in a few lyrical passages evoking, for example, the serene calm that follows a thunderstorm, many of his contemporaries were often still writing purely descriptive poems about nature.[41]

Shortly after the middle of the century, profound changes took place in Western culture. In an atmosphere of growing sentimentalism there was now an increased focus on the inner self when beholding nature and the often conflicting feelings that fought for precedence in the heart. Among members of the educated elite a veritable cult of nature soon developed in which the closely observant man relished the emotions stirred in him by his fellow beings, even the smallest worm or blade of grass,[42] while nature as a whole – at least in its harmonious aspects – was increasingly treasured as a comforter and a guide.[43] At the same time, especially thanks to Burke and also the Ossian poems (1759-1763) "translated" by James MacPherson (1736-1796), the modern version of the sublime made its appearance, accompanied by a sense of human insignificance, unease, and loneliness in the face of the violent aspects of nature – an awareness that implied an even more intense sense of the dislocating loss of an anthropocentric starting point noted by Le Brun. The old Leibnizian solution that explained away discordant natural phenomena like the devastating Lisbon earthquake (1755) by reducing them to relatively small elements in the great harmonious cosmos was less satisfactory than before. "Le mal est sur la terre" – "Evil stalks the earth" – was Voltaire's adamant response in his *Poème sur le désastre de Lisbonne* (1756).[44] But whereas at first the Enlightenment barely knew how to deal with natural phenomena such as earthquakes, volcanic eruptions, or violent thunderstorms, (pre-)Romanticism found a way round the matter by introducing the aesthetic genre of the sublime. Meanwhile, the fact that in this (pre-)Romantic sensibility nature triumphed over art[45] and accordingly the sublime pre-empted the beautiful, makes it clear that there was enough confidence to allow oneself safely to be overwhelmed by "wild" nature; an enlightened trust in man and his abilities, a trust in nature which – despite its inhospitable aspects – was nonetheless harmonious or at least enjoyable, and a trust in a loving rather than a punishing God. In short, in the course of the eighteenth century images of nature, man, and God underwent a major parallel change among members of the cultural upper class.

▶ Portrait of James MacPherson, compiler of the *Ossian* poems (1759-1763), anonymous print, eighteenth century

Although that identification of the tormented Romantic with nature was often difficult to achieve in practice,[46] the mere passionate pursuit of it offered a lead to a more transcendent experience of nature. To begin with, that transcendence meant a secular widening of the horizon, the removal of a boundary far beyond any fence, but there are also unmistakably religious aspects to it. Not infrequently, the pleasant shiver morphed into a reverent, indeed a sacred shudder, because the sublime and the religious were very often not far removed from one another – the discovery that lightning was the result of material forces did not alter that.[47] It was no coincidence that a German text extolling the lightning rod spoke of entering the "Heiligtum der Natur,"[48] nor that in 1809 the later Groningen professor of natural history Theodorus van Swinderen (1784-1851) referred to the perception of the sublime in nature as sacred.[49] Many may have felt strengthened in such a numinous interpretation since even the etymology pointed that way, for the sublime brings man to a threshold (*sub liminem*) and is therefore a borderline experience.

It had been a long road to that sacralisation, for ever since the seventeenth century nature had increasingly been read as the immanent book in which God showed his existence and qualities – albeit a reading more popular among Protestants than Catholics. We saw in Chapter 4 how a whole branch of physico-theology known as "brontotheology", the theology of thunder, had developed.[50] But now, influenced by a desire to surrender oneself to nature, the overarching edifice of physico-theology itself also changed in character. Of old, nature's astonishing order and efficiency had been the rational apologetic argument that substantiated God's existence and his providential goodness, but as of the sentimentalist climate of the last decades of the eighteenth century that role was increasingly taken over by wonder, contemplation, and introspection. By dint of an amplified awareness of the limitations of human knowledge – and, I would add, also through a lesser preoccupation with control resulting from a growing sense of safety – the active, rational study of objectified nature was increasingly replaced by a passive receptive approach, which also allowed for more "spiritual" interpretations.

But whether it was despite or because of Franklin's invention, it was precisely in the unleashed force of the thunderstorm that the artist sought to experience his direct dependence on the majestic God.[51] The aesthetic experience thus gained, the experience of the pure "wild" splendour of sublime creation, offered a new emotional underpinning to faith in God, as in the case of the preacher, poet, and politician Jacob Kantelaar (1759-1821), for example.[52] What did not change, however, was the steadfast trust that nature, even in its "untamed" aspects, was under God's providential and benevolent protection: physico-theology may therefore be regarded as the essential condition of the sublime.[53] In the view of representatives of the (Christian) Enlightenment, the sublime Creator himself thus guaranteed the ultimate enjoyment of his creation, if only by providing autonomous man with the means (such as the lightning rod) to defy its dangers.

▶ Portrait of the preacher, poet, and politician Jacob Kantelaar, print by Willem van Senus, after Jacobus Schoemaker Doyer, 1812-1851

All this was in stark contrast to the traditional perception of nature in the early modern period. In Chapter 4 we saw how naturally the storm was then associated with God's wrath at the sins of men. And that was not merely a matter of divine finger-wagging at individual sins here and there: for centuries it had been believed that ever since man's Fall the whole of Creation had lain under the curse of original sin, which had led to a general degeneration. Storms (including thunderstorms) and such barren obstacles as high mountains and unpredictable seas were the obvious proofs of it.[54] In the (Christian) enlightened discourse, on the other hand, mountains could no more be called a result of original sin than thunderstorms were proofs of God's ire. On the contrary, in the earl of Shaftesbury's view, that "wild", untouched nature was the very place in which the "pure" work of the benevolent Creator was to be found.[55] Nor was man degenerate either, for his noble disposition enabled him to devise inventions and, like a modern Prometheus, fearlessly face the threat of all sublime natural violence – an encounter that bestowed a new dignity on him.

Some went even further. For them it was not enough that under God's benevolent aegis "virtuous" man equated himself with the equally faultless Creation, or that exalted man identified himself (cautiously, of course) with the sublime aspects of immeasurable nature. Those Romantics set their goals higher, seeking nothing less than a complete absorption in nature by a cosmic expansion of feeling. For them the created world was thus no longer the indirect source of God but the direct revelation of him.[56] Although those mystics realised that nature, or at least the aestheticised version of it, to wit the landscape, was not only a sign of God's presence but at the same time a mark of his mystery,[57] tendencies such as theosophy and pantheism still crept more or less imperceptibly into the picture. In the process the personal element usually disappeared from the conception of God, for to them the numinous was rather an all-pervading energy – as we saw in Chapter 6, to many theosophists, electricity became the particular expression of God's power. And when that mysterious power was no longer regarded as divinely created but as an emanation of the Sublime One himself, the transition to pantheism was a fact.[58] Done with disguising their hesitations with a well-chosen oxymoron or intriguing chiaroscuro, a number of authors, artists, and thinkers, particularly those who espoused northern European Romanticism, deliberately got off the fence and came down pretty firmly on the side of pantheistic transcendence.[59] Whichever ideological standpoint the passionate Romantic took, though, he invariably drank in nature to her last drop, striving with mystical intimacy to incorporate her (as if she were his beloved) into his soul.[60] The master-servant relationship[61] or anthropocentric starting point had long been a thing of the past, but the Romantic, deeply aware of his own insignificance, only waited longingly for the nature he so admired to appeal to his heart.[62] This new sense of nature was not only present in high art, such as the works of Friedrich Schiller,[63] but in popular novels as well.[64] We also find the same mindset in many landscape painters of the period. They sought to render the idea of nature as a manifestation of the hand of God. A painted landscape was a microcosm related to God's macrocosm, a coherent whole that must convince the viewer of nature's divine character.[65] Perhaps that formulation had more pantheistic overtones than some artists would have liked, but in any case it can be said that for many the universe acquired an animate and emblematic character.[66]

Epilogue

With pantheism we have come as far as we can go. The circumference of the hourglass this book hopefully resembles has reached its maximum. When we finally return to our starting point and consider once again the likely consequences that the introduction of the lightning rod had for the way in which the thunderstorm was perceived, the first thing to strike us is the waning of the conviction that it was in some way the "punishment of sin". No less obvious is that the invention proved detrimental to all manner of practices and views not sanctioned by the Church and resulted in a re-interpretation of some of their ecclesiastical counterparts. In a more general and positive sense, there was a notably increased feeling of security, even if it was not always justified. That confidence-inspiring climate, in which the notion of enlightened triumph provided the basis for "sensitive" shuddering, seems in turn to have facilitated the transformation of the thunderstorm into an aesthetic and metaphorical phenomenon unprecedented in history.[67] In any case, thanks to the advent of the modern version of the sublime as a genre, thunder and lightning became an extremely popular theme in painting and music, and even more so in literature. It has rightly been remarked that from an aesthetic point of view Romanticism should be seen not as a reaction to the Enlightenment but as a corollary to it.[68] Yet it was not only in art but also in society as a whole that thunderstorms in general and the formidable and mysterious power of electricity in particular were used as metaphors for diverse types of psychological states of emergency,[69] such as sexual passion, tyranny, revolutionary or bellicose inclinations, religious conversion, and artistic creativity. In a period noted for its predilection for "internalising" weather, no class of meteorological phenomena corresponded so well with the whimsical, irrational mood of the Romantic as meteors and – mutatis mutandis – thunderstorms.[70] What scene could have greater appeal for the heart of a sensitively enlightened or early Romantic poet, pastor, painter, musician, or citizen than thunder and lightning? And that appeal also extended to an audience whose feeling for nature was not so fine-tuned, because the raw violence of the thunderstorm sometimes acted as a fairly stereotypical aesthetic tool that could easily degenerate into bombast.[71]

Eventually, thanks in part to that increased sense of safety (to which the invention of the lightning rod made a modest, but not insignificant contribution[72]), in the late eighteenth and early nineteenth century a new sensibility could emerge in which nature developed into something like a beloved friend. Fear seemed to give way to wonder, many old nervous inhibitions disappeared. In the absence of fundamental threat, fear and pleasure even intensified each other in the aesthetic experience of the sublime. Or was there real fear behind all the enthusiastic (pre-)Romantic sensitivity to nature and did the artists drown it out themselves? It does, after all, take courage to actually admit fear. Did they and their audiences willingly allow themselves to be carried away – artistically speaking – by nature's violence so that they no longer had to fight against it? Whatever the case may be, old enlightened minds probably grumbled that by doing so man gave up his dearly acquired freedom and knowingly made himself the slave of nature again. But protests of that kind were now falling on deaf ears. They had less and less effect on the enthusiasm of the grumbler's contemporary. In a striking parallel with Reverend Johannes Martinet, who in 1778 watched a thunderstorm in utter delight in the Dutch Republic,[73] the German man of letters Christoph Martin Wieland also confessed in exactly the same year that, "I do not know anything more touching than these kinds of scene."[74]

▶ *Die Dreisteine im Riesengebirge*, painting by Carl Gustav Carus, 1826. The rock formation known as the "Three Stones" is part of the *Riesengebirge* mountain range that straddles the Czech-Polish border. Carus was a physician, natural philosopher, and painter. As an artist, especially in his early period, he was strongly influenced by Caspar David Friedrich with his predilection for mountains, forests, nocturnal lunar landscapes, and ruins – "desolate and peaceful" is an apt description

Notes

Introduction: Lightning after Franklin

1 "Geen Verschynsel toch is er in de gansche Natuur treffender voor het gezigt, ontzaggelyker voor het gehoor, geweldiger voor het gevoel, en nuttiger voor de Waereld." J.F. Martinet, *Katechismus der natuur* (4 vols.; Amsterdam: J. Allart, 1777-1779), 1: 122.

2 "Als er een Onweêr opkomt, wanneer ik aan myn werk ben, werp ik alles ter neêr, om het te hooren en te zien (...). Luisterde de heilige Dichter er ook niet na, toen hy zong: de God der eeren dondert (Ps. 29:3) of wil hy, dat wy het niet zullen opmerken? Ik hoor en zie het nooit dan met groote eerbied, diepe verwondering, en met een nederig vermaak, zonder dwaaze vrees; ten zy het zeer geweldig is, en deeze zegen in een oordeel veranderd wordt." Ibid., 1: 122-23.

3 F. Bulhof, *Ma patrie est au ciel. Leven en werk van Willem Emmery de Perponcher Sedlnitzky (1741-1819)* (Hilversum: Verloren, 1993), 172 (cf. also 228).

4 Cf., e.g., J.C. Streng, *'Het is thans zeer briljant'. Aspecten van het Zwolse culturele leven tijdens de overgang van ancien régime naar moderne tijd* (Hilversum: Verloren, 1999), 155, and I. Leemans and G.-J. Johannes, *Worm en donder. Geschiedenis van de Nederlandse literatuur 1700-1800: de Republiek* (Amsterdam: Bert Bakker, 2013), 24.

5 Cf. Leemans and Johannes, *Worm en donder*, 31, 37-39 and 379.

6 See e.g., A. Reed, *Romantic Weather. The Climates of Coleridge and Baudelaire* (Hanover, NH-London: Brown University Press, 1983), 4.

7 See R. Anderson, *Lightning Conductors: Their History, Nature and Mode of Application* (London-New York: E. & F.N. Spon, 1880), 25.

8 Cf. Chr. Begemann, *Furcht und Angst im Prozeß der Aufklärung. Zu Literatur und Bewußtseinsgeschichte des 18. Jahrhunderts* (Frankfurt a.M.: Athenäum, 1987), 130 and 91 respectively.

9 R. Stichweh, *Zur Entstehung des modernen Systems wissenschaftlicher Disziplinen. Physik in Deutschland 1740-1890* (Frankfurt a. M.: Suhrkamp, 1984), 279, cited by C. Möhring in her important Ph.D. Dissertation *Eine Geschichte des Blitzableiters. Die Ableitung des Blitzes und die Neuordnung des Wissens um 1800* (Weimar 2005), 14. Available online at: https://e-pub.uni-weimar.de/opus4/frontdoor/index/index/year/2009/docId/1374 (accessed 4 November 2021).

10 Kant did so as early as 1756 in his *Fortgesetzte Betrachtung der seit einiger Zeit wahrgenommenen Erderschütterungen*. See R. Wild, "Prometheus-Franklin: Die Gestalt Benjamin Franklins in der deutschen Literatur des 18. Jahrhunderts," *Amerikastudien/American Studies* 23 (1978): 31.

11 E. Weigl, "Entzauberung der Natur durch Wissenschaft – dargestellt am Beispiel der Erfindung des Blitzableiters," in *Jahrbuch der Jean-Paul-Gesellschaft* 22 (1987): 18. Kant was initially rather sceptical about the advantages of the lightning conductor.

12 Cf. C. Grabo, *A Newton among Poets. Shelley's Use of Science in* Prometheus Unbound (Chapel Hill: University of North Carolina Press, 1930; repr. 1968), 185.

13 Mary Shelley, *Frankenstein, or the Modern Prometheus* (1818). See below, p. 87.

14 See Begemann, *Furcht und Angst*, 91.

15 Cf. J. Golinski, *British Weather and the Climate of Enlightenment* (Chicago: The University of Chicago Press, 2007), 10: "We cannot identify the Enlightenment as a whole with the extreme hubris of Victor Frankenstein."

16 "die Lehre von den Wettervorgänge", "zum ‚Wortführer' im Gespräch zwischen Theologie und Naturwissenschaft wird." M. Büttner, "Theologie und Klimatologie im 18. Jahrhundert," *Neue Zeitschrift für systematische Theologie und Religionsphilosophie* 6 (1964): 157.

17 R. van Dülmen, *Kultur und Alltag in der Frühen Neuzeit. Dritter Band: Religion, Magie, Aufklärung 16.-18. Jahrhundert* (Munich: Beck, 1994), 270.

18 A point advanced by O. Hochadel and P. Heering, "Introduction: Revisiting an Invisible Technology," *Playing with Fire. Histories of the Lightning Rod*, ed. P. Heering, O. Hochadel, and D.J. Rhees (Philadelphia: American Philosophical Society, 2009), 3-4.

19 Cited ibid., 4.

20 Cf. O. Briese, *Die Macht der Metaphern. Blitz, Erdbeben und Kometen im Gefüge der Aufklärung* (Stuttgart-Weimar: J.B. Metzler, 1998), 60.

21 J. Huizinga, "Natuurbeeld en historiebeeld in de achttiende eeuw," in idem, *Verzamelde werken* (9 vols.; Haarlem: H.D.Tjeenk Willink, 1948-1953), 4: 341-59.

22 Cf. Reed, *Romantic Weather*, 57.

23 Ibid., 54.

24 Ibid., 57.

25 Cf., e.g., J. von der Thüsen, *Het verlangen naar huivering. Over het sublieme, het wrede en het unheimliche. Essays* (Amsterdam: Querido, 1997), 7.

26 Cf. R.A. Bosch, *En nooit meer oude Psalmen zingen. Zingend geloven in een nieuwe tijd 1760-1810* (Zoetermeer: Boekencentrum, 1996), 33.

27 See above, n. 18 and O. Hochadel, *Öffentliche Wissenschaft. Elektrizität in der deutschen Aufklärung* (Göttingen: Wallstein, 2003).

28 P. Heering and O. Hochadel, "Epilogue. An Invisible Technology – What Remains to be Seen," in Heering et al., *Playing with Fire*, 271.

29 M. Beller, *Jupiter tonans. Studien zur Darstellung der Macht in der Poesie* (Heidelberg: Winter, 1979). I disregard here the popular monograph by D.M. Elsom, *Lightning. Nature and Culture* (London: Reaktion Books, 2015).

30 To mention only the most important: M. Gamper, *Elektropoetologie. Fiktionen der Elektrizität, 1740-1870* (Göttingen: Wallstein, 2009); R. Gaderer, *Poetik der Technik. Elektrizität und Optik bei E.T.A. Hoffmann*, (Freiburg: Rombach, 2009), and B. Specht, *Physik als Kunst. Die Poetisierung der Elektrizität um 1800* (Berlin-New York: De Gruyter, 2010). Somewhat comparable nineteenth-century American topics are dealt with by P. Gilmore, *Aesthetic Materialism. Electricity and American Romanticism* (Stanford: Stanford University Press, 2009). For Britain, cf.: M. Fairclough, *Literature, Electricity and Politics 1740–1840: 'Electrick Communication Every Where'* (London: Palgrave, 2017) and by the same author "Romantic Electricity: Fields and Currents in Recent Research," *Literature Compass* 15 (2018): 1-15 and the literature mentioned there. For France, cf. the Special Issue of the *Revue des Sciences Humaines* 281, no. 1 (January-March 2006): 1-255, entitled *L'Imaginaire de l'électricité dans les lettres et les arts*, edited by S. Thorel-Cailleteau and C. Jamain.

31 U. Büttner and I. Theilen, eds., *Phänomene der Atmosphäre. Ein Kompendium literarischer Meteorologie* (Stuttgart: J.B. Metzler, 2017). Similar issues are discussed in another recent volume: G. Braungart and U. Büttner, eds., *Wind und Wetter. Kultur-Wissen-Ästhetik* (Paderborn: Fink, 2018). Furthermore, the "fascinating and often unsettling connection between weather and religion" is thematised in a monograph by Peter J. Thuesen, *Tornado God. American Religion and Violent Weather* (Oxford-New York: Oxford University Press, 2020). A general survey is given by J. Hardwick and R.J. Stephens, "Acts of God: Continuities and Change in Christian Responses to Extreme Weather Events from Early Modernity to the Present," *WIREs Climate Change* 11, no. 2 (2020): 1-16.

32 *Centaurus* 57, no. 3 (2015): 129-228, edited by K. Vermeir.

33 E. Le Roy Ladurie, J. Berchtold, and J.-P. Sermain, eds., *L'Événement climatique et ses représentations (XVIIe-XIXe siècle). Histoire, littérature, musique et peinture* (Paris: Desjonquères, 2007) and D. McCallam, *Volcanoes in Eighteenth-Century Europe. An Essay in Environmental Humanities* (Oxford: Oxford University Press, 2019).

34 On this category of "electrical demonstrators", see Hochadel, *Öffentliche Wissenschaft*, 186-248. Itinerant showmen like these were also active in the Dutch Republic, see below, Chapter 2, n. 4.

35 Cf. Golinski, *British Weather*, 7.

1 - A New Invention

1 J.J.L. ten Kate, *De planeeten: een gedicht*, 2nd ed. (The Hague: Nijhoff, 1873), 70 (orig. ed. 1869).

2 Though the Leiden professor Van Musschenbroek has been credited with the invention of the "Leyden jar," the German scholar Ewald Jürgen von Kleist (1700-1748) had devised a similar apparatus some months earlier. His communication was rather inadequate, however. On the invention of the condensator, see J.L. Heilbron, *Electricity in the 17th and 18th Centuries. A Study of Early Modern Physics* (Berkeley and Los Angeles-London: University of California Press, 1979), 309-23 and J. Meya and H.O. Sibum, *Das fünfte Element. Wirkungen und Deutungen der Elektrizität* (Reinbek: Rowohlt, 1987), 64-66. On Van Musschenbroek, see C. de Pater, *Petrus van Musschenbroek (1692-1761), een newtoniaans natuuronderzoeker* (Utrecht: s.n., 1979), esp. 41-44.

3 Heilbron, *Electricity*, 339-40.

4 Ibid., 340 and W.D. Hackmann, "Instruments and Experiments. The Case of Atmospheric Electricity in Eighteenth-Century Holland," *Tijdschrift voor de geschiedenis der geneeskunde, natuurwetenschappen, wiskunde en techniek* 10 (1987): 196-97.

5 Cf. J. Riskin, *Science in an Age of Sensibility. The Sentimental Empiricists of the French Enlightenment* (Chicago: University of Chicago Press, 2002), 97.

6 I. Bernard Cohen, "Prejudice against the Introduction of Lightning Rods," *Journal of the Franklin Institute* 253 (1952): 423 and the literature listed there. For a biography of Kinnersley, see J.A.L. Lemay, *Ebenezer Kinnersley, Franklin's Friend* (Philadelphia: University of Pennsylvania Press, 1964).

7 Heilbron, *Electricity*, 348-52.

8 Ibid., 349.

9 P. Bertucci, "Enlightening Towers. Public Opinion, Local Authorities, and the Reformation of Meteorology," in Heering et al., *Playing with Fire*, 27-30. Veratti's countryman Giovanni Battista (Giambattista) Beccaria (1716-1781), a professor of physics in Turin, had built an experimental conductor on the roof of his house as early as 2 July 1752. See ibid., 32-33.

10 From the vast amount of literature on this topic I only mention here the classic publication by I. Bernard Cohen, *Benjamin Franklin's Science* (Cambridge, MA-London: Harvard University Press, 1990), which also discusses the experiments of Dalibard and others. Importantly, Cohen devotes attention to the ideas of the French lawyer Jacques de Romas (1713-1776) who, independently of Franklin, suggested using a kite to extract electric sparks from thunderclouds. Though he claimed to have carried out this experiment before Franklin, he finally had to admit that the American had been earlier in publishing. On Romas, see G. Berger and S.A. Amar, "The Noteworthy Involvement of Jacques de Romas in the Experiments on the Electric Nature of Lightning," *Journal of Electrostatics* 67 (2009): 531-35. A general overview is given by Heilbron, *Electricity*.

11 See, e.g., M.B. Schiffer, *Draw the Lightning Down. Benjamin Franklin and Electrical Technology in the Age of Enlightenment* (Berkeley and Los Angeles-London: University of California Press, 2003), 165-66 and the literature listed there. Though some of these installations will have been insulated, they do not always seem to have been properly earthed.

12 Heilbron, *Electricity*, 391.

13 Cf. Cohen, "Prejudice," 417 and, more elaborately, I. Bernard Cohen and R. Schofield, "Did Diviš erect the First European Protective Lightning Rod, and was his Invention Independent?," *Isis* 43 (1952): 358-64. On Diviš, see also Möhring, *Geschichte des Blitzableiters*, 83-100 and the literature listed there.

14 On the distinction between an innovation-centred history and a use-centred history, cf. D. Edgerton, *The Shock of the Old. Technology and Global History since 1900*, 3rd ed. (London: Profile Books, 2011), xi-xviii (orig. ed. 2006).

15 This subject is treated at greater length by Hochadel, *Öffentliche Wissenschaft*.

16 Cohen, "Prejudice," 417. Möhring, *Geschichte des Blitzableiters*, 91-92 argues that Diviš's opponents were not driven by superstition or ignorance but simply followed Diviš's own reasoning, implying that his installation would prevent the mere development of thunderstorms. However, an eighteenth-century source led Martin Scharfe to point out that Diviš was aiming at a different goal: he intended to disperse the lightning's energy in the upper strata of the air, in order to preserve only the fertilising rain. See M. Scharfe, *Menschenwerk. Erkundungen über Kultur* (Cologne-Weimar-Vienna: Böhlau, 2002), 76.

17 Hochadel, *Öffentliche Wissenschaft*, 151.

18 See Cohen, "Prejudice," 395, 414, 416, Möhring, *Geschichte des Blitzableiters*, 221, as well as Hochadel and Heering, "Introduction: Revisiting an Invisible Technology," 9. For a biography of Nollet, see J. Torlais, *Un physicien au siècle des Lumières: l'abbé Nollet, 1700-1770* (Paris: Sipuco, 1954).

19 P. Heering, "Styles of Experimentation and the Attempts to Establish the Lightning Rod in Pre-Revolutionary Paris," in Heering et al., *Playing with Fire*, 122. Importantly, electrical theory seemed to be in contradiction with the proportionality between cause and effect postulated by G.W. Leibniz (1646-1716), see Gamper, *Elektropoetologie*, 50.

20 See Heilbron, *Electricity*, 348-49.

21 Riskin, *Science in an Age of Sensibility*, 69-103. In France, Franklin's electrical research became the bone of contention between these different schools of thought.

22 Heering, "Styles of Experimentation," 122.

23 H. Meidinger, *Geschichte des Blitzableiters* (Karlsruhe: Braun, 1888), 42. Though rather dated, this book is still useful for the wealth of factual information it provides.

24 Heering, "Styles of Experimentation," 124.

25 Meidinger, *Geschichte des Blitzableiters*, 35-36.

26 Ibid., 23, 36 and Cohen, "Prejudice," 398.

27 A conclusion to the contrary, drawn by Meya and Sibum, *Das fünfte Element*, 83 seems to be incorrect.

28 Cohen, "Prejudice," 399.

29 Meidinger, *Geschichte des Blitzableiters*, 36 and T.A. Mitchell, "The Politics of Experiment in the Eighteenth Century: the Pursuit of Audience and the Manipulation of Consensus in the Debate over Lightning Rods," *Eighteenth-Century Studies* 31 (1998): 314. While relying on outdated literature, Cohen, "Prejudice," 420 n. 61 erroneously placed the disaster in 1767.

30 Meidinger, *Geschichte des Blitzableiters*, 41. Cf. also J. Wiesner, *Jan Ingen-Housz. Sein Leben und sein Wirken als Naturforscher und Arzt* (Vienna: C. Konegen, 1905), 185 and Möhring, *Geschichte des Blitzableiters*, 123 n. 345. These lightning conductors were not the first to be installed in the Habsburg monarchy, however. That honour fell to the church of Penzing near Vienna in 1770, see Hochadel en Heering, "Introduction: Revisiting an Invisible Technology," 9.

31 Meidinger, *Geschichte des Blitzableiters*, 42.

32 Ibid., 39, 41. Cf. also Möhring, *Geschichte des Blitzableiters*, 267-68.

33 Meidinger, *Geschichte des Blitzableiters*, 38. Cf. also H. Prinz, *Gewitterelektrizität* (Munich: Oldenbourg, 1979), 31 and Anderson, *Lightning Conductors*, 43. Anderson records the existence of more than 700 conductors in a 15-kilometre radius around Hamburg as early as 1774, which compares very favourably with other European regions. On the important role played by Reimarus, see Möhring, *Geschichte des Blitzableiters*, esp. 139-46. Equally important was the existence of the very old and still extant Hamburg lightning protection company D.H.W. Schultz and Sons, see https://dhw-schultz.de (accessed 3 March 2021). Recent scholarship has pointed out that to a certain extent Reimarus was able to build on his father's preliminary experiments and data, see C. Zwierlein, *Prometheus Tamed. Fire, Security, and Modernities, 1400 to 1900* (Leiden-Boston: Brill, 2021), 119.

34 See O. Hochadel, "In Nebula Nebulorum. The Dry Fog of the Summer of 1783 and the Introduction of Lightning Rods in the German Empire," in Heering et al., *Playing with Fire*, 55.

2 - The Introduction of the Lightning Rod in the Netherlands

1 L. Roberts, "Science Becomes Electric: Dutch Interaction with the Electrical Machine during the Eighteenth Century," *Isis* 90 (1999): 680-714, esp. 683-84.

2 Ibid., 688.

3 See J.H. Winkler, *Nieuwe natuurkundige ontdekkingen, aangaande de eigenschappen, werkingen, en oorzaaken der electriciteyt, benevens eene beschryvinge van twee nieuwe electrische werktuigen. Mitsgaders de aftekening der zelve.* Uit het Hoogduitsch vertaald (Amsterdam: Hendrik Vieroot, 1745) and J.H. Winkler, *De eigenschappen*

der electrische stoffe, en van het electrische vuur, uit verscheidene nieuwe proeven verklaart. Benevens eenige in 't koper gebragte nieuwe werktuigen om te electriseeren. Strekkende tot een tweede deel der Gedagten over de electriciteyt (Amsterdam: Hendrik Vieroot, 1746). The German originals were published in 1744 and 1745, respectively. On Winkler, see Meya and Sibum, *Das fünfte Element*, 48-58.

4 See M.R. Wielema, *Ketters en verlichters. De invloed van het Spinozisme en het Wolffianisme op de Verlichting in gereformeerd Nederland* ([Amsterdam:] s.n., 1999), 146 n. 17, 150 and Roberts, "Science Becomes Electric," passim. These itinerant demonstrators probably belonged to a different social category. If the comparison with Germany stands up, then they may have been mere showmen, see Hochadel, *Öffentliche Wissenschaft*, 187-220 and by the same author "The Sale of Shocks and Sparks: Itinerant Electricians in the German Enlightenment," in *Science and Spectacle in the European Enlightenment*, ed. B. Bensaude-Vincent and Chr. Blondel (Abingdon-New York: Routledge, 2008), 89-101.

5 J. Golinski, "Enlightenment Science," in *The Oxford Illustrated History of Science*, ed. I. Rhys Morus (Oxford: Oxford University Press, 2017), 195. On the sublime experience, cf. below, Chapter 7.

6 According to Th.L. Hankins, *Science and the Enlightenment* (Cambridge: Cambridge University Press, 1985), 55, early electrical research was not dominated by amateurs and was primarily aimed at developing scientific theory.

7 Cf. Hochadel and Heering, "Introduction: Revisiting an Invisible Technology," 13-14.

8 Cf. J.J. Kloek and W.W. Mijnhardt, *1800: Blauwdrukken voor een samenleving* (The Hague: Sdu, 2001), 91.

9 See J.H. Winkler, *Die Stärke der electrischen Kraft des Wassers in gläsernen Gefässen* (Leipzig: Breitkopf, 1746).

10 "Brief over den elektrikalen vlieger, geschreven door B. Franklin, October 19, 1752," *Uitgezogte verhandelingen uit de nieuwste werken van de sociëteiten der wetenschappen in Europa en van andere geleerde mannen*, vol. 1 (1757): 617-18. A Dutch translation is given there of a letter from Franklin's *New Experiments and Observations* (London: D. Henry and R. Cave, 1754).

11 See H.A.M. Snelders, "Benjamin Franklin (1706-1790), lid-correspondent van het Bataafsch Genootschap der Proefondervindelijke Wijsbegeerte te Rotterdam (1771)," *Nieuwe verhandelingen van het Bataafsch genootschap der proefondervindelijke wijsbegeerte te Rotterdam*, 3rd ser., vol. 6 (1991): 10.

12 See P.G. Hoftijzer, "An Eighteenth-Century Amateur of Books and Science in Warmond," in *Living in Posterity. Essays in Honour of Bart Westerweel*, ed. J.F. van Dijkhuizen et al. (Hilversum: Verloren, 2004), 148.

13 Ibid., 153. Edens's estate auction catalogue attests to his fascination with electrical experiments. Besides the instruments mentioned above he possessed a number of early books on this subject, e.g., Winkler's *Nieuwe natuurkundige ontdekkingen* from 1745 (see above, n. 3).

14 William Franklin to Sarah Franklin, 10 October 1761, *The Papers of Benjamin Franklin*, vol. 9, *January 1, 1760, through December 31, 1761*, ed. Leonard W. Labaree (New Haven: Yale University Press, 1966), 364–68 [online: *Founders Online*, National Archives, https://founders.archives.gov/documents/Franklin/01-09-02-0153]. See also Snelders, "Benjamin Franklin," 10 and H. Hardenberg, "Benjamin Franklin en Nederland," *Bijdragen voor de geschiedenis der Nederlanden* 5 (1950): 216. I am grateful to Lee Preedy for the first reference.

15 See Wiesner, *Jan Ingen-Housz*, 17.

16 On this subject, see S. Schaffer, "The Consuming Flame: Electrical Showmen and Tory Mystics in the World of Goods," in *Consumption and the World of Goods*, edited by J. Brewer and R. Porter (London-New York: Routledge, 1993), 494-95 and P. Bertucci, "Revealing Sparks: John Wesley and the Religious Utility of Electrical Healing," *The British Journal for the History of Science* 39 (2006): 341-62.

17 Heilbron, *Electricity*, 420.

18 J. Engelman, "Proeven over den aart der werkinge van de electriciteit," in *Verhandelingen van de Hollandsche Maatschappy van Wetenschappen*, vol. 1 (1754): 267-328. Engelman was only one of a large number of Dutch medical men in early electrical research, cf. Floris Winckel, "Medicalising Electricity in the Dutch Republic, 1745-1789," *BMGN- Low Countries Historical Review* 137, no. 3 (2022): 60-86.

19 See M. van Marum, *Verhandeling over het elektrizeeren* (Groningen: s.n., 1776), ii (Preface).

20 J.Th. Rossyn, *De tonitru et fulmine ex nova electricitatis theoria deducendis* (Franeker: Gulielmus Coulon, 1762).

21 Ibid., 33. Interestingly, in his later life Rossyn welcomed other innovations as well, such as draining by means of steam engines. See M.J. van Lieburg and H.A.M. Snelders, *'De bevordering en volmaking der proefondervindelijke wijsbegeerte'. De rol van het Bataafsch Genootschap te Rotterdam in de geschiedenis van de natuurwetenschappen, geneeskunde en techniek (1769-1988)* (Amsterdam: Bataafsch Genootschap, 1989), 73.

22 "Brief van den heer Franklin, betreffende het voorkomen van ongelukken, die het weerlicht veroorzaakt," *Vaderlandsche letter-oefeningen* 6, pt. 2 (1766): 29-31.

23 "Nieuwe waarneemingen, gedaan op het afleiden van de blixemstoffe door yzeren staaven, ter voorkominge van ongelukken; benevens eenige aanmerkingen over den koers dien de blixem houdt," *Nieuwe vaderlandsche letter-oefeningen* 3, pt. 2 (1770): 20-25. Winthrop's letter is dated 4 July 1768.

24 J.-C. Valmont de Bomare, *Algemeen en beredenerent woordenboek der natuurlyke historie* (3 vols.; Dordrecht: Abraham Blussé, 1767-1770), 1: 260.

25 J. le Francq van Berkhey, *Natuurlyke historie van Holland* (9 vols.; Amsterdam: Yntema and Tieboel, 1769-1811), 1: 356-68.

26 Wiebe E. Bijker, *Of Bicycles, Bakelites, and Bulbs. Toward a Theory of Sociotechnical Change* (Cambridge, MA-London: MIT Press, 1995), 270.

27 J.-A. Nollet, *Brieven over de elektrisiteit* (3 vols.; Utrecht-Amsterdam: S. de Waal and G. Warnars, 1773-1783), 1: xiv note (French orig. ed. 1753). Digital versions of books by Nollet and many of his contemporaries are available at http://cnum.cnam.fr (accessed 16 September 2018).

28 N. Chomel, *Algemeen huishoudelijk-, natuur-, zedekundig- en konstwoordenboek*. 2e druk geheel verbetert en meer als de helfte vermeerdert door J.A. de Chalmot (7 vols.; Leiden-Leeuwarden: J. le Mair and J.A. de Chalmot, 1778), under 'blixem' and 'donder', and *Vervolg* (9 vols.; Kampen-Amsterdam: J.A. de Chalmot and J. Yntema, 1786-1793). A similar ambivalence can be found in the dictionary of the Amsterdam polymath Egbert Buys (ca. 1725-1769). Though he acknowledged the similarity of lightning and electricity, he did not draw the firm conclusion of the lightning rod's usefulness. See Egbert Buys, *Nieuw en volkomen woordenboek van konsten en weetenschappen* (10 vols.; Amsterdam: S.J. Baalde, 1769-1778), 3: 538.

29 J. Rechter, "Beschryving van een toestel, geschikt, om met een vlieger, veilig, electrische proeven te neemen," *Algemeene vaderlandsche letter-oefeningen* 1, pt. 2 (1779): 69-81, there 81.

30 On John Cuthbertson and his brother Jonathan (who was active in Rotterdam), see W.D. Hackmann, *John and Jonathan Cuthbertson. The Invention and Development of the Eighteenth Century Plate Electrical Machine* (Leiden: Rijksmuseum voor de geschiedenis der natuurwetenschappen, 1973) and M. Rooseboom, *Bijdrage tot de geschiedenis der instrumentmakerskunst in de Noordelijke Nederlanden tot omstreeks 1840* (Leiden: s.n., 1950), 49-52, as well as Roberts, "Science Becomes Electric," 696-99.

31 See M. Keyser, "Het intekenboek van Benjamin Bosma – natuurwetenschappelijk en wijsgerig onderwijs te Amsterdam, 1752- 1790. Een verkenning," in *Jaarverslagen Koninklijk Oudheidkundig Genootschap, uitgebracht door de voorzitters in de 124e, de 125e, de 126e en de 127e jaarvergadering* (1986): 70-71. Both brothers attended Bosma's lectures in 1769 and 1771-1772. Bosma was one of the founders of Amsterdam's famous learned society *Felix Meritis*. He inaugurated its physics department on 11 November 1777. He gave a lecture on electricity only a week later, see ibid., 73.

32 Cf. A. Wiechmann and L.C. Palm, eds., *Een elektriserend geleerde. Martinus van Marum 1750-1837* (Haarlem: Joh. Enschedé, 1987), 20.

33 John Cuthbertson included a catalogue of instruments and objects for sale in his shop at the back of the second volume of the Dutch translation of Joseph Priestley's *The History and Present State of Electricity* (1767): J. Priestley, *Geschiedenis en tegenwoordige staat der proefondervindelijke wijsbegeerte* (2 vols.; Amsterdam: M. Magerus, 1773), 2: 319-27, there 322. Priestley's book was translated and brought out by the Amsterdam publisher Martinus Magerus (d. before 1784), with the assistance of and additions by John Cuthbertson.

34 See J. Cuthbertson, *Algemeene eigenschappen van de electriciteit, onderrichting van de werktuigen en het neemen van proeven in dezelve*, 2nd ed. (2 vols.; Amsterdam: Pieter Hayman, 1782), 2: 219 (orig. ed. Amsterdam 1776). An account of the experiments of 18 and 23 July 1772 is given, e.g., in the *'s Hertogenbossche courant*, 11 August 1772. On the experimenters mentioned here, in a somewhat different context, cf. H.A.M. Snelders, *Het gezelschap der Hollandsche scheikundigen. Amsterdamse chemici uit het einde van de achttiende eeuw* (Amsterdam: Rodopi, 1980).

35 Cf. Prinz, *Gewitterelektrizität*, 27-28 (with illustration).

36 According to J.A. Gruys and C. de Wolf, *Thesaurus 1473-1800* (Nieuwkoop: B. de Graaf, 1989), 117.

37 See above, n. 33.

38 Priestley, *Geschiedenis*, 2: 60-61 n.

39 See D.A. von Gallitzin, *Sendschreiben an die Kaiserliche Akademie der Wissenschaften zu St. Petersburg über einige Gegenstände der Electricität* (Münster-Leipzig: Ph.H. Perrenon, 1780). Cf. also H. Prinz,"Fulminantes über Wolkenelektrizität," *Bulletin des schweizerischen elektrotechnischen Vereins* 64, no. 1 (1973): 8-9. On Gallitzin, who was appointed a director of the *Hollandsche Maatschappij der Wetenschappen* in 1777, see L. de Gou, *Biografische bijdragen over achttiende-eeuwers* (Haarlem: Joh. Enschedé, 1988), 55-79. He had his own electrostatic generator built around 1777, see ibid., 71.

40 Hardenberg, "Benjamin Franklin en Nederland," 217 and Snelders, "Benjamin Franklin," 10-11. The appointment took place in 1771. As appears from surviving correspondence, Franklin appreciated receiving the transactions of this learned society.

41 See Van Lieburg and Snelders, "De bevordering," 48-49 and M. van Marum, "Verhandeling over de electrische luchtverhevelingen," *Verhandelingen van het Bataafsch Genootschap der Proefondervindelyke Wysbegeerte*, vol. 6 (1781): 1-74.

42 W. May, "Verhaal der uitwerkingen van eenen enkelden blixem slag, voorgevallen op één van 's lands oorlogschepen," *Verhandelingen uitgegeeven door de Hollandsche Maatschappye der Weetenschappen*, vol. 12 (1770): 391-412.

43 See L. Jensen, "Disaster upon Disaster Inflicted on the Dutch," *BMGN- Low Countries Historical Review* 134, no. 2 (2019): 45-70, there 57.

44 W. Watson, "Raadgeeving tot voorkoming van de ongelukken, die de schepen dikwyl door het weerlicht overkomen," *Vaderlandsche letter-oefeningen* 4, pt. 2 (1764): 455-58.

45 D. Radermacher, "Berigt wegens de uitwerkingen van den donder op een Oost-Indisch schip," *Verhandelingen van het Bataafsch Genootschap der Proefondervindelyke Wysbegeerte*, vol. 8 (1787): 173-80, there 174 and 178 (citation). On p. lviii a correction is made, because this paper had initially been linked to the wrong author: J.C.M. Radermacher, who died in December 1783.

46 *Captain Cook's Journal during his First Voyage round the World, made in H.M. Bark Endeavour, 1768-71: a Literal Transcription of the Original Mss*, ed. W.J.L. Wharton (Cambridge: Cambridge University Press, 2014), 354. The incident occurred on 12 October 1770 and prompted Cook to comment that every ship should have a lightning conductor.

47 A preliminary survey of archival materials held at the National Archives in The Hague failed to yield any indication of either discussions or actions in 1770 or 1771.

48 W. van Barneveld, *Het onweersvuur in zyne richting en uitwerksels nagespoort* (Amsterdam: P. Hayman, 1780), 26.

49 Ibid. Van Barneveld (1747-1826), an erudite Amsterdam apothecary and a member of multiple learned societies, was a friend of Cuthbertson's.

50 Meidinger, *Geschichte des Blitzableiters*, 42.

51 Cuthbertson, *Algemeene eigenschappen* 2: 160 and 177-78.

52 (M. van Marum), "Bijvoegsel over de bliksem-afleiders, door den secretaris der maatschappij," *Natuurkundige verhandelingen van de Hollandsche Maatschappij der Wetenschappen*, vol. 9 (1819): 319-44, there 333. For the year 1817 this learned society had offered a prize for the best essay on the following issue: "To which defects of lightning conductors the sometimes imperfect protection of buildings and ships can be attributed?" ("Aan welke gebreken der afleiders is het toe te schrijven dat zij in enkele gevallen de gebouwen of schepen niet geheel beveiligd hebben?"). On this topic, see J.G. de Bruijn, *Inventaris van de prijsvragen uitgeschreven door de Hollandsche Maatschappij der Wetenschappen 1753-1917* (Haarlem-Groningen: H.D. Tjeenk Willink, 1977), no. 194.

53 On this company, see I. de Graaf, "Als de bliksem...," *Nijvertijd* 7, no. 12 (December 2003): n.p. Coppersmith Hendrik Jan Helderman (1744-1825) founded the company in Deventer in the eastern part of the Dutch Republic in 1779. His nephew P.O. Schaap subsequently took it over.

54 Gallitzin's unpublished letter, entitled "Description du conducteur électrique placé sur le château de Rosendael en Gueldre," is referred to only in *Mémoires de l'Académie impériale et royale des sciences et belles-lettres de Bruxelles*, vol. 3 (1780), xi. The memorial itself, however, has been printed in the volume just mentioned (pp. 1-12). It bears the title: "Mémoire sur la forme la plus propre et la plus efficace pour terminer les conducteurs électriques qu'on place sur les bâtimens." On page 8 it becomes clear that Gallitzin placed the conductor on Rosendael Castle around June 1778 and that he was so agreeably surprised about its costs that in his opinion these should not form an obstacle for its introduction.

55 See J.C. Bierens de Haan, *Rosendael, Groen Hemeltjen op Aerd. Kasteel, tuinen en bewoners sedert 1579* (Zutphen: Walburg Pers, 1994), 179. The installation of the conductor may have been prompted by a lightning strike, see ibid., 183. Possibly Gallitzin's conductor still had a rather experimental character.

56 L.S. Jacquet, *Proef eener elektrische natuurkunde*, na 't Fransch van den abt Jacquet. Vertaald en uitgebreid door C.R.Th. Kraijenhoff (Utrecht-Harderwijk: B. Wild and J. van Kasteel, 1783), 297-310 (appendix no. 3). Jacquet's *Précis de l'électricité* was published in Vienna in 1775. In a more general context, Kraijenhoff revisited the theme of his Doesburg installation in an article in 1785: "Onzydige beproeving en verdediging der metaalen blixem-afleideren," *Algemeen magazyn van wetenschap, konst en smaak* 1, pt. 2 (1785): 952ff. On Kraijenhoff, see also W. Uitterhoeve, *Cornelis Kraijenhoff 1758-1840. Een loopbaan onder vijf regeervormen* (Nijmegen: Vantilt, 2009), 59-61.

57 Jacquet, *Proef*, 297, 300 and C.R.Th. Kraijenhoff, *Handleiding tot het stellen van bliksem-afleiders* (Nijmegen: Vieweg, 1836), 26-27. Kraijenhoff knew Van Leenhof's sons since he had been a student at the University of Harderwijk, see Uitterhoeve, *Cornelis Kraijenhoff*, 46.

58 Kraijenhoff, *Handleiding*, 3.

59 G. Moll, "Over afleiders en het onweder van 12 februarij 1836," *Algemeene konst- en letterbode* 1836, no. 21 (6 May): 325.

60 C.O.A. Schimmelpenninck van der Oije, "Een reysje na Gelderland en de electrike vloeistof," *Gelders oudheidkundig contactbericht* 47 (1970): 12-17, there 14.

61 *Oprechte Haarlemse courant*, 4 October 1800. Several of Lugt's popularising publications are included in the bibliography at the end of this book.

62 His name was mentioned by Kraijenhoff in Jacquet, *Proef*, 269 n.

63 See below, Chapter 5. On the British socio-cultural context, cf. Golinski, *British Weather*, 41-76, esp. 67.

64 The same holds true for Germany, cf. Möhring, *Geschichte des Blitzableiters*, 123.

65 See above, n. 54.

66 Hochadel and Heering, "Introduction: Revisiting an Invisible Technology," 9 and 16. On nineteenth-century developments, see A. Mohun, "Lightning Rods and the Commodification of Risk in Nineteenth Century America," in Heering et al., *Playing with Fire*, 167-80.

67 Kraijenhoff, *Handleiding*, 41 and 52 n. (costs of tower lightning conductors).

68 C.R.Th. Kraijenhoff, *Voorschriften wegens het plaatsen van bliksemafleiders aan 's rijks gebouwen, behoorende onder de administratie van het Departement van Oorlog, en wel bijzonder de buskruid-magazijnen*, (The Hague: s.n., 1819), 3 (citation).

69 Kraijenhoff, *Handleiding*, 3.

70 On the significant role played by physicians and apothecaries in Dutch natural philosophy, cf. H.A.M. Snelders, "Romantiek en natuurwetenschappers in Nederland (1800-1850)," *De negentiende eeuw* 8 (1984): 73. Strikingly, Snelders considers eighteenth-century Dutch original scientific research to have been of a higher standard than that of the first half of the nineteenth century.

71 Scale models were used as a means of facilitating the introduction of the lightning conductor. See Hochadel and Heering, "Introduction: Revisiting an Invisible Technology," 12-13. I only mention in passing here that motley world of thunder houses, churches, and ships. See W.D. Hackmann, "The Lightning Rod. A Case Study of Eighteenth Century Model Experiments," in Heering et al., *Playing with Fire*, 209-29 and Hackmann, "Instruments and Experiments," 198. "Profitable" and "pleasant" sociability was fostered by learned societies by organising lectures and doing experiments. See e.g. A.G. Schulte, ed., *Arnhems Historisch Genootschap 'Prodesse Conamur' 1792-1992* (Zutphen: Walburg Pers, 1992), 43, 47, 49 and 200.

72 See G. Hesselink, "Bericht wegens het onweder van 24 february 1804," *Algemeene vaderlandsche letter-oefeningen* (1804), pt. 2: 168-73. Cf. also J. Buisman, *Duizend jaar weer, wind en water in de Lage Landen* (7 vols.; Franeker: Van Wijnen, 1995-2019), 7: 138.

73 Cf. F. Westra, *Martinitoren* (Groningen: Passage, 2009), 88-89. Only after a subsequent lightning strike in 1836 was a conductor installed.

74 Perhaps symptomatically, those members of *Felix Meritis* who were interested in electrical research, conducted most of their experiments outside the learned society. See Roberts, "Science Becomes Electric," 695-96.

75 Hochadel and Heering, "Introduction: Revisiting an Invisible Technology," 10. Cf. also Hochadel, *Öffentliche Wissenschaft*, 167 and Meidinger, *Geschichte des Blitzableiters*, 44.

76 For a detailed account, see Hochadel, "In Nebula Nebulorum," 45-70.

3 - Eighteenth-Century Physical Theories on Thunderstorms

1 Hochadel, "In Nebula Nebulorum," 45-70 and McCallam, *Volcanoes in Eighteenth-Century Europe*, 213-31. On the Laki volcano eruption, see K. Kleemann, *A Mist Connection. An Environmental History of the Laki Eruption of 1783 and Its Legacy* (Berlin-Boston: De Gruyter, 2023). On meteorological observations in the Austrian Netherlands, see G.R. Demarée, "'De grote droge nevel' van 1783 in de Zuidelijke Nederlanden: een historisch-klimatologische studie," *Tijdschrift voor ecologische geschiedenis* 2 (1997): 27-36. On the British situation, see J. Grattan and M. Brayshay, "An Amazing and Portentous Summer. Environmental and Social Responses in Britain to the 1783 Eruption of an Iceland Volcano," *The Geographical Journal* 161 (1995): 125-34. A general meteorological survey is given by J.A. Kington, *The Weather of the 1780s over Europe* (Cambridge: Cambridge University Press, 1988).

2 See M. Engammare, "Tonnerre de Dieu et 'courses d'exhalations encloses es nuées'. Controverses autour de la foudre et du tonnerre au soir de la Renaissance," in *Sciences et religions de Copernic à Galilée (1540-1610)*, ed. C. Brise and A. Romano (Rome: École Française de Rome, 1999), 168 and R. Vermij, "Erschütterung und Bewältigung. Erdbebenkatastrophen in der Frühen Neuzeit," in *Um Himmels Willen. Religion in Katastrophenzeiten*, ed. M. Jakubowski-Tiessen and H. Lehmann (Göttingen: Vandenhoeck & Ruprecht, 2003), 237-38, as well as R. Vermij, *Thinking on Earthquakes in Early Modern Europe* (London-New York: Routledge, 2021), 26. Cf. also K.-H. Hentschel, "Man erhitze sich daher zur Zeit der Ungewitter nicht. Dunst und Dämpfe, Kult und Kämpfe: Menschen, Blitz und Donner," *Hierzuland* 8, no. 15 (1993): 6-23, esp. 7, and H.H. Frisinger, "Early Theories on the Cause of Thunder and Lightning," *Bulletin of the American Meteorological Society* 46 (1965): 785-87.

3 See De Pater, *Petrus van Musschenbroek*, 329. Sodom's destruction is described in Genesis 19. Later on, eighteenth-century perception proved to be false, however, because a lightning strike produces a smell of ozone, not sulphur.

4 This phrase was used, for example, by the English scientist John Woodward (1665-1728) in *An Essay toward a Natural History of the Earth and Terrestrial Bodies* (London: R. Wilkin, 1695). Cited by V. Janković, *Reading the Skies. A Cultural History of English Weather, 1650-1820* (Manchester: Manchester University Press, 2000), 27. On the complex way Aristotle linked earthquakes to thunderstorms, see E. Oeser, *Historische Erdbebentheorien von der Antike bis zum Ende des 19. Jahrhunderts* (Vienna, n.p., 2003), 14-15. This book is available online at: https://docplayer.org/41470765-Abhandlungen-geologische-bundesanstalt-historische-erdbebentheorien-von-der-antike-bis-zum-ende-des-19-jahrhunderts-band-58-erhard-oeser.html (accessed 15 March 2021).

5 In this vein, Franz von Beroldingen (1740-1798), a representative of the Catholic Enlightenment, typically argued that the dry fog consisted of "inflammable air", see Hochadel, "In Nebula Nebulorum," 49.

6 Oeser, *Historische Erdbebentheorien*, 16.

7 Vermij, *Thinking on Earthquakes*, 62.

8 Cohen, *Benjamin Franklin's Science*, 146.

9 Oeser, *Historische Erdbebentheorien*, 37-38.

10 Ibid., 38. On Bertholon, see J.-P. Poirier, *L'abbé Bertholon. Un électricien des Lumières en province* (Paris: Hermann, 2008). By the same author, cf. also "Electrical Earthquakes: A Short-Lived Theory in the 18th Century," *Earth Sciences History* 35 (2016): 283-302. On Bertholon's similar ideas on volcanic eruptions, see McCallam, *Volcanoes in Eighteenth-Century Europe*, 67ff.

11 Oeser, *Historische Erdbebentheorien*, 38 and Heering, "Styles of Experimentation," 125-26. Bertholon was the first to install lightning conductors in Paris in 1782. The chemist Joseph Priestley, who also regarded earthquakes as electrical phenomena, preferred to restore the balance between atmospheric and earth electricity by means of flying "electrical kites", see Oeser, *Historische Erdbebentheorien*, 86.

12 G. de Santillana, "Volta," *Scientific American* 212 (January 1965): 85 and F.L. Holmes, "Phlogiston in the Air," in *Nuova Voltiana. Studies on Volta and his Times*, ed. F. Bevilacqua and L. Fregonese (5 vols.; Milan: Hoepli, 2000-2003), 2: 89.

13 See, e.g., Cohen, "Prejudice," 426. The same idea was found in Germany, see Hochadel, "In Nebula Nebulorum," 50-51. On the Boston earthquake, see W.B. Robles, "Atlantic Disaster: Boston Responds to the Cape Ann Earthquake of 1755," *The New England Quarterly* 90 (2017): 7-35, esp. 23-24 (on Prince) and D.G. Hart, *Benjamin Franklin. Cultural Protestant* (Oxford-New York 2021: Oxford University Press), 163-64, as well as Thuesen, *Tornado God*, 63. Thuesen only has an eye for Prince's interpretation of the earthquake as potential divine punishment for the "presumption" of using lightning conductors, while disregarding the minister's physical reflections.

14 Cited by Hochadel, *Öffentliche Wissenschaft*, 152.

15 Ibid. and Hochadel, "In Nebula Nebulorum," 50-51.

16 Cited by W. Breidert, "Das Erdbeben von Lissabon und die Erschütterung seiner Zeitgenossen," in *Lichtenberg Jahrbuch* (1994): 60. Cf. also Briese, *Macht der Metaphern*, 147-48. In the early modern period and even during the Napoleonic Era, pyramids were often depicted as more pointed structures than they really are, see, e.g., the engravings in *Description de l'Égypte, ou recueil des observations et des recherches qui ont été faites en Egypte pendant l'expédition de l'armée Française* (23 vols.; Paris: Imprimerie impériale, 1809-1828).

17 W.M. Sudduth, "Eighteenth-Century Identifications of Electricity with Phlogiston," *Ambix* 25 (1978): 131-47.

18 Cf. Hochadel, *Öffentliche Wissenschaft*, 71. Some regarded phlogiston as the substance that made bodies conductors of electricity. Therefore, Volta thought that the kindling of fires and the subsequent saturation of the air with phlogiston would carry off atmospheric electricity even more efficiently than the lightning conductor, see A. Volta, *Meteorologische Briefe nebst Beschreibung seines Eudiometers* (Leipzig: J.G. Müller, 1793), 189 and 192, cited by Briese, *Macht der Metaphern*, 24. Cf. also Holmes, "Phlogiston in the Air," 88 (Volta was building on Priestley's views).

19 Janković, *Reading the Skies*, 147-48 wrongly gives this impression.

20 On the battle over the existence of one or two electrical fluids, see Heilbron, *Electricity*, 431-48.

21 Hochadel, "In Nebula Nebulorum," 51 and Hochadel, *Öffentliche Wissenschaft*, 152.

22 On writings of this kind (for the most part published anonymously), see Hochadel, "In Nebula Nebulorum," 50.

23 See above, Chapter 1, n. 16. Cf. also E. Benz, *Theologie der Elektrizität. Zur Begegnung und Auseinandersetzung von Theologie und Naturwissenschaft im 17. und 18. Jahrhundert* (Mainz-Wiesbaden: Steiner, 1971), 36.

24 Cf. Hochadel, *Öffentliche Wissenschaft*, 152.

25 P. Bertholon, *De l'électricité des végétaux* (Lyon: Bernuset, 1783), 393-404. Cf. also Möhring, *Geschichte des Blitzableiters*, 281-82.

26 Hochadel, *Öffentliche Wissenschaft*, 171.

27 Hackmann, "The Lightning Rod," 213 and Hochadel and Heering, "Introduction: Revisiting an Invisible Technology," 9.

28 Cf. B.F.J. Schonland, *The Flight of Thunderbolts*, 2nd ed. (Oxford: Oxford University Press, 1964), 24 (orig. ed. Oxford 1950).

29 Hochadel and Heering, "Introduction: Revisiting an Invisible Technology," 9 and Hochadel, *Öffentliche Wissenschaft*, 163. Cf. also Cohen, "Prejudice," 411.

30 Hochadel, "In Nebula Nebulorum," 53 and 62. A more detailed account is given by F. Dross, "Gottes elektrischer Wille? Zum Düsseldorfer 'Blitzableiter-Aufruhr' 1782/83," in *Landes- und Reichsgeschichte. Festschrift für Hansgeorg Molitor zum 65. Geburtstag*, ed. J. Engelbrecht and S. Laux (Bielefeld: Verlag für Regionalgeschichte, 2004), 281-302. Many of those who were in favour of the lightning conductor rather thoughtlessly stigmatised these objections as mere superstitions.

31 On this dispute, see R.W. Home, "Points or Knobs. Lightning Rods and the Basis of Decision Making in Late Eighteenth Century British Science," in Heering et al., *Playing with Fire*, 97-120 and Heilbron, *Electricity*, 380-82.

32 See Cohen, *Benjamin Franklin's Science*, 9 and 30, Hochadel, *Öffentliche Wissenschaft*, 160-61 and (in more detail) A.O. Aldridge, *Franklin and his Contemporaries* (New York: New York University Press, 1957), 124-38. The French statesman and economist Anne Robert Jacques Turgot (1727-1781) first applied this epigram (freely derived from a hexameter line by the first-century Roman poet Marcus Manilius, *Astronomica*) to Franklin. It was inscribed on a marble bust of Franklin made by the sculptor Jean-Antoine Houdon (1741-1828) in 1778. The epigram was published in the *Journal de Paris* on 15 November 1778. Religious and political protest went hand in hand here, see Wild, "Prometheus-Franklin," 33-34.

33 J. Ingenhousz, "Overweegingen, betreffende het geschil, of scherpgepunte afleiders te verkiezen zyn boven zulken, die in een bol eindigen," in J. Ingenhousz, *Verzameling van verhandelingen over verschillende natuurkundige onderwerpen*. Translated by J. van Breda (2 vols.; The Hague: Isaac van Cleef, 1785), 1: 50-104.

34 Hankins, *Science and the Enlightenment*, 63.

35 Remarkably enough, the outcomes of modern research seem rather to point to a higher efficacy of blunt lightning conductors, see C.B. Moore, G.D. Aulich, and W. Rison, "A Modern Assessment of Benjamin Franklin's Lightning Rods," in Heering et al., *Playing with Fire*, 256-68, esp. 264-65 and 267.

36 See above, p. 28, 40.

37 Hackmann, "The Lightning Rod," 215. Van Marum and Cuthbertson are referred to as experts here.

38 On the Dutch situation, see J.W. Buisman, *Tussen vroomheid en Verlichting. Een cultuurhistorisch en -sociologisch onderzoek naar enkele aspecten van de Verlichting in Nederland (1755-1810)* (2 vols.; Zwolle: Waanders, 1992), 1: 198-99 and the literature listed there. See also the weather chronicle by my namesake Buisman, *Duizend jaar weer* 6: 608-17.

39 See above, p. 42.

40 See M. van Hattum, *Lezingen en verhandelingen in 'Concordia et libertate' (1769-1806) en 'Felix meritis' (Dep. letterk.) (1779-1808, 1810-1832, 1865-1873)* (Amstelveen: s.n., 1983), 14 (typescript). The lecture in question was held before the Amsterdam society *Concordia et libertate* on 12 December 1786.

41 Cf. H. Geurts, "De uitzonderlijke zomer van 1783 met de warmste juli sinds het begin der waarnemingen," *Weerspiegel* 7 (1982): 355-59. At the start of August, thunderstorms caused severe damage in Brabant, in the southern part of the Dutch Republic. Cf. also Buisman, *Duizend jaar weer* 6: 608ff.

42 The Groningen magistracy considered the installation of a lightning rod on St Martin's Tower in 1783, but only in 1836 did it actually happen, see above, Chapter 2, n. 73.

43 Kraijenhoff, "Onzydige beproeving," 953.

44 See, e.g., above, n. 33 and Chapter 2, n. 45. The medical uses of electricity continued to receive a good deal of contemporary publicity. These are not dealt with here, however.

45 Kraijenhoff, "Onzydige beproeving," 964-67 and 976. Kraijenhoff held a lecture advocating the installation of lightning rods in *Concordia et libertate* on 5 April 1785. It is briefly mentioned by Van Hattum, *Lezingen*, 13.

46 Kraijenhoff, "Onzydige beproeving," 967-68. Cf. also above, pp. 56-57.

47 Hesselink, "Bericht wegens het onweder van 24 february 1804," 172. Hesselink was already engaged with volcanism and earthquakes in his doctoral dissertation (1778), see G.J. Hesselink, *Dissertatio physica inauguralis De montium ignivomorum, terraeque motuum origine, et cognatione, ex stupendis illorum effectibus colligenda* (Groningen: Jacob Dikema, 1778). Later, in his lectures at the Mennonite Seminary at Amsterdam, Hesselink criticised the use of Franklin's terminology of positive and negative electricity. In his opinion, these terms did not have an explicative meaning, but only descriptive significance. See R.K. Koopmans, *Hulde aan Gerrit Hesselink* (Amsterdam: J.W. Yntema, 1812), 101-03.

48 Van Marum, *Verhandeling over het elektrizeeren*, iii and "Proefneemingen ter aanwyzing, hoe de aardbeeving en waterberoering zomwylen door eene electrische ontlaading veroorzaakt kunnen worden," included as Chapter 8 in *Eerste vervolg der proefneemingen met Teyler's electrizeer-machine in 't werk gesteld* (Haarlem: Joh. Enschedé and Jan van Walré, 1787), 171-79 (*Verhandelingen uitgegeeven door Teyler's tweede genootschap*, vol. 4).

49 See above, Chapter 2, n. 33. On Magerus's kite experiments in the summer of 1772, see Priestley, *Geschiedenis en tegenwoordige staat* 2: 60-61 n.

50 Le Francq van Berkhey, *Natuurlyke historie* 1: 364-65, and 356, respectively.

51 Chomel, *Algemeen huishoudelijk-, natuur-, zedekundig- en konstwoordenboek* 1: 196. A similar argument is made by "Kort begrip der natuurkunde," *Vaderlandsche letter-oefeningen* 7, pt. 2 (1767): 658. Remarkably, not a word is said about Franklin there.

52 As remarked in n. 3 above, it must have been ozone, not sulphur, that Keuchenius could smell. This pungent gas has only been discovered in 1840. Its name originates from the Greek verb ὄζειν, to stink.

53 J.F. Martinet, "Verbaazende uitwerksels van een onweder," in *Verhandelingen uitgegeeven door de Hollandsche Maatschappye der Weetenschappen*, vol. 14, section Communications (Berichten) (1773): 33-45. Remarkably enough, Martinet did not yet make a plea here for the installation of a lightning rod.

54 Martinet, *Katechismus der natuur* 1: 124. On p. 126 Martinet returned to the issue of the lightning strike at the Alem church. Subsequently, on p. 129, he deemed it advisable to install lightning conductors.

55 Martinet's friend and colleague the Reverend E.M. Engelberts (1731-1807), minister of the Reformed church at Hoorn in North Holland, rejoiced that after a thunderstorm the grass looked as if it was "filled with new saltpetre particles", see E.M. Engelberts, *Bespiegelingen over de vier getijden des jaars* (Amsterdam: Yntema and Tieboel, 1769), 144. De Perponcher was of a slightly different opinion, because he felt experiments had demonstrated that lightning electricity had "an expanding potential, which accelerated the circulation of the liquids" and stimulated plant growth accordingly. See W.E. de Perponcher Sedlnitzky, *De Zuid-Bevelandse dorps-leeraar* (Utrecht: J. van Schoonhoven, 1799), 217-18.

56 Kraijenhoff in Jacquet, *Proef*, 191 n. It should be noted that notwithstanding his advocacy of the lightning rod, Kraijenhoff still counselled the use of traditional precautionary measures inspired by the sulphur discourse.

57 Pauline B., *De vrouw om en in de pastorie* (Amsterdam: Johannes van der Hey, 1853), 173 (cf. also 181).

58 Chomel, *Algemeen huishoudelijk-, natuur-, zedekundig- en konstwoordenboek* 1: 522.

59 Martinet, *Katechismus der natuur* 3: 227 and, by the same author, *Kleine katechismus der natuur voor kinderen* (Amsterdam: J. Allart, 1779), 24. Apparently some folk locked themselves in their cellars during thunderstorms.

60 J.F. Martinet, *Huisboek voor vaderlandsche huisgezinnen* (Amsterdam: J. Allart, 1793), 124. In Martinet's opinion, thunderstorms could be used for religious education.

61 This warning was given as late as the beginning of the nineteenth century by the German instrument maker J.C. Gütle, *Allgemeine Sicherheitsregeln bei Gewittern* (Nuremberg: Schneider and Weigel, 1805). Cf. also F.L. Supprian, *Vernünftige Gedanken von den Ursachen des Blitzes und dessen wunderbaren Wirkungen* (Potsdam: C.F. Voss, 1746). Both publications are cited by Hentschel, "Man erhitze sich daher zur Zeit der Ungewitter nicht," 7-9 and 20-21.

62 J.A.E. Goeze, "Beredeneerde, stichtlijke en proefondervindlijke bedenkingen over de vrees voor het onweder van donder en bliksem," *Nieuwe vaderlandsche bibliotheek van wetenschap, kunst en smaak* 5, pt. 2 (1801): 197. J.A.E. Goeze was the younger brother of the famous Hamburg Lutheran minister Johann Melchior Goeze (1717-1786), the orthodox opponent of the German writer Gotthold Ephraim Lessing (1729-1781).

63 O. Briese, "Der abgeleitete Blitz. Metapherngeschichte als Mentalitätengeschichte," *Euphorion. Zeitschrift für Literaturgeschichte* 92 (1998): 415 and the literature listed there.

64 J. Kraus, *Gewitterkatechismus, oder Unterricht über Blitz und Donner* (Augsburg: N. Doll, 1797), 25-26 (orig. ed. 1794). On the importance and meaning of the fireplace, cf. below, p. 146.

65 Hentschel, "Man erhitze sich daher zur Zeit der Ungewitter nicht," 11.

66 Advice given by the Greifswald professor Peter Ahlwardt, *Bronto-theologie, oder vernünftige und theologische Betrachtungen über den Blitz und Donner* (Greifswald-Leipzig: J.J. Weitbrecht, 1745), 315-16 and 335. Cited by H.D. Kittsteiner, *Die Entstehung des modernen Gewissens* (Frankfurt a.M.: Insel Verlag, 1991), 77.

67 Ibid. On the role played by human and animal faeces, cf. also below, p. 155.

68 See above, n. 61.

69 Heilbron, *Electricity*, 341 and Cohen, "Prejudice," 396 n. 9. In 1753 a serious discussion of the effectiveness of bell-ringing during thunderstorms was also held at the philosophical faculty of Prague university.

4 - Official Religion

1 Small parts of this chapter were published in an earlier article entitled: "Op zoek naar het juiste midden: C.R.Th. Kraijenhoff en S.J. van de Wijnpersse over het onweer," in *Verlichting in Nederland 1650-1850. Vrede tussen rede en religie?*, ed. J.W. Buisman (Nijmegen: Vantilt, 2013), 191-206.

2 Cf. K.H. Miskotte, *Antwoord uit het onweer. Een verhandeling over het boek Job*, 2nd ed. (Amsterdam: Uitgeversmaatschappij Holland, 1936) (orig. ed. Amsterdam 1929). On weather conceptions of the Andaman Islanders, see L. Boia, *The Weather in the Imagination* (London: Reaktion Books, 2005), 120.

3 Begemann, *Furcht und Angst*, 73.

4 Sometimes only the expression "the dear thunderstorm" was deemed acceptable, see A. de Cock, "Eerbied voor de natuur en haar verschijnsels," *Volkskunde* 17 (1905): 48.

5 As a child, the Dutch theologian Hendrikus Berkhof (1914-1995) was urged to silence by his grandfather during a thunderstorm, by saying: "shut up, the voice of God!" See *Waar is God in deze tijd? De betekenis van de geschiedenis in de theologie van dr. H. Berkhof*, ed. A. van de Beek et al. (Nijkerk: Callenbach, 1994), 202.

6 K. Heckscher, *Die Volkskunde des germanischen Kulturkreises* (Hamburg: M. Riegel, 1925), 124.

7 E. Kück, *Wetterglaube in der Lüneburger Heide* (Hamburg: R. Hermes, 1915), 145.

8 F. Hempler, *Das Gewitter im Volksglauben unserer Heimat* (Danzig: A.W. Kafemann, 1928), 7, 22 (*Heimatblätter des deutschen Heimatbundes Danzig* 5 [1928], Issue 5/6).

9 P. Sébillot, *Le folk-lore de France* (4 vols.; Paris: Librairie Orientale et Americaine, 1904-1907), 1: 93. On the broader German context of these popular beliefs, see H. Freudenthal, *Das Feuer im deutschen Glauben und Brauch* (Berlin: De Gruyter, 1931), 38, 41.

10 Cf. A. Rieth, *Der Blitz in der bildenden Kunst* (Munich: Heimeran, 1953), 40 and Golinski, *British Weather*, 81-82. The interpretation of thunderstorms as announcements of the Day of Judgement was rather self-evident in seventeenth-century England, see A. Walsham, "Deciphering Divine Wrath and Displaying Godly Sorrow: Providentialism and Emotion in Early Modern England," in *Disaster, Death and the Emotions in the Shadow of the Apocalypse, 1400-1700*, ed. J. Spinks and C. Zika (London: Palgrave Macmillan, 2016), 26. But even as late as 1783, that year's memorable summer occasioned comparable eschatological speculations (letter of William Cowper [1731-1800] to John Newton [1725-1807], dated 29 June 1783), see Grattan and Brayshay, "Amazing and Portentous Summer," 130.

11 A. Winter-Fritzsche, *'Naturgetreue' Blitzdarstellungen in der venezianischen Malerei des 16. Jahrhunderts mit einer Einführung in Deutung und Darstellung des Blitzes* (Frankfurt a.M.: Peter Lang, 1990), 32.

12 Kittsteiner, *Die Entstehung des modernen Gewissens*, 47.

13 F.M. Phayer, *Religion und das gewöhnliche Volk in Bayern in der Zeit von 1750 bis 1850* (Munich: Stadtarchiv, 1970), 46.

14 Jacques-Bénigne Bossuet, *Œuvres complètes*, ed. F. Lachat (31 vols.; Paris: Vivès, 1862-1879), 10: 120, cited by S. Hache, "Thunder or Celestial Harmony: French Theological Debates on the Sonorous Sublime," in *Music and the Sonorous Sublime in European Culture, 1680-1880*, ed. S. Hibberd and M. Stanyon (Cambridge-New York: Cambridge University Press, 2020), 34. Hache points out, though, that Bossuet also had an eye for what he considered to be Christ's silent voice in the heart, which strikes the faithful sometimes like a flash of lightning, see ibid., 31.

15 On *Wetterpredigt* and *Wettergebet*, see S. Wodianka, *Betrachtungen des Todes. Formen und Funktionen der 'meditatio mortis' in der europäischen Literatur des 17. Jahrhunderts* (Tübingen: Niemeyer, 2004), 133-37 and H.D. Kittsteiner's monograph mentioned in n. 12 above. *Gewitterbänke* are shortly mentioned by R. Alewyn, *Probleme und Gestalten* (Frankfurt a.M.: Insel, 1974), 310. In Catholic regions home prayers were said in the so-called *Herrgottswinkel*, a little corner with, e.g., a crucifix; consecrated candles (*Wetterkerze*) were lit there as well, see F. Fabich, *Bauernmedizin* (Rosenheim: Rosenheimer, 1991), 96.

16 On these booklets, see Kittsteiner, *Die Entstehung des modernen Gewissens*, 31-100.

17 Begemann, *Furcht und Angst*, 73 and, more elaborately, A. Bernd-Brinkmann, "Wetterlieder im 17. und 18. Jahrhundert," *Lied und populäre Kultur* 45 (2000), 89-108, esp. 95ff. and 104.

18 Thuesen, *Tornado God*, 56. The hymn in question is "God the Thunderer, or the Last Judgement and Hell," see I. Watts, *Hymns and Spiritual Songs. In Three Books* (London: printed by J. Humfreys, for John Lawrence, 1707), 2: hymn LXII.

19 On these special or immediate providences, cf. K. Thomas, *Religion and the Decline of Magic. Studies in Popular Beliefs in Sixteenth- and Seventeenth-Century England,* 2nd ed. (Harmondsworth: Penguin, 1973), 97-98 (orig. ed. London: Weidenfeld & Nicolson, 1971).
20 Freudenthal, *Feuer,* 18.
21 Cf. H.D. Kittsteiner, "Das Gewissen im Gewitter," in *Jahrbuch für Volkskunde der Görres-Gesellschaft,* n.s. 10 (1987): 17.
22 Wodianka, *Betrachtungen des Todes,* 135.
23 Mitchell, "The Politics of Experiment," 312. On the dual aspect of the Protestant thunder prayer (doing penance after experiencing the signs of God's punishing wrath), see also Kittsteiner, *Die Entstehung des modernen Gewissens,* 38.
24 This was the interpretation of the Saxon Lutheran minister M. Bapst von Rochlitz, *Wetterspiegel, oder nützliche unnd eigentliche Erzelung* (Leipzig: Berwaldt, 1589), cv. Cited by Kittsteiner, *Die Entstehung des modernen Gewissens,* 47.
25 Kittsteiner, "Das Gewissen im Gewitter," 14.
26 Kittsteiner, *Die Entstehung des modernen Gewissens,* 49.
27 Thuesen, *Tornado God,* 36-38.
28 Ibid., 61-62.
29 W.A.T. Kruythoff, *Geschiedverhaal van de vuur en watersnood, de stad Edam meermaals, dog byzonder op den 24 february 1602 overgekomen. Nevens eene korte beschryving van dezelve stad. Met aantekeningen en byvoegsels vermeerderd en verbeterd door Mr. Jan Leoninus, oud schepen der gemelde stad* (Edam: H. Rolff, 1781), Voorreeden (preface), 5.
30 G. Westmijse, "Protestantsche bede-, dank- en vastendagen in de Republiek der Zeven Vereenigde Nederlanden," in *Kerkhistorische opstellen van het gezelschap S.S.S. Nieuwe bundel,* ed. R. Riphaagen Jr. et al. (The Hague: Nijhoff, 1914), 114. Cf. also E. Geudeke, *De classis Edam, 1572-1650. Opbouw van een nieuwe kerk in een verdeelde samenleving* (Gouda: Vereniging voor Nederlandse Kerkgeschiedenis, 2010), 368-69 n. 55 and the literature listed there.
31 On Tuinman's and Janssen's reactions, see J.W. Buisman, "Opvattingen over het onweer in de overgang van Verlichting naar Romantiek," *Zeeland* 3 (1994): 70.
32 W. Mengels, "Chronijk van het dorp Opcanne bij Maastricht, 1740-1778," ed. V. de Stuers, in *Publications de la Société historique et archéologique dans le Duché de Limbourg,* vol. 24, n.s. vol. 4 (Roermond: Romen, 1887): 297.
33 J. van Eijk, *Het dorp Eerenhoef, uit het diepst verval tot voorige welvaard hersteld* (Rotterdam: Cornel, 1816), 53.
34 "Hoe vreeslyk loeit en kraakt en dreunt de donderslag | (...) 't Is of 't een voorspel was van 's waerelds jongsten dag. || (...) Bedenk eens, zondaar, kan dit hier uw hert doen beven | Daar zyn [God's] langkmoedigheid u duld in het leven | Wat zal het zyn, als u zyn eeuwge wraak eens vind? || Val dan in tyds hem door bekering in de roede; | Of hy noch mogelyk aan u dacht ten goede. | Draal, draal niet, opdat u zyn bliksem niet verslind'!" "Ter gelegenheid van een vreeslyk onweder," in P. Boddaert Czn., *Stichtelyke gedichten,* 5th ed. (2 vols.; Leiden: Luchtmans, 1765), 1: 515 (orig. ed. 1726).
35 De Perponcher, *Zuid-Bevelandse dorps-leeraar,* 210-11.
36 Ibid., 212-14, 220-21.
37 See, e.g., Scharfe, *Menschenwerk,* 73-74.
38 J.-P. Schneider, "De l'orage châtiment au chaos maîtrisé," in Le Roy Ladurie et al., *L'Événement climatique,* 127, 136.
39 Engammare, "Tonnerre de Dieu," 173-75 (cf. also 177). In my opinion, however, Engammare has a tendency to make the Renaissance and Baroque figures he cites look a bit more modern than they actually were.
40 More elaborately on this: Thuesen, *Tornado God,* 44ff.
41 Ibid., 60. The article in question is entitled "Of Superstitious Fears, and Their Causes Natural and Accidental," *American Magazine and Historical Chronicle* (May 1744), 372-75. Probably it has been written by Jeremiah Gridley (1702-1767), who was Grand Master of the Masons in North America as of 1754.
42 (J.-A. Courtonne?), *L'Observateur hollandois* (Leeuwarden: A. Ferwerda, 1750-1751): 273-80 (Issue no. 35). With thanks to Dr. J. Spaans for this reference. On the probable participation of Jacques-André Courtonne in this journal, see W.J. Kooiman, "Philippus Ludovicus Statius Muller," in *Earebondel ta de tachtichste jierdei fan dr. G.A. Wumkes op 4 septimber 1949. Oanbean troch de Fryske Akademy* (Bolsward: Osinga, 1950), 80.

43 "würde ich (...) [die Menschen] etwa durch Blitze und Donnerkeile vernünftiger machen?", C.M. Wieland, *Sämtliche Werke* (39 + 6 vols.; Leipzig: Göschen, 1794-1811), 25: 208, cited by Weigl, "Entzauberung der Natur," 29.
44 Wieland, *Sämtliche Werke* 25: 148ff, cited by Weigl, "Entzauberung," 29.
45 Thuesen, *Tornado God*, 50.
46 On the religious trends depicted here, see Kittsteiner, "Das Gewissen im Gewitter," 7-26 and his book *Die Entstehung des modernen Gewissens*, 31-100.
47 James Delbourgo, *A Most Amazing Scene of Wonders. Electricity and Enlightenment in Early America* (Cambridge, MA-London: Harvard University Press, 2006), 51-52. Cf. also Begemann, *Furcht und Angst*, 68 and above, p. 14.
48 Thuesen, *Tornado God*, 63. Cf. also Hart, *Benjamin Franklin*, 105.
49 Briese, *Macht der Metaphern*, 25. Briese does not seem to realise that declining the divine punishment interpretation does not imply supporting the idea of blind chance. Rather, more stress was put on the inscrutability of God's will.
50 See above, Chapter 1, n. 13.
51 Cohen, "Prejudice," 420.
52 Ibid., 418.
53 Cohen, *Benjamin Franklin's Science*, 139-40 and Mitchell, "The politics of experiment," 312.
54 Goeze, "Beredeneerde, stichtlijke en proefondervindlijke bedenkingen," 198. However, his elder brother Johann Melchior Goeze, a Lutheran minister in Hamburg and a well-known antagonist of Lessing, was an opponent of the lightning conductor, see Möhring, *Geschichte des Blitzableiters*, 144.
55 Kittsteiner, *Die Entstehung des modernen Gewissens*, 93.
56 Mitchell, "The Politics of Experiment," 313. Franklin pointed out, however, that lightning flashes actually moved from the earth to the sky, see Cohen, "Prejudice," 421-22.
57 It was with this term that German agricultural communities rejected the lightning rod in the late eighteenth century. See Chr. Fuhrmeister, "'Eripuit Caelo Fulmen Sceptrumque Tyrannis.' The Political Iconography of Lightning in Europe and North America, 1750-1800," in Heering et al., *Playing with Fire*, 148.
58 W. Ruppert, "Volksaufklärung im späten 18. Jahrhundert," in *Deutsche Aufklärung bis zur Französischen Revolution 1680-1789*, ed. R. Grimminger (Munich: Hanser, 1980), 349.
59 Cohen, *Benjamin Franklin's Science*, 145-53 and Cohen, "Prejudice," 425-40. Möhring, *Geschichte des Blitzableiters*, 127 erroneously supposes that Calvinist dogma cannot object to the introduction of the lightning conductor. Partly for this reason, she underestimates the extent of the opposition in Protestant British America. On Prince's opinions, see above, Chapter 3, p. 57, and on his controversy with Winthrop: Thuesen, *Tornado God*, 63-64.
60 Cohen, *Benjamin Franklin's Science*, 158.
61 On this affair, see Cohen, "Prejudice," 436-37 and, in more detail, J. Riskin, "The Lawyer and the Lightning Rod," *Science in Context* 12 (1999): 61-99.
62 Rieth, *Blitz*, 43 and Meidinger, *Geschichte des Blitzableiters*, 28.
63 "die gottesleugnerischen Franzosen"; "um unserm Herrgott die Augen auszustechen", E.L. Rocholz, *Deutscher Glaube und Brauch im Spiegel der heidnischen Vorzeit* (2 vols.; Berlin: Dümmler, 1867), 2: 113.
64 L. Trénard, *Lyon. De l'Encyclopédie au Préromantisme* (2 vols.; Paris: P.U.F., 1958), 202-03. Bertholon successfully installed the first conductors in Lyon in the late 1770s.
65 See above, Introduction, p. 13.
66 Hochadel, *Öffentliche Wissenschaft*, 152.
67 Wild, "Prometheus-Franklin," 36. On the importance of the self-made man or autodidact as a distinct type in the Western Enlightenment, see Buisman, *Tussen vroomheid en Verlichting* 1: 258-62 and 2: 383-420, esp. 403.
68 Wild, "Prometheus-Franklin," 38. The most comprehensive study on the figure of Prometheus is R. Trousson, *Le thème de Prométhée dans la littérature européenne* (2 vols.; Geneva: Droz, 1964). Even in its third, revised edition (Geneva: Droz, 2001) the European reception of Franklin as the modern Prometheus is ignored, however.
69 Cited by Hochadel, "In Nebula Nebulorum," 59.

70 G. Goubier-Robert, "De la fulgurisation sadienne aux foudres républicaines," in Le Roy Ladurie et al., *L'Événement climatique*, 418-20, 423.

71 Begemann, *Furcht und Angst*, 91.

72 On this topic, cf. esp. Trousson, *Le thème de Prométhée* 1: 260ff. As early as 1771 the famous German poet Johann Wolfgang von Goethe (1749-1832) considered that his poetical genius put him on an equal footing with the gods, see C. Thacker, *The Wildness Pleases. The Origins of Romanticism* (London: Croom Helm, 1983), 218.

73 Cf. Hackmann, "The Lightning Rod," 218.

74 Hochadel, "In Nebula Nebulorum," 59.

75 See above, Introduction, p. 13.

76 Mary Shelley, *Frankenstein*, Chapter 3.

77 See below, Chapter 7.

78 Wild, "Prometheus-Franklin," 34.

79 See above, Chapter 3, p. 62. On lightning as a political metaphor, cf. below, Chapter 7, pp. 186-91.

80 Cf. Delbourgo, *Amazing Scene*, 132 (on electricity) and passim.

81 See above, p. 84 and Cohen, *Benjamin Franklin's Science*, 157. According to some, Richmann's death in 1753 illustrated the actuality of Nollet's fear, see above, p. 27.

82 "Befreiung vom Aberglauben heißt Aufklärung", I. Kant, *Kritik der Urteilskraft* (1790). In: *Werke*, ed. W. Weischedel (12 vols.; Frankfurt a.M.: Suhrkamp, 1974-1977), 10: 226. Cited by M. Pott, *Aufklärung und Aberglaube. Die deutsche Frühaufklärung im Spiegel ihrer Aberglaubenskritik* (Tübingen: Niemeyer, 1992), 3.

83 Kant, *Kritik der Urteilskraft* 10: 164-207 (§§ 23-29). Cf. H. Böhme, *Natur und Subjekt* (Frankfurt a.M.: Suhrkamp, 1988), 393.

84 Schneider, "De l'orage châtiment," 136.

85 H.-B. de Saussure, *Exposition abrégée de l'utilité des conducteurs électriques* (Geneva: n.p., 1771). On De Saussure's actions, see also Möhring, *Geschichte des Blitzableiters*, 143. Genevans tried to pull down the device, see Anderson, *Lightning Conductors*, 43.

86 Cf. also Cohen, "Prejudice," 425.

87 In turn, however, Franklin's opponents regarded his invention as a presumptuous human effort to "improve" nature.

88 "Subjektzentrismus in der Erfahrung der Natur", Begemann, *Furcht und Angst*, 76.

89 Though he pays ample attention to Pietism, Kittsteiner hardly detects the subtle influence of the Enlightenment here.

90 "Hij, die met God en godsdienst spot | De booze beef voor zulk een' God! | Maar hij, die voor den Heere leeft | En Jesus tot zijn' Voorspraak heeft | Mag altoos weezen welgemoed | Dewijl Gods almagt hem behoedt." J. van Eijk, *Liederen voor den landman* (3 pieces; Amsterdam: J. Allart, 1794-1796), 3: 33. See on Van Eijk D. Nauta et al., *Biografisch lexicon voor de geschiedenis van het Nederlandse protestantisme* (6 vols.; Kampen: Kok, 1978-2006), 5: 178-79.

91 S.J. van de Wijnpersse, *Het loeijend onweder godvruchtig beschouwd, of opwekking tot een godvereerend gemoedsbestaan, onder het rollen der donderslagen en het klateren der bliksemen, in eene leerrede over Psalm XXIX* (Amsterdam: H. Diederiks and H. Larde, 1799), 193 (citation).

92 J. van Loo, "Nuttige en godvrugtige beschouwing van het onweder," in J. van Loo, *Leerredenen* (4 vols.; Utrecht: A. van Paddenburg, 1784-1789), 3: 47-84. Translations from the German: J.P. Gerricken (= J.P. Gericke), *De Heere in 't vuur of een leerreden gedaan op palm sondag 1750. Na het afbranden van (...) de kerke St. Michaels tot Hamburg (...).* Uit het Hoogduits vertaald (Amsterdam: D. Weege, 1750) and on the same accident H. Hoek, *Het heilzaam oogmerk van Godt in zyne oordeel-gerichten door vuur (...). Uitgesproken op den plegtigen boet- vast- en beededag den 19 Maart 1750.* Uit het Hoogduits vertaald (Haarlem: J. Nieuwenhuyzen and J. Huytink, 1750), as well as J.F. Stapfer, *Een dertig-tal van predikatien.* Uit het Hoogduits vertaalt en met eenige aantekeningen uitgegeven door Antonius de Stoppelaar (2 vols.; Utrecht: G.T. and A. van Paddenburg, 1768-1771), 1: 263-81. Thunderstorms were not only a foreshadowing of the Day of Judgement for Van de Wijnpersse and Van Loo but for Stapfer as well (1: 279-81). For a more detailed account of the German sermons on the occasion of the lightning strike at the Hamburg St. Michael's Church in 1750, see Zwierlein, *Prometheus Tamed*, 114-27.

93 "De christen bij het onweder," *Nieuwe vaderlandsche bibliotheek van wetenschap, kunst en smaak* 2, pt. 2 (1798): 432.

94 De Perponcher, *Zuid-Bevelandse dorps-leeraar*, 213.

95 Cf. m.m. A. Clerc, "Tempêtes libertins et naufrages spirituels au XVIIe siècle," in Le Roy Ladurie et al., *L'Événement climatique*, 348 and Briese, *Macht der Metaphern*, 32 (lightning strikes regarded as punishments for sexual sin).

96 "Waar is uw moed nu, Ongodist! | Zie, zie hem zich verschuilen! | Die 's Hoogsten aanzijn stout betwist | Kruipt bloode in diepe kuilen. | Koom, sterke geest! bepleit uw zaak | Bij 't klaatren van den donder! | Ontken hier 's Hemels gunst en wraak! (...) | Maar hoe! gij zwijgt (...) geen wonder!" G.P. (Gerrit Paape), "Het onweder," in *Proeven van poëtische mengelstoffen, door het dichtlievend kunstgenootschap onder de spreuk: Kunstliefde spaart geen vlijt* (13 vols.; Leiden: C. van Hoogeveen Jr., 1774-1789), 5: 165-66. On this poem, see also P. Altena, *Gerrit Paape (1752-1803). Levens en werken* (Nijmegen: Vantilt, 2012), 86-89. In a similar traditional manner Paape uses thunderstorms as a "penitential trumpet" in some of his other early poems, see Altena, *Gerrit Paape*, 106-07.

97 Van de Wijnpersse, *Het loeijend onweder*, 204, 205 and 208, respectively.

98 Ibid., 201, 204-05.

99 *Almanach der natuur, voor iedere maand van het jaar* (Amsterdam: J. de Jongh, 1789), June, 36.

100 Cf. m.m. Thomas, *Religion*, 128.

101 C.S. Ulber, *Der christliche Creutzträger* (Liegnitz-Hamburg: R. Beneke, 1760), 11. Cited by Kittsteiner, "Das Gewissen im Gewitter," 20.

102 C.S. Ulber, *De christelyke kruis-drager*. Uit het Hoogduitsch vertaald door Antoni Fredrik Klenke (2 vols., Amsterdam: A. Eichhorn, 1772-1780).

103 "De rampen, die Hij zendt, zijn liefdebanden; | De straf is zegen in zijn handen: | Hemelblijdschap is haar vrucht." R. Feith, "Het onweder. Eene cantate," in *Oden en gedichten* (3 vols.; Zwolle: H. Doyer, 1824), 2: 48 and 51 (originally published in 1787 in *Tael- en dichtlievende oefeningen van 'Kunst wordt door arbeid verkreegen'* [7 vols.; Leiden: C. van Hoogeveen Jr., 1775-1790], 5: 167-68). On the reinterpretation of divine punishments as fatherly chastisements or trials, cf. also above, p. 80.

104 "Gedagten by het eindigen eens zomer-onweders," *Hedendaagsche vaderlandsche letter-oefeningen* 7, pt. 2 (1778): 317 (by J.V.P.).

105 "Gods grootheid bij het onweder," in *Almanach der musen voor het jaar 1795* (Amsterdam: W. Brave, n.d.): 12-14 (by Hn.).

106 At the end of the eighteenth century interpreting disasters as divine punishments had lost a considerable amount of its self-evident character in most of the Western world. On the Dutch context, see Buisman, *Tussen vroomheid en Verlichting* vol. 1, and on the German situation Zwierlein, *Prometheus Tamed*, 109-32, as well as Chr. Weber, *Vom Gottesgericht zur verhängnisvollen Natur. Darstellung und Bewältigung von Naturkatastrophen im 18. Jahrhundert* (Hamburg: F. Meiner, 2015). Similar French and other European developments are described in *L'invention de la catastrophe au XVIIIe siècle. Du châtiment divin au désastre naturel*, ed. A.-M. Mercier-Faivre and C. Thomas (Geneva: Droz, 2008).

107 See above, n. 42.

108 See "Beschouwing der natuur na eene zwaare dondervlaag," *De menschenvriend* 9 (1796): 253-54 and *Zamenspraak tusschen een landheer en zijn' tuinman over het onweder* (Amsterdam: De Vries, Van Munster, and Van der Hey, 1805), 20-21, as well as *Het onweêr. Eene gemeenzame zamenspraak* (Amsterdam: n.p., 1805), esp. 12-16. Both booklets were published with a common title: *Zamenspraken over de oorzaken van het onweder en het nut der afleiders* (Amsterdam: De Vries, Van Munster, and Van der Hey, 1805).

109 P. Altena, "Die holländische Himmelsstürmer. Sturm und Drang in der niederländischen Literatur 1770-1800," in *Sturm und Drang. Geistiger Aufbruch 1770-1790 im Spiegel der Literatur*, ed. B. Plachta and W. Woesler (Tübingen: Niemeyer, 1997), 191.

110 Rossyn, *De tonitru*, 33. Moreover, he deems the divine punishment interpretation stupid and impious there.

111 P.W. Feenstra, *Leerrede en gebed, uitgesproken op den dank- en bededag 22 februari 1809, in de kerk der doopsgezinden op het Cingel te Sneek* (Sneek: C. van Gorcum, 1809). Cf. Streng, *'Het is thans zeer briljant'*, 160-61.

112 Van de Wijnpersse, *Het loeijend onweder*, 198.

113 "De onmaatige vrees voor 't onweder tegengegaan," *De vaderlander* 1 (1776): 235.

114 Ibid., 248. Remarkably, even here thunderstorms are still interpreted as the expression of God's voice. On the other hand the use of the lightning rod is recommended (ibid., 247).

115 Cf. M. Nieuwenhuyzen, *Lettergeschenk voor de Nederlandsche jeugd*, 3rd ed. (Haarlem: A. Loosjes, 1814), 103-06 (orig. ed. 1790).

116 "Over de vrees by een onweder," *De menschenvriend* 4 (1791): 264.

117 Ibid., 263.

118 Ibid., 260. In enlightened circles the ancient focus on sin and penitence was also considered an impediment to resolute action, because it would result in fatalism and passivity. Cf. Kittsteiner, "Das Gewissen im Gewitter," 17 and by the same author *Die Entstehung des modernen Gewissens*, 78.

119 "Over de vrees by een onweder," 261 (cf. also 258 and 263).

120 "De vyanden des lichts. Een vertelzel," *Algemeene vaderlandsche letter-oefeningen* (1791), pt. 2: 536-37. Obviously Franklin was not a Quaker. Having received a Puritan education he developed deistic leanings. See Hart, *Benjamin Franklin*, Chapters 1 and 2.

121 (H. van Alphen), *De christelyke spectator. In twaalf vertoogen* (The Hague: J. Thierry and C. Mensing, 1799): 135.

122 Ibid. In Latin: *Eripuit morti miseros, sceptrumque tyranno.*

123 *Rapport over den staat en de middelen ter verbetering der laagere schoolen; door eene daar toe benoemde commissie, bij den raad der gemeente van Leyden ingeleverd 28 Junij 1796* (Leiden: Herdingh and Du Mortier, 1796), 107 note c.

124 "Het dondert! – Vrienden! leêgt uw glazen | Drinkt lustig! laat de orkanen razen! | We ontschuilen 't Veiligst bij den wijn | De ontzagbre wraak van god Jupijn | Wanneer zijn schorre donders klaatren: | Drinkt leêg – Wat siddert ge als een riet! | Drinkt leêg – hij bliksemt langs de waatren | Hij bliksemt in zijn' nectar niet." J. Immerzeel jr., "Bij een onweder," in *Nederlandsche muzen-almanak 1819* (Rotterdam: J. Immerzeel jr., n.d.): 46. Cf. also Streng, *'Het is thans zeer briljant'*, 156. See on Immerzeel: B.P.M. Dongelmans, *Johannes Immerzeel junior (1776-1841). Het bedrijf van een uitgever-boekhandelaar in de eerste helft van de negentiende eeuw* (Amstelveen: Ernst, 1992). Young Lessing's drinking song was entitled *Der Donner* (The Thunder) and dates from 1751. It is cited in full by Beller, *Jupiter tonans*, 160.

125 Letter written by Abraham Capadose to Dirk van Hogendorp (1797-1845), dating from 29 May 1824, cited by D. Kalmijn, *Abraham Capadose* (The Hague: Boekencentrum, 1955), 341-42.

126 See Num. 21:4-9. I did not come across this comparison in any eighteenth-century Dutch source.

127 C. Geertz, *Local Knowledge. Further Essays in Interpretative Anthropology* (New York: Basic Books, 1983). To a certain extent, Ten Kate's line of argument is an exception to Zwierlein's surmise that enlightened advocacy of the introduction of the lightning conductor followed a straightforward top-down path, see Zwierlein, *Prometheus Tamed*, 240.

128 "Toch was dat waagstuk geen schennige waan | Koortsig in 't brein van een Titan ontstaan. | Denken en trachten | Studie en strijd | Slaaplooze nachten | Hadden 't doen rijpen, gebeden 't gewijd. | Ziet, daar verschijnt hij, de Ontdekker! Zijn oog | Blikt naar omhoog: | 'God der Natuur! O bekroon mijn verlangen! | Gij die het weet, geen zelfzuchtige wensch | 't Heil van den Mensch | Prikkelt mijn streven.' – En 't doodsbleek der wangen | Wordt van den vuurgloed der geestdrift vervangen. | 'Herwaards de stang!' | En daar verschijnt ze – als de Koperen Slang | Voor wie gelooven | Teeken des heils – op den Tempel, daarboven! (...) | Daar leeft een God, die den heldenmoed loont: | F r a n k l i n ! uw naam is voor eeuwig gekroond." Ten Kate, *De planeeten*, 69-70, 72. In other cases the lightning rod did not replace the religious symbol but rather seems to have detracted from the latter's supposed efficacy. When, e.g., one of the first summit crosses was erected on the Erzberg in Styria (Austria) in 1823, the ancient protective power of the crucifix was not though great enough to render the installation of a lightning rod superfluous, see M. Scharfe, "Kruzifix mit Blitzableiter," *Österreichische Zeitschrift für Volkskunde* 102 (1999): 289-336.

129 Cf. "De onmaatige vrees," 237. On Dutch physico-theology, see J. Bots, *Tussen Descartes en Darwin. Geloof en natuurwetenschap in de achttiende eeuw in Nederland* (Assen: Van Gorcum, 1972).

130 Cf. "De onmaatige vrees," 240. Physico-theology sometimes came close to the divine punishment interpretation, also in Germany. Cf. Zwierlein, *Prometheus Tamed*, 126.

131 H.K. Poot, "Zomeronweer," in idem, *Gedichten*, 2nd ed. (3 vols.; Delft: R. Boitet, 1726-1735), 3: 99-105, there 105 (orig. ed. Rotterdam: A. Willis, 1716-1722).

132 "de Albestierder [weet] uit de duisternisse het licht te doen voortkomen, en de zwaarste gevallen ons tot zegeningen te doen gedyen. Zoo wil hy ontzag en eere voor zyne gedugte Majesteit in het hert van reukelooze stervelingen inboezemen." Engelberts, *Bespiegelingen over de vier getijden des jaars*, 143-44.

133 E.M. Post, *Het land, in brieven* (Amsterdam: J. Allart, 1788), 263-64.

134 "Het onweder," *Nieuwe vaderlandsche bibliotheek van wetenschap, kunst en smaak* 4, pt. 2 (1800): 318 (probably a translation from the German). A patrician poet also humbly acknowledged God's majesty and omnipotence in a thunderstorm in 1783, see C. Onderdewyngaart, "Het onweder," in *Proeven van poëtische mengelstoffen* 9: 133-34.

135 I. van Haastert, "Ode aan de wolken," in *Proeven van poëtische mengelstoffen* 11: 40-41.

136 See above, p. 91. Cf. also a poem written in 1755 by Bernardus Elikink (1701-1767), a Reformed minister from Papendrecht near Rotterdam: B. Elikink, "Bij gelegenheid van een onweer," *Stichtelijke gezangen* (Amsterdam: Johannes Covens Jr., 1769), 171-74. In this poem the Lord's "terrifying thunder word" and the eternal punishments were impressed upon the godless.

137 "Wat is (...) wonderlyker dan honderd Blixempylen uit de donkere wolk, nu naar boven, dan naar beneden of naar de zyden uitgeschoten, in de hemelsblaauwe donker geworden lucht te zien schitteren, en oogenblikkelyk daarin verdwynen? (...) Heft uwe oogen op om hooge en ziet, wie deeze dingen geschapen heeft (...) wat kan aangenaamer zyn dan Hem, wiens Wysheid en Goedheid zo onbeperkt voor my is als zyne Almagt, in myne tegenwoordigheid te zien werken?" "Godvrugtige bespiegelingen over de wolken," *De denker* 10 (1773): 247-48 (Issue no. 501, 3 August 1772).

138 Martinet, *Katechismus der natuur* 3: 228.

139 Goeze, "Beredeneerde, stichtlijke en proefondervindlijke bedenkingen," 198.

140 "Over de vrees by een onweder," 257, 259-60. The Dutch poet J.F. Helmers (1767-1813) gives another instance of stressing the Lord's omnipotence on one hand while retaining a mild view of him on the other. See "De nacht. Ode," in *Kleine dichterlijke handschriften* (20 vols.; Amsterdam: Uylenbroek, 1788-1809), 1: 23-32. See also Streng, *'Het is thans zeer briljant'*, 158-59.

141 J. van Os, "Bij een onweder," in *Dichtstukken van het Haagsch genootschap onder de spreuk: Kunstliefde spaart geen vlijt, en bekroonde keurstoffen* (3 vols.; The Hague: J. Thierry and C. Mensing, 1791-1799), 2: 138.

142 "De onmaatige vrees," 245. Cf. also Goeze, "Beredeneerde, stichtlijke en proefondervindlijke bedenkingen," 195.

143 "Bedenkingen over de vrees die veele menschen voor onweders hebben," *Hedendaagsche vaderlandsche letter-oefeningen* 4, pt. 2 (1775): 243 (by E.).

144 "Over de vrees by een onweder," 262. Cf. ook Goeze, "Beredeneerde, stichtlijke en proefondervindlijke bedenkingen," 195.

145 "Mag 't stookhuis zo vol stof door 't vuur der Zon vergaârd | Die vreemde mengsels dan niet scheiden | Niet lout'ren, zuiv'ren en bereiden | Om ze als een vruchtbaar vocht te storten op deeze Aard?" A. Veldhoven, *Dichtkundige bespiegelingen over de aarde, den mensch en den starrenhemel* (Dordrecht: H. de Haas, 1774), 40.

146 See Nieuwenhuyzen, *Lettergeschenk*, 105-06. Even without a physico-theological context this threefold use of thunderstorms was in accordance with the eighteenth-century scientific state of the art, see, e.g., Valmont de Bomare, *Algemeen en beredenerent woordenboek* 3: 108.

147 H. van Alphen, "Het onweder," in *Kleine gedigten voor kinderen*, 2e vervolg (Utrecht: wed. J. van Terveen, 1782), 76.

148 A. Blussé, *De beste opvoeding der jeugd, ten nutte van dit gemeenebest; en bespiegelingen bij een onweder* (Dordrecht: Blussé, 1776), 31.

149 "Over den donder en den blixem, en hunnen oorsprong. Weinig daarvan te vrezen. Voorzorgen evenwel daartegen," *De opmerker* 3 (1774-1775): 339 (Issue no. 147, 14 August 1775).

150 J.S. Goslin, "Bij een onweder," in *Dichtoefeningen van het genootschap onder de zinspreuk: Kunstliefde spaart geen vlyt* (The Hague: Van Cleef, 1813), 177.

151 Chomel, *Algemeen huishoudelijk-, natuur-, zedekundig- en konstwoordenboek* 1: 522. Cf. (in a physico-theological context) also "Over de vrees by een onweder," 262.

152 Martinet, *Katechismus der natuur* 1: 132.

153 Ibid., 135.

154 Ibid., 130.

155 In Dutch: *onweer*. See W.E. de Perponcher Sedlnitzky, *De wijsgeer der natuur* (2 vols.; Utrecht: J. van Schoonhoven, 1797), 1: 197.
156 Van de Wijnpersse, *Het loeijend onweder*, 204-05.
157 Feith, *Oden en gedichten* 2: 50.
158 W.E. de Perponcher Sedlnitzky, *Wijsgeerige gronden* (Utrecht: J. van Schoonhoven, 1817), 116.
159 "Wat vreest gij slaafsch! Aanbidt den Heer | Dank Hem voor dit gezegend weêr | Het welk de heete lucht bekoelt | De onreinheid uit den dampkring spoelt | Aan mensch en beest verfrissching schenkt | En 't dorstig aardrijk laaft en drenkt." Van Eijk, *Liederen voor den landman* 3: 32.
160 "Over de vrees by een onweder," 261.
161 W. Derham, *Physico-Theology, or a Demonstration of the Being and Attributes of God, from His Works of Creation* (2 vols.; London: A. Strahan; T. Cadell Jr. and W. Davies, 1798), 2: 61 n. (orig. ed. 1713).
162 S. Hottinger, *Brontologia physico-sacra* (Basel: n.p., 1698).
163 A number of "bronto-theologians" are listed by W. Philipp, *Das Werden der Aufklärung in theologiegeschichtlicher Sicht* (Göttingen: Vandenhoeck & Ruprecht, 1957), 22. On physico-theology in general, see the recent volume by A. Blair and K. von Greyerz, eds., *Physico-Theology. Religion and Science in Europe, 1650-1750*, (Baltimore: Johns Hopkins University Press, 2020).
164 See above, Chapter 3, n. 66. The Dutch translation even went through a second edition: P. Ahlwardt, *Bronto-theologie, of redelyke en theologische bespiegelingen over het onweder*, 2nd ed. (Rotterdam: Ph. Losel, 1760) (orig. ed. Delft: A. Voorstad, 1750). On Ahlwardt's book, see also Kittsteiner, *Die Entstehung des modernen Gewissens*, 71, 74-80, and Möhring, *Geschichte des Blitzableiters*, 211-14.
165 "Gesang zur Zeit des Ungewitters," in B.H. Brockes, *Raths-Herrn der Stadt Hamburg, verdeutschte Grund-Sätze der Welt-Weisheit, des Herrn Abts Genest : nebst verschiedenen eigenen theils Physicalischen theils Moralischen Gedichten, als des Irdischen Vergnügens in Gott Dritter Theil* (Hamburg: König and Richter, 1728), 598-601.
166 "Gedanken über Vernunft, Aberglauben und Unglauben," a poem written in 1729 and included in A. von Haller, *Die Alpen und andere Gedichte* (Stuttgart: Reclam, 1994), 3-22.
167 A modern edition of this poem is included in F.G. Klopstock, *Oden* (Stuttgart: Reclam, 1994), 59-67. Cf. also Thacker, *The Wildness Pleases*, 218 and O. Briese, "Blitze: Klopstocks Ode Die Frühlingsfeyer," in Büttner and Theilen, *Phänomene der Atmosphäre*, 133-42.
168 Cf. Philipp, *Das Werden der Aufklärung*, 165.
169 J.W. von Goethe, *Maximen und Reflexionen* no. 9 (Hamburg Edition, vol. 12, Hamburg: C. Wegner, 1953, 365ff.), cited by P. Michel, *Physikotheologie. Ursprünge, Leistung und Niedergang einer Denkform* (Zürich: Beer, 2008), 201 n. 329.
170 Michel, *Physikotheologie*, 148.
171 N.-A. Pluche, *Le spectacle de la nature* (9 vols.; Paris: Estienne, 1732-1742), 3: 266-68 (sequel to the second part, 1737). This work was translated many times, including into Dutch, see N.-A. Pluche, *Schouwtoneel der natuur of samenspraaken over de bysonderheden der natuurlyke historie* (16 vols.; The Hague-Amsterdam: J. Neaulme et al., 1737-1749).
172 Bertucci, "Enlightening Towers," 40.
173 Cf. Gamper, *Elektropoetologie*, 36.
174 Specht, *Physik als Kunst*, 31 wrongly gives this impression.
175 See, e.g., A. Schmidt, "Gewitter und Blitzableiter. Historische Deutungsmuster von Gewitter und deren Umschlag in Technik," in *Natur-Bilder. Wahrnehmungen von Natur und Umwelt in der Geschichte*, ed. R.P. Sieferle and H. Breuninger (Frankfurt a.M.-New York: Campus Verlag, 1999), 286-90.

5 - Marginal and Marginalised Religious Reactions

1 Parts of this chapter were previously published in J.W. Buisman, "Thunderstorms: the Devil's Work?," in *Religion und Aufklärung. Akten der Ersten Internationalen Kongresses zur Erforschung der Aufklärungstheologie (Münster, 30. März bis 2. April 2014)*, ed. A. Beutel and M. Nooke (Tübingen: Mohr Siebeck, 2016), 639-46.
2 "Wetterprophet": Drei Wolken am Himmel | Was soll dies bedeuten? | Der Meßner soll heimgeh'n | Soll Wetter läuten! in A. von Arnim and C. Brentano, *Des Knaben Wunderhorn: alte deutsche Lieder* (3 vols.;

Stuttgart: Reclam, 1987), 3: Chapter 201 (orig. ed. Heidelberg: Mohr and Zimmer, 1806).

3 A.D. White, *A History of the Warfare of Science with Theology in Christendom* (2 vols.; London: Macmillan, 1896), Introduction. Cited by Hochadel, "In Nebula Nebulorum," 51. A special focus on the battle between (dogmatic) theology and meteorological science is given in Chapter 11 of White's book.

4 So, e.g., an episcopal document from the Perpignan diocese in 1663, see J. Delumeau, *La peur en Occident (XIVe-XVIIIe siècles). Une cité assiégée* (Paris: Fayard, 1978), 93.

5 This approbation was given by the Council of Milan in 1576, for example. Cf. E.H. Landon, *A Manual of Councils of the Holy Catholic Church*, rev. ed. (2 vols.; Edinburgh: Grant, 1909), 1: 380 (orig. ed. London: Rivington, 1846).

6 Thomas, *Religion*, 85.

7 See James C. Ungureanu, *Science, Religion, and the Protestant Tradition: Retracing the Origins of Conflict* (Pittsburgh: University of Pittsburgh Press, 2019), esp. Chapter 2.

8 W. Frijhoff, "Vraagtekens bij het vroegmoderne kersteningsoffensief," in *Religieuze volkscultuur. De spanning tussen de voorgeschreven orde en de geleefde praktijk*, ed. G. Rooijakkers and Th. van der Zee (Nijmegen: SUN, 1986), 78.

9 Boia, *Weather*, 123 points out that the idea of demons inhabiting the skies is even older, being already mooted by Plato and his followers.

10 White, *A History of the Warfare* 1: 344, cited by Cohen, *Benjamin Franklin's Science*, 244 n. 3. Cf. also W.H.Th. Knippenberg, *Devotionalia. Religieuze voorwerpen uit het katholieke leven*, 2nd ed. (2 vols.; Eindhoven: Bura, 1985), 2: 33 (orig. ed. 1980). Some specimens of apotropaic bell-ringing can even be found as early as the seventh century, see A. Lehr, *Magie en klokken. Het occult gebruik van klokken en bellen, vroeger en nu – hier en elders. Een verkenning in de literatuur* (Asten: Nationaal Beiaardmuseum, 2000), 87. Charlemagne himself turned against this practice, however, see ibid., 100.

11 Lehr, *Magie en klokken*, 88.

12 A. Lehr, *Fulgura frango, ik breek de bliksem. Een middeleeuws klokopschrift* (Asten: Nationaal Beiaardmuseum, 1983), 24-25. On the demonological interpretation of thunderstorms, see also Sébillot, *Folk-lore* 1: 72.

13 White, *A History of the Warfare* 1: 348.

14 J. van Nuys Klinkenberg, *Onderwys in den godsdienst* (11 vols.; Amsterdam: J. Allart, 1780-1794), 7: 644.

15 G. Rooijakkers, *Rituele repertoires. Volkscultuur in oostelijk Noord-Brabant 1559-1853* (Nijmegen: SUN, 1994), 470. Rooijakkers's informant here is the enlightened Calvinist minister S. Hanewinkel (1766-1856). According to a more positive but exceptional idea (found in Wallonia), in principle thunderstorms enabled mortal man to peer through a crevice into the abode of the blessed, if not this curiosity would inevitably result in blindness. See Sébillot, *Folk-lore* 1: 73. Cf. also H. Bächtold-Stäubli, ed., *Handwörterbuch des deutschen Aberglaubens* (10 vols.; Berlin-Leipzig: De Gruyter, 1927-1942), 1: 1416 (e.g., on Bohemia).

16 Cf. J.G. Frazer, *The Golden Bough*, abr. ed. (Mineola, N.Y.: Dover, 2002), Chapter 3 (orig. ed. 12 vols.; London: Macmillan, 1906-1915).

17 Lehr, *Fulgura frango*, 26.

18 W. de Fonvielle, *Éclairs et tonnerre*, 2nd ed. (Paris: Hachette, 1869), 172 (orig. ed. 1867).

19 R. von Strele, "Wetterläuten und Wetterschiessen. Eine culturgeschichtliche Studie," in *Jahrbuch des Deutschen und Österreichischen Alpenvereins* 29 (1898): 123-24.

20 S. Wilson, *The Magical Universe. Everyday Ritual and Magic in Pre-Modern Europe* (London: Hambledon and London, 2000), 71-72 and Sébillot, *Folk-lore* 1: 108.

21 K. Adrian, "Wind und Wetter im Glauben und Brauchtum unseres Volkes," in *Mitteilungen der Gesellschaft für Salzburger Landeskunde* 84-85 (1944-1945): 4.

22 J. Schwarz, "Aachen- oder Wetterhörner," *Westdeutsche Zeitschrift für Volkskunde* 32 (1935), Appendix 'Heimatmuseum und Volkskunde': 13, 15 (citation on 6).

23 J. Delumeau, *Rassurer et protéger. Le sentiment de sécurité dans l'Occident d'autrefois* (Paris: Fayard, 1989), 83. On Thiers cf. also A. Corbin, *Les cloches de la terre. Paysage sonore et culture sensible dans les campagnes au XIXe siècle* (Paris: Flammarion, 1994), 104. Remarkably enough Thiers was a sceptic on the issue of exorcism, see F. Young, *A History of Exorcism in Catholic Christianity* (London-New York: Palgrave Macmillan, 2016), 164.

24 S. Zak, *Musik als 'Ehr und Zier'* (Neuss: Päffgen, 1979), 37-38, cited by Lehr, *Magie en klokken*, 33 (cf. also 101). Cf. Num. 10: 1-10 and Josh. 6: 5.

25 Rooijakkers, *Rituele repertoires*, 471.
26 Von Strele, "Wetterläuten," 124 and, in more detail, H. Otte, *Glockenkunde*, 2nd ed. (Leipzig: Weigel, 1884), 17-18 and Lehr, *Magie en klokken*, 103-05. A contemporary enlightened account of a 'bell baptism' is given by *Algemeene vaderlandsche letter-oefeningen* (1796), pt. 1: 538-41. On some seventeenth-century Ligurian texts for the exorcism of thunderstorms, see J. Jensen, *Kirchliche Rituale als Waffen gegen Dämonenwirken und Zauberei* (Berlin: LIT, 2007), 134-37.
27 Thomas, *Religion*, 34.
28 Delumeau, *Rassurer et protéger*, 82.
29 Ibid. The ritual for the consecration of bells in Paris likewise makes clear that exorcism is at issue here, because in that ceremony evil spirits are explicitly mentioned as originators of thunderstorms and other misfortune, see Cohen, *Benjamin Franklin's Science*, 119.
30 Cf., e.g., C. van de Wiel, *Het recht van de Kerk met betrekking tot handelingen van goddelijke eredienst, gewijde plaatsen en heilige tijden* (Louvain: Peeters, 2000), 56. Remarkably, in the Austrian village of Kaprun baptismal robes were thrown outside the house and collected afterwards, as a prophylactic ritual against thunderstorms, see Adrian, "Wind und Wetter," 4.
31 Von Strele, "Wetterläuten," 125. Dog names were popular in Austria, though in Tirol the biggest bells were also called "the bull." See Adrian, "Wind und Wetter," 10 and F. Hager and H. Heyn, *Drudenhax und Allelujawasser. Volksbrauch im Jahreslauf* (Rosenheim: Förg, 1975), 205, respectively.
32 Lehr, *Fulgura frango*, 13.
33 Otte, *Glockenkunde*, 23. On the attendance of a godfather and godmother at the consecration ceremony of bells, see also Lehr, *Magie en klokken*, 100-01.
34 E. Gleichmann, *Die Glocke im Volksglauben, Sage und Dichtung* (Kulmbach: R. Rehm, 1930), 58 and Otte, *Glockenkunde*, 173-74. Otte refers, e.g., to Th. Fontane, *Wanderungen durch die Mark Brandenburg* (Berlin: Hertz, 1862-1882).
35 Cf. Aby M. Warburg, *Schlangenritual. Ein Reisebericht*, 5th ed. (Berlin: K. Wagenbach, 2011), a revision of a lecture held in 1923. Cf. also Elsom, *Lightning*, 198-200. This ritual aimed at triggering rain showers.
36 See above, p. 100. However, Fontane's book in question does not appear in the catalogue of Ten Kate's private library, as is proved by *Catalogus van de bibliotheken van J.J.L. ten Kate en van een rechtsgeleerde* (Amsterdam: Van Looy, 1890).
37 Von Strele, "Wetterläuten," 125.
38 A. Brittinger, *Die bayerische Verwaltung und das volksfromme Brauchtum im Zeitalter der Aufklärung* (Munich: Mössl, 1938), 14.
39 Adrian, "Wind und Wetter," 9 (passing remark). Consecration by a well known cleric was also an important factor here. For that reason a bell consecrated in the German town of Meissen in the eleventh century by Bishop Benno himself was considered to be highly effective against thunderstorms, see Lehr, *Fulgura frango*, 16.
40 Von Strele, "Wetterläuten," 133.
41 Cf. Corbin, *Cloches de la terre*, 186.
42 G. Bouchard, *Le village immobile. Sennely-en-Sologne au XVIIIe siècle* (Paris: Plon, 1972), 319.
43 See above, n. 38 and Fabich, *Bauernmedizin*, 98-99.
44 Von Strele, "Wetterläuten," 139-40.
45 Ibid., 133.
46 Kittsteiner, *Die Entstehung des modernen Gewissens*, 92. On French priests credited with the same qualities, see Wilson, *Magical Universe*, 76. Cf. also Sébillot, *Folk-lore* 1: 108-09. Even as late as the nineteenth century also in Pietism a special local protective power against thunderstorms was ascribed to some pious inhabitants, cf. M. Scharfe, *Die Religion des Volkes. Kleine Kultur- und Sozialgeschichte des Pietismus* (Gütersloh: Mohn, 1980), 36-37 (on the devout shoemaker J. A. Straub, 1776-1857).
47 See Von Strele, "Wetterläuten," 130-31. Usually the church went no further than prohibiting the priest from leaving the village during thunderstorms, because he was believed to dispose of the rites to propitiate God and to dispel the demons. This prohibition was promulgated, e.g., in the Perpignan diocese in 1663, see Delumeau, *Rassurer et protéger*, 84.
48 Adrian, "Wind und Wetter," 11.

49 On the distinction between different categories of thunderstorms, see Brittinger, *Bayerische Verwaltung*, 22. Cf. also Von Strele, "Wetterläuten," 123 and (in passing) Adrian, "Wind und Wetter," 9 and 21.

50 The distinction between the two types of thunderstorm is part of a more general differentiation between "natural magic" and "demonic magic." Cf. J. Ehrard, *L'idée de nature en France dans la première moitié du XVIIIe siècle* (2 vols.; Paris: S.E.V.P.E.N., 1963), 1: 37.

51 Adrian, "Wind und Wetter," 20ff. In Heinrich Kramer's well-known *Malleus Maleficarum* (1486) a separate chapter (Vol. 2, Question 1, Chapter 15) was already devoted to witches, who caused thunder- and hailstorms with the help of Satan: cf. R. Vervoort, *'Vrouwen op den besem en dergelijck gespoock'. Pieter Bruegel en de traditie van hekserijvoorstellingen in de Nederlanden tussen 1450 en 1700* (Nijmegen: Vantilt, 2011), 82.

52 Adrian, "Wind und Wetter," 8.

53 Von Strele, "Wetterläuten," 136-37.

54 Adrian, "Wind und Wetter," 8. Adrian is quoting here from a Salzburg student's diary. The custom of *Hexenläuten* was abolished only by the Salzburger *Hirtenbrief* (pastoral letter) of 1782, see S. Merkle, *Die kirchliche Aufklärung im katholischen Deutschland* (Berlin: Reichl, 1910), 106.

55 On "witch- ringing" in Sennely, see Bouchard, *Le village immobile*, 319.

56 Adrian, "Wind und Wetter," 21 (a story about an ancient "thunder bell" from Wald im Pinzgau, a village in the Salzburger Land).

57 Von Strele, "Wetterläuten," 124.

58 Lehr, *Fulgura frango*, 29.

59 See Gleichmann, *Die Glocke im Volksglauben*, 48 and Von Strele, "Wetterläuten," 125.

60 Adrian, "Wind und Wetter," 10.

61 Rieth, *Blitz*, 31.

62 Fuhrmeister, "'Eripuit Caelo Fulmen,'" 144 and 158 n. 1.

63 Corbin, *Cloches de la terre*, 104-05. It remains unclear, however, in how far these inscriptions had become a dead letter by then.

64 In my opinion, the Catholic priest N. Kyll (1904-1973) wrongly gives this impression, see Kyll, "Glocke," 172.

65 Gleichmann, *Die Glocke im Volksglauben*, 58. Characteristically, the trolls at the end of the sixth scene of the second act of H. Ibsen's play *Peer Gynt* (1867) disappear as soon as the church bells start ringing.

66 Lehr, *Fulgura frango*, 17 and idem, *Magie en klokken*, 110.

67 Von Strele, "Wetterläuten," 133.

68 Ibid., 132-33 and Adrian, "Wind und Wetter," 7.

69 Knippenberg, *Devotionalia* 1: 128 and Lehr, *Magie en klokken*, 87.

70 M. Gari, *Le vinaigre et le fiel* (Paris: Plon, 1983), 21, cited by Delumeau, *Rassurer et protéger*, 84.

71 Cf. Fonvielle, *Éclairs et tonnerre*, 173.

72 Von Strele, "Wetterläuten," 132.

73 Lehr, *Fulgura frango*, 35. In some German territories bells were rung with three interruptions, allegedly because it was precisely during those intervals that the demon was suffocated, see F. Steffan, "Wetterabweisende Heiltümer. Fund in der Turmkugel von Altenhohenau bei Wasserburg am Inn," Volkskunst. *Zeitschrift für volkstümliche Sachkultur*, Issue 4 (1985): 23.

74 Cf. Delumeau, *Rassurer et protéger*, 82.

75 Corbin, *Cloches de la terre*, 104, 109.

76 Augustinus de Angelis, *Lectiones meteorologicae* (Rome: Corbi, 1663), 75. Cited by White, *A History of the Warfare* 1: n. 248.

77 Lehr, *Fulgura frango*, 45.

78 Hochadel, "In Nebula Nebulorum," 51.

79 Even beyond the field of meteorology! See Corbin, *Cloches de la terre*, 110 (nineteenth-century examples from the French Moselle and Charente areas).

80 Lehr, *Fulgura frango*, 44.

81 Ibid., 45.

82 Enlightened eighteenth-century Protestants were definitely more sceptical. Cf. Hochadel, "In Nebula Nebulorum," 51: in their opinion, apotropaic bell-ringing was simply a Roman Catholic custom, whereas Protestants had only prayers to resort to. And even there they proved to be critical, because they felt uneasy with the view of a punishing God inherent in many thunder prayers.

83 Hentschel, "Man erhitze sich daher zur Zeit der Ungewitter nicht," 11.

84 Cf. Lehr, *Fulgura frango*, 61. Typically, the enlightened Italian priest L.A. Muratori (1672-1750) regarded objects of devotion as mere prayer aids, rather than as items that were powerful in themselves, see Ulrich L. Lehner, *The Catholic Enlightenment. The Forgotten History of a Global Movement* (Oxford-New York: Oxford University Press, 2016), 132.

85 See above, p. 114.

86 Cohen, "Prejudice," 432.

87 Otte, *Glockenkunde*, 24.

88 R.F.H. Praamsma, *Zacharias Webber (1644-1696), irenisch lutheraan – verlicht protestant. Kerk en theologie in het denken van een zeventiende-eeuwse kunstschilder* (Delft: Eburon, 2013), 245.

89 Otte, *Glockenkunde*, 27. Even at the end of the twentieth century this collective prayer was still in vogue in some foundries in the Netherlands when important bells were cast, see Lehr, *Magie en klokken*, 13.

90 Lehr, *Fulgura frango*, 40 and idem, *Magie en klokken*, 92.

91 By royal injunction in 1547, the first year of Edward VI's reign. See Thomas, *Religion*, 60.

92 H.W. Hielkema, "It kloklieden as folksbrûkme yn Fryslân," in *Frysk studinte-almenak* 1937: 44. Cf. also Lehr, *Fulgura frango*, 34.

93 G. Rooijakkers, *Eer en schande. Volksgebruiken in het oude Brabant* (Nijmegen: SUN, 1995), 129-30, cited by Lehr, *Magie en klokken*, 92.

94 Thomas, *Religion*, 60.

95 Rooijakkers, *Rituele repertoires*, 471 and Lehr, *Magie en klokken*, 81.

96 Cf. Kittsteiner, *Die Entstehung des modernen Gewissens*, 56.

97 J.J. van Deinse, *Uit het land van katoen en heide*, 2nd ed. (Enschede: Van der Loeff, 1925), 106 and Heckscher, *Volkskunde*, 372.

98 M. Broeckhove, "Devotie in het Vlaamssprekend gedeelte van België tot de H. Donatus, biezondere patroon tegen schadelijke natuur- en vernielende kunstmatige luchtelementen," *Oostvlaamsche zanten* 34 (1959): 128, and 35 (1960): 48. See also Lehr, *Fulgura frango*, 14-16 and Buisman, "Opvattingen over het onweer," 72. In a similar way small silver St Anthony's bells were used, see Fabich, *Bauernmedizin*, 97 and Hager and Heyn, *Drudenhax und Allelujawasser*, 201.

99 E. Kreissl, ed., *Aberglauben – Aberwissen. Welt ohne Zufall* (Graz: Universalmuseum Joanneum, 2014), 45.

100 Wodianka, *Betrachtungen des Todes*, 133-34 and Kittsteiner, *Die Entstehung des modernen Gewissens*, 57-58.

101 Cf. Delumeau, *Rassurer et protéger*, 436.

102 See above, Chapter 4, p. 74.

103 Lehr, *Magie en klokken*, 107-08.

104 Brittinger, *Bayerische Verwaltung*, 19.

105 *Volledige verzameling van alle stukken, betreffende de afscheiding der kerk van den staat, door de Nationaale Vergadering, representeerende het volk van Nederland.* Met eene voorrede van C. Rogge (Leiden: Du Mortier, 1796), 421. Not only was the principle of full equality between all denominations applied here but also the transfer of the property rights of church towers to the civic authorities was involved.

106 E.g., H.L. Quesnel (1752-1804) and J.H. Midderigh (1753-1800), see ibid., 399-400. This conclusion tallies with Edwina Hagen's remark that no anti-papism can be traced in the First National Assembly, see E. Hagen, "Een zaal van staatsmannen, niet van godgeleerden. Godsdienstige sentimenten in de Nationale Vergadering," in *Het Bataafse experiment. Politiek en cultuur rond 1800*, ed. F. Grijzenhout, N. van Sas, and W. Velema (Nijmegen: Vantilt, 2013), 137.

107 Bonifacius Stöltzlin, *Geistliches Donner- und Wetter-Büchlein: Das ist: Einfältige Erinnerung vom Donner, Blitz, Straal, Hagel und schädlichen Wettern* (Ulm: Vogt, 1650, many reprints), cited by Kittsteiner, *Die Entstehung des modernen Gewissens*, 61.

108 Cf. Kittsteiner, *Die Entstehung des modernen Gewissens*, 61, 63.

109 Ibid., 91.

110 Cf. ibid., 60 (Kittsteiner's terms).

111 Cf. ibid., 68.

112 Ibid., 91. On this change of attitude, cf. also Bernd-Brinkmann, "Wetterlieder," 99-100.

113 Otte, *Glockenkunde*, 24.

114 See above, p. 112.

115 See above, Chapter 3, pp. 53ff.

116 Hochadel, "In Nebula Nebulorum," 52. No fewer than 103 bell ringers were killed in Germany between 1750 and 1783 (ibid.). Allegedly, Emperor Joseph II called church bells "the clergy's artillery," see E. Roloff, *Göttliche Geistesblitze. Pfarrer und Priester als Erfinder und Entdecker* (Weinheim: Wiley-VCH, 2010), 40. When the English music historian Charles Burney (1726-1814) visited the Bavarian town of Freising in 1772, a thunderstorm raged over the city. During the night, Burney was very frightened, because not a single lightning rod had been installed there, the inhabitants relying solely on bell-ringing instead, see Chr. Kürzeder, "Wetterläuten in Freising und Mr. Burneys Todesangst. Phänomene der Frömmigkeit im Rokoko," in *Mit Leib und Seele. Münchner Rokoko von Asam bis Günther*, ed. R. Diederen and Chr. Kürzeder (Munich: Sieveking, 2014), 211.

117 Lehr, *Fulgura frango*, 56 and 52, respectively.

118 Cf. Hentschel, "Man erhitze sich daher zur Zeit der Ungewitter nicht," 17 (a specimen of such a utilitarian motivation in the Bavarian ban on bell-ringing).

119 H. Bausinger, "Aufklärung und Aberglaube," *Deutsches Vierteljahrsschrift für Literaturwissenschaft und Geistesgeschichte* 37 (1963): 349-50. Cf. also Brittinger, *Bayerische Verwaltung*, 15. Bausinger's interpretation seems somewhat anachronistic, because contemporary scholars such as Van Musschenbroek also supposed that moving bells and clappers would attract electricity, see Chomel, *Algemeen huishoudelijk-, natuur-, zedekundig- en konstwoordenboek* 1: 522. It was only modern scientific research that demonstrated convincingly that apotropaic bell-ringing does not protect against lightning strikes but rather increases the risk of their occurrence. However, the increased risk does not operate via the electrification of thunderclouds by the tolling of bells, cf. Heering and Hochadel, "Epilogue," 275 n. 2.

120 Heering, "Styles of Experimentation," 139 n. 26 and Cohen, "Prejudice," 421. Nonetheless, even Franklin's little bells retained their thunder-announcing function! See also Elsom, *Lightning*, 113. Nollet also used such a set of small warning bells in his research, see Prinz, "Fulminantes über Wolkenelektrizität," 8. German and Dutch amateurs could construct their own warning device, see the manual in *Onderrigtingen omtrent de wyze, op welke men zich te gedragen heeft by het opkomen van onweder, benevens de middelen, om zich voor de nadeelige uitwerkselen van den blixem te beveiligen. Uitgegeeven tot onderrigting van onkundigen.* Na den derden vermeerderden druk uit het Hoogduitsch vertaald (The Hague: I. van Cleef, 1779), 90-93 (appendix).

121 Cohen, "Prejudice," 395.

122 Hentschel, "Man erhitze sich daher zur Zeit der Ungewitter nicht," 17-18.

123 Corbin, *Cloches de la terre*, 40.

124 Ibid., 105-06. Corbin refers here to a *cahier de doléances* from 1789.

125 See above, Chapter 3, p. 67.

126 Cited by Heilbron, *Electricity*, 341.

127 See above, Chapter 3, p. 67.

128 Valmont de Bomare, *Algemeen en beredenerent woordenboek* 3: 108.

129 Corbin, *Cloches de la terre*, 107.

130 Cf. Engammare, "Tonnerre de Dieu," 178.

131 According to Abbé Pluche, as phrased by Corbin, *Cloches de la terre*, 106.

132 A. van der Woud, *De Bataafse hut. Denken over het oudste Nederland (1750-1850)*, 2nd ed. (Amsterdam-Antwerp: Contact, 1998), 47 (orig. ed. Amsterdam: Meulenhoff, 1990).

133 Cohen, "Prejudice," 396 n. 9 and Lehr, *Fulgura frango*, 58.

134 See, e.g., J. van Lier, *Oudheidkundige brieven* (The Hague: Van Thol, 1760), 112-13.

135 Ibid., 109-10, 112-13. According to an ancient popular belief, streaks of lightning caused stones to fall from heaven. These "thunderstones" or "thunder chisels" were in great demand afterwards, being regarded as curative and prophylactic amulets. Later generations would identify them as belemnites.

136 Engammare, "Tonnerre de Dieu," 178.

137 Brittinger, Bayerische Verwaltung, 20.
138 Adrian, "Wind und Wetter," 16-17.
139 See above, p. 115.
140 Cited by Adrian, "Wind und Wetter," 17. See Herodotus, *The Histories*, Book 4, para. 94.
141 Lehr, *Fulgura frango*, 26. Cf. also Lehr, *Magie en klokken*, 95. Unfortunately, there is no reference to when this occurred. On the Japanese example, see B. Vogel, Transmission and Performance of Taiko in Edo Bayashi, Hachijo, and Modern Kumi-daiko Styles (unpublished PhD thesis, Houston 2009), 13 (see https://scholarship.rice.edu/bitstream/handle/1911/61883/3362425.pdf?sequence=1&isAllowed=y (accessed 28 November 2018)). Sounding the taiko drums had other functions as well, e.g., giving a storm warning.
142 Cohen, "Prejudice," 395 n. 7.
143 The enlightened Salzburg archbishop Hieronymus von Colloredo (1732-1812), e.g., put a ban on both apotropaic bell-ringing and *Wetterschießen* in 1771 and 1786 respectively. See Adrian, "Wind und Wetter," 17.
144 On the use of the cloud seeding technique to interrupt the formation of hailstones, see Harry Geurts, *Weerspiegeld. Het weer nader verklaard* (Franeker: Van Wijnen, 2021), 192. Cf. also Cynthia Barnett, *Rain. A Natural and Cultural History* (New York: Crown, 2015), Chapter 8. However, the context there is the fight against drought. Remarkably, even well into nineteenth-century America some people thought that hard noise, produced by instruments of war or otherwise, could provoke rain showers (ibid.).
145 Cf. Cohen, "Prejudice," 396 n. 9 and also above, p. 129.
146 Cf., respectively, F. Ferré, *Hellfire and Lightning Rods. Liberating Science, Technology, and Religion* (Maryknoll: Orbis Books, 1993), 27, and the eschatological views held by the Boston clergyman Mather Byles (1706-1782) in 1755, cited by Robles, "Atlantic Disaster," 15. Cf. 2 Peter 3: 10. On ancient ideas on sulphur and hell, see also the monograph by S. Kroonenberg, *Why Hell Stinks of Sulfur. Mythology and Geology of the Underworld* (London: Reaktion Books, 2013).
147 Cf. Lehr, *Fulgura frango*, 59. Even as late as 1906 the English cleric and campanologist J.J. Raven (1833-1906) believed in a natural explanation of the effectivity of apotropaic bell-ringing, see J.J. Raven, *The Bells of England* (London: Methuen, 1906), 280, cited by Cohen, "Prejudice," 401.
148 Adrian, "Wind und Wetter," 17.
149 H. Filipetti and J. Trotereau, *Symboles et pratiques rituelles dans la maison paysanne traditionnelle* (Paris: Berger-Levrault, 1978), 102. Cf. also Sébillot, *Folk-lore* 1: 109-10.
150 P. Sartori, *Das Buch von deutschen Glocken* (Berlin: De Gruyter, 1932), 208.
151 See https://gtb.ivdnt.org/search/, *Woordenboek der Nederlandsche taal*, s.v. "Pieter" and "kegelen" (accessed 4 May 2021).
152 Sébillot, *Folk-lore* 1: 74 (south-west France and Franche-Comté) and De Cock, "Eerbied voor de natuur," 52 (Italy and Flanders). A similar idea can be found in the French Alps, see Wilson, *The Magical Universe*, 59. Typically, the inhabitants of the Salzburger Land feared the thunder rather than the lightning, see Adrian, "Wind und Wetter," 18. This may be deemed remarkable, because it is lightning with which the devil is explicitly associated in Luke 10: 18.
153 Broeckhove, "Devotie" (1959), 133.
154 Ibid.(1959), 189.
155 Ibid.(1960), 48 and (1959), 131.
156 A. Feldmann, "'Donnerkeil, Feuer, und ein entsetzlicher Hagel...' Der Heilige Donatus von Münstereifel," *Brand. Heimatkundliche Blätter* (2002): 66.
157 Knippenberg, *Devotionalia* 1: 128.
158 On the information in this paragraph, see Gooren, "Over Sint Donatus," 33, *Bedevaartplaatsen in Nederland*, ed. P.J. Margry et al. (4 vols.; Amsterdam-Hilversum: Meertens Instituut and Verloren, 1997-2004), 2: 737, and Broeckhove, "Devotie" (1959), 128, 132.
159 Cf. M. English, "De donderheilige Sint-Donatus en zijn verering in West-Vlaanderen," *Biekorf* 60 (1959), 72.
160 P. Hildebrand, *De Kapucijnen in de Nederlanden en het prinsbisdom Luik* (10 vols.; Antwerp: Archief der Kapucijnen, 1945-1956), 9: 742.
161 In this connection, Bernd-Brinkmann points to the parallel between Roman Catholic litanies of the saints and Protestant thunderstorm hymns, see Bernd-Brinkmann, "Wetterlieder," 98: both aimed at bringing about a comforting feeling among the members of the family.

162 Eusebius, drawing on Tertullian, already mentioned these events in his *Historia ecclesiastica* (Book 5, Chapter 5). In his opinion, they were the outcome of the Christians' prayers. On this, see in more detail, Ido Israelowich, "The Rain Miracle of Marcus Aurelius: (Re-)Construction of Consensus," *Greece and Rome*, 2nd ser. 55 (2008), 83-102. Cassius Dio (ca. 155-235), a non-Christian author, related the same events, but he identified the Egyptian magician Arnuphis as an instigator of them. It seems unlikely, however, that the Twelfth Legion was involved in this battle and that large numbers of Christians took part in it, see Israelowich, "Rain Miracle," 90. Moreover, no evidence exists for Donatus's presence in the Roman legion, far less for his involvement in the "rain miracle".

163 See, f.i., Broeckhove, "Devotie" (1959), 45-46 and Feldmann, "Donnerkeil," 70-71.

164 Broeckhove, "Devotie" (1960), 50-51 and (1959), 56.

165 See on this H. Achermann, *Die Katakombenheiligen und ihre Translationen in der schweizerischen Quart des Bistums Konstanz* (Stans: Historischer Verein Nidwalden, 1979), cited by Feldmann, "Donnerkeil," 72.

166 These kinds of objects were inspired by a thirteenth-century Spanish "miracle cross" and were worn as amulets to prevent the harm caused by thunderstorms. They were banned in the course of the eighteenth century. Nevertheless, they continued to be represented on house blessings.

167 Fact and fiction are quite often hard to separate in the various descriptions of these events. The "pious" version presented here is based on annual reports of the Jesuits in Münstereifel, see Feldmann, "Donnerkeil," 62-64 and in more detail B. Plück, Die Verehrung des hl. Donatus von Münstereifel in den Bistümern Köln und Trier 1652-1773 (unpublished Master Thesis, Bonn 1980), 18-21 and 27 n. 2 (on the little Caravaca cross). Cf. also Broeckhove, "Devotie" (1959), 43-44. A summary of the Master Thesis just mentioned has been given in B. Plück, "Der Kult des Katakombenheiligen Donatus von Münstereifel," in *Jahrbuch für Volkskunde*, n.s. 4 (1981): 112-26.

168 Feldmann, "Donnerkeil," 76 and F. Pohle, 'Mera ossa et cadavera.' Studien zum Jesuitentheater in Jülich-Berg, Ravenstein und Aachen (1601-1817) (unpublished PhD thesis, Aachen 2006), 318 and the literature listed there: https://publications.rwth-aachen.de/record/62507 (accessed 11 May 2021). Even today, pilgrimages are still made to the relics in Bad Münstereifel, see Feldmann, "Donnerkeil," 86.

169 English, "De donderheilige Sint-Donatus," 72. As shown above, objects of this kind were considered defences or even amulets against thunderstorms.

170 Hildebrand, *Kapucijnen* 9: 741-42, 749 and Broeckhove, "Devotie" (1959), 50, 55. When I attended the religious service in St Donatus's Church in Arlon on 11 July 1999 (i.e. the saint's day, the second Sunday of that month), a double Mass was celebrated, a statue of Donatus was placed in the front of the church building, and his relics were exposed for veneration.

171 See Margry et al., *Bedevaartplaatsen* 2: 735-40 and Gooren, "Over Sint Donatus," 34, respectively.

172 Margry et al., *Bedevaartplaatsen* 2: 735.

173 Ibid. 2: 736.

174 Ibid. 2: 737. The fraternity of St Donatus published a booklet on the occasion of its bicentennial in 1979: *H. Donatus Martelaar: 200-jarige viering St. Donatusbroederschap Reek* (s.l., s.n., 1979).

175 Plück, Die Verehrung des hl. Donatus, 30.

176 Feldmann, "Donnerkeil," 67, referring to N. Kyll, "Die Hagelfeier im alten Erzbistum Trier und seinen Randgebieten," in *Rheinisches Jahrbuch für Volkskunde* 13-14 (1962-1963): 113-71.

177 Margry et al., *Bedevaartplaatsen* 2: 556 (Loosbroek) and Gooren, "Over Sint Donatus," 11 (Vragender).

178 Besides Reek, another early Donatus pilgrimage site in Brabant is the village of Handel (1848), see Margry et al., *Bedevaartplaatsen* 2: 652. However, the institution of most Donatus pilgrimages can only be established from the second half of the nineteenth century onwards.

179 On these objects in general, see G.C.M. Egelie, *Gietijzeren wegkruisen in Limburg* (Zutphen: Walburg Pers, 1983), 99. Feldmann reports that such crosses were already erected in the Eifel and Ardennes regions during the eighteenth century and that they frequently bore the image of Donatus, see Feldmann, "Donnerkeil," 75.

180 Cf. Pohle, 'Mera ossa et cadavera,' 319-20, Hildebrand, *Kapucijnen* 9: 745, and Broeckhove, "Devotie" (1959), 221.

181 An example of such a song can be found in Feldmann, "Donnerkeil," 91. Unfortunately, it is undated. Cf. also Margry et al., *Bedevaartplaatsen* 2: 370 (late twentieth century).

182 Plück, Die Verehrung des hl. Donatus, 39.

183 See G. Rooijakkers, "De dynamiek van devotionalia. De materiële cultuur van het geleefde geloof in oostelijk Noord-Brabant," in *De dynamiek van religie en cultuur*, ed. M. Monteiro, G. Rooijakkers, and J. Rosendaal (Kampen: Kok, 1993), 95-96 and by the same author *Rituele repertoires*, 607-09. Variants on these "papiertjes met heiligdom" (little papers with a sacred part) can be found also in Germany: see Steffan, "Wetterabweisende Heiltümer," 23 and 25 n. 3, as well as Kück, *Wetterglaube*, 149-50 (on the somewhat different *Himmelsbriefe*).

184 See Feldmann, "Donnerkeil," 79.

185 See above, p. 125.

186 F. Hunolt, *Christliche Sitten-Lehr über die Evangelische Wahrheiten* (6 vols.; Cologne: Huisch, 1740-1748), 2: 282, cited by Feldmann, "Donnerkeil," 79. Similar views (albeit with a stronger focus on penance and conversion) were held by certain eighteenth-century Bavarian preachers, see Phayer, *Religion und das gewöhnliche Volk*, 46.

187 Cf. Lehner, *Catholic Enlightenment*, 132. Lehner never tires of stressing how deeply the Catholic Enlightenment was rooted in the Tridentine efforts at reform. However, it is important to distinguish carefully between intentions and actions here.

188 See above, p. 100. Delbourgo, *Amazing Scene*, 83 even refers to belief in the lightning rod among the enlightened as "talismanic": "what was remarkable was not that it worked, but that is should ever fail."

189 See above, Chapter 2, n. 53.

190 Cf. Gooren, "Over Sint Donatus," 33.

191 Feldmann, "Donnerkeil," 79-80.

192 See above, Chapter 4, n. 128.

193 Hildebrand, *Kapucijnen* 9: 746.

194 Cf. Rieth, *Blitz*, 39. As late as 1770, when the spire of St James's Church in Hamburg was restored and equipped with the first lightning rod in the German-speaking lands, an ancient tradition was honoured by the principal Lutheran minister Christian Samuel Ulber (cf. above, Chapter 4, n. 101 and 102). He placed a variety of sacred and secular objects in the spire's new lantern, including a Bible, a catechism, a memorial of his own, and some coins – a curious Protestant counterpart of the ways Catholics traditionally treated relics and other objects of devotion, see Möhring, *Geschichte des Blitzableiters*, 142 and Steffan, "Wetterabweisende Heiltümer," 23-25, respectively.

195 Pohle, 'Mera ossa et cadavera,' 319 n. 4. The parish acquired this relic in 1783.

196 Plück, Die Verehrung des hl. Donatus, 57.

197 See P.G.J. Post, "'Het suizen van een zachte koelte.' Het weer en christelijke rituelen: terreinverkenning en pleidooi voor wetenschappelijk grensverkeer gethematiseerd via palmzondagriten," *Volkskundig bulletin* 15 (1989): 198 n. 26 and Filipetti and Trotereau, *Symboles et pratiques rituelles*, 101. Cf. also Sébillot, *Folk-lore* 1: 106-07 and Thuesen, *Tornado God*, 26.

198 Hentschel, "Man erhitze sich daher zur Zeit der Ungewitter nicht," 12.

199 Freudenthal, *Feuer*, 371-72.

200 Broeckhove,"Devotie" (1959), 181 and Kittsteiner, *Die Entstehung des modernen Gewissens*, 56.

201 Feldmann, "Donnerkeil," 73 and Plück, Die Verehrung des hl. Donatus, 57. Other weather patron saints were, e.g., John the Baptist, Michael (leader of the war against Satan in the celestial sphere!), Anne, Margaret, Ulrich, and Medard.

202 Plück, Die Verehrung des hl. Donatus, 58.

203 Plück assumes a coincidence here, see ibid., 76.

204 H. Jedin, ed., *Handbuch der Kirchengeschichte* (10 vols.; Freiburg-Basel-Vienna: Herder, 1962-1979), 3: 597-605, cited by Plück, Die Verehrung des hl. Donatus, 75.

205 See J.S. Friedman, *Out of the Blue. A History of Lightning: Science, Superstition, and Amazing Stories of Survival* (New York: Delacorte Press, 2008), 37 and Sébillot, *Folk-lore* 1: 118 and 104. Cf. above, n. 135.

206 S. Ruhe, *Visserstruien uit 40 Nederlandse vissersplaatsen* (Baarn: Forte, 2013), 23 and 42 and idem, *Visserstruien 2. 65 historische truien met breischema's uit 55 Nederlandse vissersplaatsen* (Baarn: Forte, 2015), 50. Sometimes the zigzag pattern is also associated with conjugal discord.

207 Scharfe, *Religion des Volkes*, 37.

208 B. Hoggard, "The Archaeology of Counter-Witchcraft and Popular Magic," in *Beyond the Witch Trials. Witchcraft and Magic in Enlightenment Europe*, ed. O. Davies and W. de Blécourt (Manchester-New York: Manchester University Press, 2004), 170-73. Sharp metal objects and acid fluids were supposed to keep demons at bay, see below, p. 147 and pp. 155-57.

209 Filipetti and Trotereau, *Symboles et pratiques rituelles*, 79. On the central importance of the fireside, see also Wilson, *The Magical Universe*, 6-7.

210 Hentschel, "Man erhitze sich daher zur Zeit der Ungewitter nicht," 7. Interestingly, in Germany and England the chimney sweep is still regarded as a bringer of good fortune.

211 See above, Chapter 3, p. 66.

212 Filipetti and Trotereau, *Symboles et pratiques rituelles*, 79 and 102 (examples from Provence and Brittany).

213 See above, p. 132.

214 Kück, *Wetterglaube*, 142. On this ambivalence, see also Bächtold-Stäubli, *Handwörterbuch* 3: 824.

215 As said in Chapter 3, n. 18, Volta considered the kindling of fires a more efficient means of protection than the installation of a lightning rod, because the fires would fully saturate the air with phlogiston and subsequently carry off atmospheric electricity. In this connection the Italian scholar referred to the wisdom of the ancients, who lit sacrificial fires for Jupiter on mountains, see Volta, *Meteorologische Briefe*, 192-93. On the idea that smoke dispelled thunderstorms, see also Bächtold-Stäubli, *Handwörterbuch* 3: 828.

216 Cf. M. Eliade, *The Forge and the Crucible*, 2nd ed. (Chicago: University of Chicago Press, 1978), 27ff. (orig. French ed. Paris: Flammarion, 1956).

217 Freudenthal, *Feuer*, 39.

218 Hempler, *Das Gewitter*, 26.

219 See above, n. 217 and Sébillot, *Folk-lore* 1: 105 (on Basque Country). In the Alps, Provence, and Italy, for instance, such objects were swung in a terrifying manner, see Wilson, *The Magical Universe*, 70. Sometimes sickles or knives were hurled at the witches who were believed to be hiding in the thunderclouds, see Hager and Heyn, *Drudenhax und Allelujawasser*, 204. Bächtold-Stäubli, *Handwörterbuch* 3: 828 thinks a primitive kind of lightning rod is involved. However, this interpretation can only be reconciled with the aggression becoming evident here, if the demons swooping down from heaven are supposed to hurt themselves on the exposed objects. The display of sharpened metal objects such as swords and scythes to ward off thunder demons can also be found among the indigenous inhabitants of Canada and Russia, see Elsom, *Lightning*, 80.

220 Consequently, in other places all metal objects were carefully stored away as they were thought to attract lightning. Cf., e.g., R. van de Weijer, "Een huis als herinneringsmachine. Van agrarisch bedrijf Sigmans naar museumboerderij," in *De musealisering van het platteland: de historie van een Brabants boerenhuis*, ed. G. Rooijakkers, A. van Lierop, and R. van de Weijer (Nijmegen: SUN, [2002]), 136-69, there 153.

221 Filipetti and Trotereau, *Symboles et pratiques rituelles*, 122.

222 See above, n. 217.

223 See above, p. 139. Sometimes shoes were hidden in house openings to ward off witches, see Hoggard, "Archaeology," 179. Cf. the behaviour of some French priests, who hurled their shoes at heaven when their prayers against thunderstorms did not help, see above, p. 132.

224 See G.H. Kuijt, *De schaduw van drie sterren* (Goes: Oosterbaan & Le Cointre, 1994), 15-16. I am grateful to Dr. J. Zwemer at Serooskerke (Zeeland) for this reference. Cf. Exodus 12: 23.

225 Kuijt, *Schaduw*, 16.

226 Chr. Hill, *Society and Puritanism in Pre-Revolutionary England* (New York: Schocken Books, 1964), 458-62. The concept of the family as a little church can be found in Roman Catholic circles as well, see, e.g., N. Govers, *Conferenties over de heiliging van het huisgezin* (Roermond-Maaseik: Romen, 1946), 171-78.

227 Cf. Hill, *Society and Puritanism*, Chapter 13.

228 Cf. above, Chapter 4, p. 74.

229 Bernd-Brinkmann, "Wetterlieder," 95-96.

230 Van Deinse, *Uit het land van katoen en heide*, 108. Sometimes the figure of these gable decorations was included in the masonry of the house as well, not coincidentally in the brickwork next to the stable door (ibid.).

231 Heckscher, *Volkskunde*, 345 and Freudenthal, *Feuer*, 33.

232 See above, n. 135.

233 In France the use of "thunderstones" still occurred on the eve of the Second World War, see Filipetti and Trotereau, *Symboles et pratiques rituelles*, 290.

234 For instance, as an antidote against diseases of man and animals, see M.A. van Andel, "De ceraunia of dondersteen in de volksgeneeskunst," *Volkskunde* 38 (1933): 3-13. On the concept of "contagious sympathy", see below, pp. 154-55.

235 Filipetti and Trotereau, *Symboles et pratiques rituelles*, 121. Throughout France thunderstones were traditionally buried under the stable's threshold as well, see ibid., 273.

236 Chr. Blinkenberg, *The Thunderweapon in Religion and Folklore. A Study in Comparative Archaeology* (Cambridge: The University Press, 1911), 1-2. Belief in thunderstones can also be found in non-European cultures, see ibid., 6.

237 Filipetti and Trotereau, *Symboles et pratiques rituelles*, 73, 187.

238 Prinz, *Gewitterelektrizität*, 15, Sébillot, *Folk-lore* 1: 105 and Elsom, *Lightning*, 53. Perhaps the local clergy supposed the peasants invoked St Peter.

239 Filipetti and Trotereau, *Symboles et pratiques rituelles*, 122.

240 Ibid., 73.

241 Rooijakkers, *Rituele repertoires*, 613. To the same end, white crosses were also chalked on the outer walls of farmhouses.

242 Filipetti and Trotereau, *Symboles et pratiques rituelles*, 114-16. The little plant was supposed to protect against disease as well. On Jupiter's beard, see also Bächtold-Stäubli, *Handwörterbuch* 1: 1412.

243 Kück, *Wetterglaube*, 145-46.

244 So, e.g., the learned pharmaceutical chemist Van Barneveld. See H.A.M. Snelders, "The Amsterdam Pharmacist Willem van Barneveld (1747-1826) and the Discovery of Photosynthesis (1778)," *Janus* 68 (1981): 7.

245 Bertucci, "Enlightening Towers," 39. Such leaves and branches were put there also on beds and fields.

246 Prinz, *Gewitterelektrizität*, 17.

247 See above, p. 134.

248 Due to an erroneous association palm twigs in Nordic regions are frequently replaced with those of the box tree. In Central Europe sometimes *Palmkätzschen* (willow catkins) are used, see Kreissl, *Aberglauben*, 16.

249 Called in Dutch: "palmsteken".

250 Post, "Het suizen van een zachte koelte," 185-86. Cf. also W. de Blécourt, *Volksverhalen uit Noord-Brabant* (Utrecht-Antwerp: Spectrum, 1980), 122 (short mention of the making of the sign of the cross with a consecrated box tree twig and holy water at the corners of houses in Son near Eindhoven, November 1888).

251 Van de Weijer, "Een huis als herinneringsmachine," 154.

252 Knippenberg, *Devotionalia* 1 : 205-06 (cf. 208).

253 Filipetti and Trotereau, *Symboles et pratiques rituelles*, 211. On the St John's bunch, cf. also Knippenberg, *Devotionalia* 1: 209.

254 Filipetti and Trotereau, *Symboles et pratiques rituelles*, 248 and Freudenthal, *Feuer*, 27-28 (on the double protective power). See also H. Kleijn, *Planten en hun naam: een botanisch lexicon voor de Lage Landen*, 3rd ed. (Amsterdam: Meulenhoff, 1980), 251 and Knippenberg, *Devotionalia* 1: 206. Of course all kinds of herbs were thrown also in the traditional St John's fires on the 24th of June, but this ritual had not the specific purpose in mind to ward off the harmful consequences of thunderstorms.

255 Kleijn, *Planten*, 159 and Knippenberg, *Devotionalia* 1: 208.

256 See, e.g., Freudenthal, *Feuer*, 32 and Adrian, "Wind und Wetter," 6.

257 Knippenberg, *Devotionalia* 1: 209 (cf. 206).

258 D. Schuth, *Die Farbe Blau. Versuch einer Charakteristik* (Münster: LIT, 1995), 139 and 145. The colour of this plant, dedicated to Donar by the Germanic peoples, was associated with the bluish light of the "sulphuric" lightning. Typically, an ancient German name for ground ivy is *Donnerrebe*.

259 Kleijn, *Planten*, 290.

260 Freudenthal, *Feuer*, 32.

261 Hempler, *Das Gewitter*, 25.

262 On this popular belief, see B. van Meurs, *Verhandelingen over neus, niezen, bliksem, anagram en slaapwandelen* (Culemborg: Blom & Olivierse, 1890), 105. Cf. also Wilson, *The Magical Universe*, 78 (on the same idea in several French regions).
263 See above, n. 261.
264 Freudenthal, *Feuer*, 30, Adrian, "Wind und Wetter," 46-47, Heckscher, *Volkskunde*, 386, and Filipetti and Trotereau, *Symboles et pratiques rituelles*, 216 (leaf of a walnut tree). Branches of a hazelnut tree had a reputation for protecting against a lot of other misfortune as well.
265 Cf. Adrian, "Wind und Wetter," 5-6.
266 Filipetti and Trotereau, *Symboles et pratiques rituelles*, 215.
267 Ibid., 23.
268 W.E. de Perponcher Sedlnitzky, "Het onweer," in idem, *Gedichten* (3 vols.; Utrecht: J. van Schoonhoven, 1800-1802), 1: 120. As shown below, the lofty oak and the sublime thunderstorm formed a rather fixed combination in contemporary art, see Chapter 7, p. 211, 215.
269 Heering and Hochadel, "Epilogue," 270 and Freudenthal, *Feuer*, 45 (also protection against disease, 46).
270 Freudenthal, *Feuer*, 45.
271 Ibid., 46-47. Cf. also De Cock, "Eerbied voor de natuur," 49. Even as late as 1811 this idea was expressed by a peasant from the neighbourhood of Leipzig, see Bächtold-Stäubli, *Handwörterbuch* 1: 1420.
272 Hempler, *Das Gewitter*, 23 and De Cock, "Eerbied voor de natuur," 49.
273 Freudenthal, *Feuer*, 48 and De Cock, "Eerbied voor de natuur," 49.
274 Freudenthal, *Feuer*, 42-43. Reversely, powers of protection against lightning strikes, fires, and (cattle) diseases were ascribed to pieces of wood, carbonised in the great St John's fire on the eve of 24 June and taken home afterwards, see Buisman, *Tussen vroomheid en Verlichting* 1: 148-49. Cf. also Sébillot, *Folk-lore* 1: 106.
275 Freudenthal, *Feuer*, 48.
276 So Kraijenhoff in an explicative note in Jacquet, Proef, 193. On the data mentioned in this paragraph, cf. also J.W. Buisman, "Het onweer in Nederland in godsdienst, wetenschap en kunst, 1752-1830," *Nederlands archief voor kerkgeschiedenis* 78 (1998): 228-42, there 237.
277 Martinet, *Katechismus der natuur* 4: 419 and J. Toaldo, *Verhandeling over het weeder met betrekking tot den landbouw* (...). Vertaald en daarenboven voorzien met eene voorrede van M. van Geuns (Amsterdam-Utrecht: Esveldt and Holtrop, 1778), 73. On Toaldo, see above, Chapter 4, p. 83. According to Martinet, oaks were often vulnerable also because of their height.
278 Heilbron, *Electricity*, 250-60, esp. 257.
279 Cf. Bertucci, "Enlightening Towers," 39: "Enlightened philosophers like Toaldo did not confront such traditions with sarcasm. On the contrary, Toaldo engaged in the reformulation of popular beliefs in terms of Franklinian electrical science."
280 H. Aaftink, *Leven met bliksem*, 4th ed. (Amsterdam: Winterthur Verzekeringen, 1991), par. 30.
281 Even at the end of the eighteenth century an enlightened Catholic priest still deemed it necessary to put up a vigorous fight against the popular interpretation of thunderstorms as *Hexenwetter* (witches' thunderstorms), see Begemann, *Furcht und Angst*, 71.
282 Corbin, *Cloches de la terre*, 111. Sometimes it was thought that witches stole the milk, see Amber J. Rose, "The Witch on the Wall; the Milk-Stealing Witch in Scandinavian Iconography," in *Arv: Nordic Yearbook of Folklore* 71 (2015): 27–43.
283 Chomel, *Algemeen huishoudelijk-, natuur-, zedekundig- en konstwoordenboek* 1: 522.
284 Hempler, *Das Gewitter*, 22. Cf. also Bächtold-Stäubli, *Handwörterbuch* 1: 1421.
285 Kleijn, *Planten*, 290.
286 Freudenthal, *Feuer*, 383. As is known, the ammonia coming from manure generates nitric acid.
287 At its best, water was considered helpful if it had fallen during a thunderstorm. Sometimes such rainwater was given to children in order to diminish their fear of thunder and lightning. This happened, e.g., in several villages on the Lüneburg Heath, see Kück, *Wetterglaube*, 144. In Saxony this kind of rainwater was given to children in order to stimulate the development of speech, see Bächtold-Stäubli, *Handwörterbuch* 3: 830. Both examples are fine specimens of what Frazer called "contagious sympathy," see Frazer, *The Golden Bough*, Chapter 3.

288 See e.g., Adrian, "Wind und Wetter," 4. When the Church of our Lady (*Frauenkirche*) in the Bavarian town of Wasserburg was struck by lightning in 1679 large quantities of milk were used to extinguish the resulting fire. What was left over, was given to the priest. See Steffan, "Wetterabweisende Heiltümer," 25 n. 4. Likewise, the inhabitants of the Zeeland town of Domburg offered all their available milk to fight the fire when the church tower was struck twice on 25 August 1783, see Buisman, *Duizend jaar* 6: 613-14.
289 The Bavarian General Fire Regulations, which simultaneously recommended the installation of lightning rods, are cited by Möhring, *Geschichte des Blitzableiters*, 117-18, albeit without "sympathetic" interpretation. Sometimes salt was also used as a preventive. In the area around Liège, for instance, it was scattered in the four corners of the room, see Sébillot, *Folk-lore* 1: 108. As is known, salt is also added to holy water. The Lüneburg preference for vinegar to salt spring water is mentioned by Zwierlein, *Prometheus Tamed*, 234. In the same manner even as late as 1865 a few domestic servants took vinegar as a prophylactic when a thunderstorm was brewing near Arnhem, see A. Ver Huell, *Het dagboek van Alexander Ver Huell 1860-1865*, ed. Jan Bervoets and Rody Chamuleau (Zutphen: Walburg Pers, 1985), 199. I am grateful to A. Erftemeijer for this reference.
290 Kruythoff, *Geschiedverhaal van de vuur en watersnood*, 61-62.
291 Cf. above, Chapter 3, p. 66.
292 Kittsteiner, *Die Entstehung des modernen Gewissens*, 56 and Wodianka, *Betrachtungen des Todes*, 133.
293 Filipetti and Trotereau, *Symboles et pratiques rituelles*, 108. Cf. also Bächtold-Stäubli, *Handwörterbuch* 1: 1413.
294 Though eggs laid on Good Friday and Holy Saturday were highly valued as well, those laid on Maundy Thursday were favourite. These eggs were called *Antlaßeier* in Germany and Austria. See Freudenthal, *Feuer*, 35 and Kreissl, *Aberglauben*, 31 (an example from 2003!). The word is derived from *Ablass*, indulgence.
295 Freudenthal, *Feuer*, 35.
296 So, e.g., in the Lungau region in central Austria, see Adrian, "Wind und Wetter," 4.
297 Filipetti and Trotereau, *Symboles et pratiques rituelles*, 108. Even as late as 1977 this custom could be found in Normandy (ibid.).
298 Ibid.
299 Freudenthal, *Feuer*, 35.
300 Hager and Heyn, *Drudenhax und Allelujawasser*, 160.
301 Filipetti and Trotereau, *Symboles et pratiques rituelles*, 108.
302 Freudenthal, *Feuer*, 34.
303 Broeckhove, "Devotie" (1959), 181. Cf. above, p. 142.
304 Filipetti and Trotereau, *Symboles et pratiques rituelles*, 129.
305 Pliny, *Naturalis historia* 10: 152, cited by Hempler, *Das Gewitter*, 23.
306 On this meat offering or *Elementefüttern*, see Freudenthal, *Feuer*, 39 and 370-71. The custom was in use in Switzerland, for example.
307 Hentschel, "Man erhitze sich daher zur Zeit der Ungewitter nicht," 11 (ox skulls).
308 Freudenthal, *Feuer*, 26, Kück, *Wetterglaube*, 148-49, and Heckscher, *Volkskunde*, 389.
309 Heering and Hochadel, "Epilogue," 270. The antlers remained attached to the belfry even after the installation of a lightning rod in 1810. Probably they were only removed in 1839. See Chr. Aspalter and A. Tantner, "Ironieverlust und verleugnete Rezeption: Kontroversen um Romantik in Wiener Zeitschriften," in *Paradoxien der Romantik. Gesellschaft, Kultur und Wissenschaft in Wien im frühen 19. Jahrhundert*, ed. Chr. Aspalter et al. (Vienna: WUV, 2006), 65.
310 Freudenthal, *Feuer*, 37.
311 Cf. ibid., 360.
312 Steffan, "Wetterabweisende Heiltümer," 23.
313 Kück, *Wetterglaube*, 148 and Heckscher, *Volkskunde*, 132. Cf. also Bächtold-Stäubli, *Handwörterbuch* 1: 1412-13.
314 Freudenthal, *Feuer*, 21 and 24, respectively.
315 Ibid., 23 and Kück, *Wetterglaube*, 148. Cf. also Bächtold-Stäubli, *Handwörterbuch* 1: 1413.
316 Heckscher, *Volkskunde*, 349.
317 Adrian, "Wind und Wetter," 43 and Freudenthal, *Feuer*, 26.
318 Hempler, *Das Gewitter*, 12.

319 Ibid., 25-26.

320 Ibid. Sometimes it was thought that such a *Blitzkatze* rather prevented a lightning strike, see Bächtold-Stäubli, *Handwörterbuch* 1: 1413. Perhaps a case of imitative magic?

6 - Intermezzo: Electrical Nature? The Animated Nature of Theosophy

1 H. Mandos, "De strijkster van Middelbeers," *Brabants heem* 16 (1964): 13.

2 Cf. P. Fara, "An Attractive Therapy: Animal Magnetism in Eighteenth-Century England," *History of Science* 33 (1995): 127-77, there 135.

3 Obviously the concept of "sympathy" ("com-passion") is used here in a literal, not a metaphorical sense. On mesmerism, see R. Darnton, *Mesmerism and the End of the Enlightenment in France*, 4th ed. (Cambridge, MA: Harvard University Press, 2009; orig. ed. 1968) and the volume *Mesmer et les mesmérismes. Le magnétisme animal en contexte*, ed. B. Belhoste and N. Edelman (Paris: Montreuil Omniscience, 2015), as well as F. Zanetti, *L'Électricité médicale dans la France des Lumières* (Oxford: Voltaire Foundation, 2017), 75-80. Cf. also H. Schott, *Magie der Natur. Historische Variationen über ein Motiv der Heilkunst* (2 vols.; Aachen: Shaker, 2014), esp. 1: 465-94, 519-48 and 602-07.

4 Benz, *Theologie der Elektrizität*, 20. The relationship between magnetism and electricity was confirmed by a publication by the German natural scientist F.U.T. Aepinus (1724-1802), *Tentamen theoriae electricitatis et magnetismi* (St. Petersburg: Typis Academiae Scientiarum, 1759). Cf. also J.H. van Swinden, *Recueil des mémoires sur l'analogie de l'électricité et du magnétisme* (3 vols.; The Hague: Libraires Associés, 1784).

5 Delbourgo, *Amazing Scene*, 165-99, esp. 167, 177-79 and the literature listed there; as well as J. Vijselaar, *De magnetische geest. Het dierlijk magnetisme 1770-1830* (Nijmegen: SUN, 2001), 54.

6 Delbourgo, *Amazing Scene*, 206, Möhring, *Geschichte des Blitzableiters*, 286-87, and Darnton, *Mesmerism*, 62-64.

7 See Vijselaar, *Magnetische geest*, 35 and Heilbron, *Electricity*, 420, respectively.

8 In England mesmerism became the fashion only after it had been discredited in France. It went through a short-lived period of success, primarily in London from 1785 to 1790. See Fara, "An Attractive Therapy," 127 and 134. On the activities of James Graham (1745-1794), the controversial Scottish healer who used electricity as an instrument of sexual therapy, see below, Chapter 7, p. 181.

9 Benz, *Theologie der Elektrizität*, 68. Oetinger had a good grounding in medical matters as well.

10 Ibid., 45-46, 48. See also N. Goodrick-Clarke, "The Esoteric Uses of Electricity: Theologies of Electricity from Swabian Pietism to Ariosophy," *Aries* 4 (2004): 70-73.

11 Benz, *Theologie der Elektrizität*, 37 and Möhring, *Geschichte des Blitzableiters*, 96ff.

12 *Procopii Diviš Theologiae Doctoris et Pastoris zu Prendiz bey Znaim in Mähren längst verlangte Theorie von der meteorologischen Electricité, welche er selbst Magiam naturalem benamt. Sammt einem Anhang vom Gebrauch der elektrischen Gründe zur Chemie.* Zum Druck befördert durch Vorsorge des Württembergischen Superintendenten in Herrenberg, Fridr. Christoph Oetinger (Tübingen: Schramm, 1765). Cf. also J. Haubelt, "Václav Prokop Diviš und Johann Ludwig Fricker," in *Mathesis, Naturphilosophie und Arkanwissenschaft im Umkreis Friedrich Christoph Oetingers (1702-1782)*, ed. S. Holtz, G. Betsch, and E. Zwink (Stuttgart: F. Steiner, 2005), 158ff. Interestingly, Oetinger typified Diviš as the "magician from the East," Emanuel Swedenborg (1688-1772) as the "magician from the North," and himself as the "magician from the South".

13 Briefly mentioned by, e.g., Möhring, *Geschichte des Blitzableiters*, 277 n. 805.

14 On Diviš's "electrical theology", see also Scharfe, *Menschenwerk*, 78-80.

15 Augustin von Balthasar, *Programma quo rector Academiae Gryphiswaldensis Augustinus Balthasar,... sacra pentecostalia rite ac pie celebranda civibus academicis indicit et in hunc usum pias quasdam meditationes de electrisatione spirituali cum iisdem communicat* (Greifswald: Struck, 1745). Cf. also Benz, *Theologie der Elektrizität*, 43, who gives only the shortened title and an incorrect first name. Possibly Von Balthasar's interpretation of Acts 2 was inspired by the so-called "beatification" experiments carried out by the German physicist Georg Matthias Bose (1710-1761): the subject of that experiment was isolated and electrified, so that in the dark a halo became visible around his head, cf. Specht, *Physik als Kunst*, 66.

16 R. Barton, *The Analogy of Divine Wisdom, in the Material, Sensitive, Moral, Civil, and Spiritual System of Things, in Eight Parts* (Dublin: Printed for the author, and sold by George and Alexander Ewing, 1750), 55-66, cited by G.N. Cantor, "The Theological Significance of Ethers," in *Conceptions of Ether. Studies in the History of Ether Theories 1740-1900*, ed. G.N. Cantor and M.J.S. Hodge (Cambridge: Cambridge University Press, 1981), 149.

17 The same text can be found in Matthew 3: 11.

18 Darnton, *Mesmerism*, 36.

19 Delbourgo, *Amazing Scene*, 133 and 142. Importantly, Delbourgo notes that in the American Enlightenment the electrical metaphor had a political rather than a religious character, see ibid., 134.

20 Ibid., 216.

21 Ibid., 133 (on a former slave's account of his conversion experience).

22 Ibid., 159 on "the link between machines and illusions (...) in sceptical accounts of evangelical awakening during the Enlightenment."

23 Cf. ibid., 132 (on the same dilemma in a political sense: can man take his destiny in his own hands or should he obediently accept it?).

24 Ibid., 44.

25 On the rivalry between Gassner and Mesmer, see H.C.E. Midelfort, *Exorcism and Enlightenment. Johann Joseph Gassner and the Demons of Eighteenth-Century Germany* (New Haven: Yale University Press, 2005), 17-20. Gassner's exorcism practices also aroused interest among Protestants, especially among theologians of a pietist orientation such as Oetinger and Johann Kaspar Lavater (1741-1801), see ibid., 102-03.

26 Delbourgo, *Amazing Scene*, 218.

27 Cf. ibid., 137, 280. Apparently, the Protestant separation of nature and grace, creating "conditions that freed natural philosophy as a sphere of inquiry distinct from theology," was not an absolute prerequisite of making scientific progress. Cf. Hart, *Benjamin Franklin*, 166.

28 On this term, see J.W. Buisman, "Inleiding: vreemde Verlichting," in Buisman, *Verlichting in Nederland*, 9-13. Sometimes the term "illuminism" is used as well, to describe a growing interest in the mystical and the occult, see R.H. Kielman, *In het laatste der dagen. Eindtijdverwachting in Nederland op de drempel van de moderne* tijd *(1790-1880)* (Delft: Eburon, 2017), 107.

29 Cf. the title Diviš gave to his manuscript: *Magia naturalis*, and the similar name modern theosophical author Schott chose for his books: *Magie der Natur* (see above, n. 3).

30 G.H.M. Delprat, "Het dierlijk magnetisme toegepast op de wonderen en voorspellingen in de H. Schrift voorkomende," *Bijdragen tot de beoefening en geschiedenis der godgeleerde wetenschappen* 5 (1817): 323-48, cited by Vijselaar, *Magnetische geest*, 454-55.

31 Cf. Vijselaar, *Magnetische geest*, 63-64.

32 W. Bernardi, "The Controversy on Animal Electricity in Eighteenth-Century Italy: Galvani, Volta and Others," *Revue d'histoire des sciences* 54 (2001): 53.

33 Cf. Specht, *Physik als Kunst*, 6. As is known, Volta made corrections to Galvani's ideas: the convulsions of the dead frog proved to be not the result of animal electricity but of the electromotoric force of a variety of metals or other heterogeneous conductors. See ibid., 79ff.

34 J. van Eijnatten, *Hogere sferen. De ideeënwereld van Willem Bilderdijk (1756-1831)* (Hilversum: Verloren, 1998), 391. Otherwise Bilderdijk was sufficiently a theosophist to occasionally accept intermediarity at the same time: in his conception, "ether" was something like the instrument of the Holy Spirit, see ibid. Interestingly, Bilderdijk's ideas bear a striking resemblance to those of the American medical practitioner T. Gale mentioned above, see Delbourgo, *Amazing Scene*, esp. 200-32. On the contemporary discourse on analogy, cf. also Barton's book mentioned in n. 16 above.

35 Even if it often remains unclear at which point the drawing of a "magnetic" parallel merges into a downright metaphorical interpretation of biblical miracles, the latter mindset can unmistakably be found with the enlightened Protestant minister J.A. Uilkens (1772-1825), see Vijselaar, *Magnetische geest*, 454. As a typical member of the Society for Public Welfare and a proponent of the popularisation of the Enlightenment, Uilkens also published a treatise on thunderstorms, now lost, unfortunately: *Verhandeling over de vooroordeelen omtrent het onweder* (Groningen, 1808). In his mind, more or less theosophical and physico-theological ideas went together. See on him IJ. Botke, *Jacobus Albertus Uilkens, 1772-1825* (Groningen: Universiteitsmuseum, 1984).

36 Cf. Gamper, *Elektropoetologie*, 8. Cf. also Fairclough, "Romantic Electricity," 4. The most striking example is of course Mary Shelley's well-known novel *Frankenstein, or the Modern Prometheus* (1818), in which Victor Frankenstein used galvanism to animate his monster.

7 - Thunderstorms and Electricity in Poetry, Music, and Painting

1 Cf. Delbourgo, *Amazing Scene*, 133. Obviously, the electrical metaphor was used for preachers as well as for their audience. Ever since about 1600, thunderstorms have been the symbol of ministry, see J. Schopff, *Wetterglöcklin* (Tübingen: Erhardo Cellio, 1599), cited by Kittsteiner, *Die Entstehung des modernen Gewissens*, 47. Cf. above, Chapter 4, p. 74.

2 Delbourgo, *Amazing Scene*, 234.

3 On the ideas of Shebbeare and Rabiqueau, see ibid., 117. The versatile English physician William Stukeley also stressed the importance of electricity in human sexual life, see Fairclough, *Literature, Electricity and Politics*, 81-82, 85. Fairclough's book gives a fine overview of electricity and eroticism in eighteenth-century England (77-119), including the activity of James Graham (83-103) (cf. above, Chapter 6, n. 8). On the sexualisation of electricity in France, see F. Zanetti, *L'Électricité médicale*, 57 and M. Delon, "Electriser, un mot d'ordre au siècle des Lumières," *Revue des sciences humaines* 281 (2006): 39-51.

4 This play was commented upon at length by B. Peperkamp, "'Bald wird komen das Feur'. Over de representatie van natuurwetenschappelijke en medische kennis in de klucht *De electriciteit; of Pefroen, met het schaepshoofd ge-electriseerd* (1746)," *Gewina* 29 (2006): 269-92, there 283-84. Some scientific terms acquired an ambiguous, erotic connotation here (289).

5 Well before Mesmer, Albrecht von Haller had already published his thoughts on the amazing and important role the finger purportedly played in the transfer of "electricity". In April 1745, in an article translated into English, he wrote: "Could one believe that a lady's finger (...) should send forth flashes of true lightening?" See A. von Haller, "An Historical Account of the Wonderful Discoveries Made in Germany etc. Concerning Electricity," *The Gentleman's Magazine* 15 (1745): 194. Cf. also Delbourgo, *Amazing Scene*, 117 and 313 n. 35. In Paape's story the finger seems to be more than a simple phallic symbol, just because of the addition of the phrase "a beatific electrical shock."

6 (G. Paape), *Vrolijke reis van een Engelschman, door Holland* (The Hague: J.C. Leeuwestyn, 1796), 49, 50 and 53, respectively. Cf. also Altena, *Gerrit Paape*, 537. Possibly Paape is referring here to the idea of Franklin's colleague Kinnersley, who suggested that the piercing look of an attractive lady contained real "electric fire," see Fairclough, *Literature, Electricity and Politics*, 58, 80.

7 Frictional electricity charged with a comparable erotic tension can also be found in "Klingsohrs Märchen," part of the unfinished novel *Heinrich von Ofterdingen*, written by Novalis (1772-1801) in 1800, as well as in the debut novel of Achim von Arnim (1781-1831), *Hollins Liebeleben* (Göttingen: Dieterich, 1802). See Gamper, *Elektropoetologie*, 147 and 236, respectively. Mesmerist influence of some kind is unmistakably perceptible in the oeuvres of both Novalis and Von Arnim.

8 See P. Bertucci, "Domestic Spectacles: Electrical Instruments between Business and Conversation," in Bensaude-Vincent and Blondel, *Science and Spectacle*, 82-83.

9 Delbourgo, *Amazing Scene*, 117.

10 Chr. Martin, "Femmes, orages, tempêtes dans le roman français du XVIIIe siècle," in Le Roy Ladurie et al., *L'Événement climatique*, 372. Thunder and lightning were already used as a sexual metaphor in Antiquity, see Beller, *Jupiter tonans*, 132ff.

11 On De Sade's use of atmospheric electricity as a metaphor for sexual arousal, cf. Goubier-Robert, "De la fulgurisation sadienne," 424.

12 Martin, "Femmes, orages, tempêtes," 369 and Goubier-Robert, "De la fulgurisation sadienne," 420.

13 Martin, "Femmes, orages, tempêtes," 364.

14 Van Eijnatten, *Hogere sferen*, 394-96, 398 and Vijselaar, *Magnetische geest*, 452.

15 *The Letters and Works of Lady Mary Wortley Montagu*, ed. Lord Wharncliffe (2 vols.; Paris: A. and W. Galignani, 1837), 1: 325-26.

16 W. Bilderdijk, "Op het huwelijksbed van een bruid en bruidegom, in hun eersten huwlijksnacht door den bliksem getroffen" (1796), in W. Bilderdijk, *Dichtwerken*, ed. I. da Costa (16 vols.; Haarlem: A.C. Kruseman, 1856-1859), 13: 66. Cf. also Van Eijnatten, *Hogere sferen*, 197.

17 A fitting example in this context is the story of Dido and Aeneas, to be found in the fourth book of Virgil's *Aeneid*. The lovers take shelter from a thunderstorm in a cave and seal their alliance in that safe spot.

18 M. Thalmann, *Der Trivialroman des 18. Jahrhunderts und der romantische Roman. Ein Beitrag zur Entwicklungsgeschichte der Geheimbundmystik* (Berlin: E. Ebering, 1923), 15 and 33, respectively.

19 See the poem "Bruiloftsbrand" written in 1822 (after the Danish of A.W. Schack Staffeldt [1769-1826]), published in *Dichtwerken* 1: 303-08. Bilderdijk imitated another, similar epitaph of Lady Montagu in 1808 or 1809: "Grafschrift op twee gelieven, door den donder getroffen," ibid., 13: 186. The theme made itself felt long after, for even as late as 1963 the Belgian writer Marnix Gijsen (1899-1984) wrote a similar poem on the consummation of a marriage during a nocturnal thunderstorm, albeit with more explicit sexual metaphorical overtones, see M. Gijsen, "The House by the Leaning Tree" (strophe VII), in G. Komrij, *De Nederlandse poëzie van de negentiende en twintigste eeuw in duizend en enige gedichten* (Amsterdam: Bert Bakker, 1996), 660.

20 Kittsteiner, *Die Entstehung des modernen Gewissens*, 54. According to the *Wetterpredigt* cited here the bride also died in this accident.

21 R. Feith, *Julia*. Opnieuw uitgegeven en van inleiding, aantekeningen en bijlagen voorzien door J.J. Kloek en A.N. Paasman (The Hague: Nijhoff, 1982), 99 and Leemans and Johannes, *Worm en donder*, 378. J. Bellamy, "Het onweeder," in J. Bellamy, *Proeven voor het verstand, den smaak en het hart* (Dordrecht: De Leeuw and Krap, 1790), 174-76.

22 "symphonische Begleiterscheinung der Gefühlshöhen." With these words Thalmann characterises the relationship between thunderstorms and the theme of love in Romantic literature. In her opinion, a direct reference to the sexual is absent here. See Thalmann, *Der Trivialroman*, 184.

23 See Hochadel, *Öffentliche Wissenschaft*, 47. Cf. also the title of D. Sturkenboom's book, *De elektrieke kus. Over vrouwen, fysica en vriendschap in de 18de en 19de eeuw. Het verhaal van het Natuurkundig genootschap der dames in Middelburg* (Amsterdam: Augustus, 2004). That this gallant experiment was really conducted in Middelburg cannot be proved beyond doubt, however. On the other hand, the "electric kiss" obviously kept its metaphorical meaning in a literary context, e.g., in a passage from Von Arnim's novel *Hollins Liebeleben*. The principal character kisses his love Maria goodnight there, strikingly enough against a pane of glass – an element that both separates (because electricity insulating) and connects (because transparent). See Gamper, *Elektropoetologie*, 239. The "electric kiss" experiment also occurs in the burlesque mentioned in n. 4 above.

24 Bose did not, however, mention that the woman suffered pain because of this experiment, too. On the other hand it was the lovelorn man who was punished here, dependent as he was on the woman. Cf. Specht, *Physik als Kunst*, 65 n. 170.

25 See above, Chapter 3, n. 32.

26 Goubier-Robert, "De la fulgurisation sadienne," 427.

27 As shown by the watercolour on p. 187, electricity, sexuality, and politics were combined here in a quite remarkable way.

28 Delbourgo, *Amazing Scene*, 160.

29 "Urbild und Patron der Monarchen." H.-W. Jäger, *Politische Metaphorik im Jakobinismus und im Vormärz* (Stuttgart: J.B. Metzler, 1971), 58. Cf. also Beller, *Jupiter tonans*, 67-131.

30 Cf. Fuhrmeister, "Eripuit Caelo Fulmen," 145 and 149, as well as A. Demandt, *Metaphern für Geschichte. Sprachbilder und Gleichnisse im historisch-politischen Denken* (Munich: Beck, 1978), 136. It was by a very mixed company of politicians that the "electric fluid" was used in Great Britain as a metaphor for very quick, "contagious" communication, that was nearly imperceptible and therefore almost uncontrollable. See Fairclough, "Romantic Electricity" and, in more detail, her *Literature, Electricity and Politics*, 121-69. Atmospheric electricity sometimes stood symbol for revolution, see ibid., 207.

31 J. von der Thüsen, *Het verlangen naar huivering. Over het sublieme, het wrede en het unheimliche. Essays* (Amsterdam: Querido, 1997), 50.

32 Fuhrmeister, "Eripuit Caelo Fulmen," 158.

33 Jäger, *Politische Metaphorik*, 32.

34 In this context Von der Thüsen alludes to the German revolutionary Georg Forster (1754-1794), who considered revolution a stream of lava against which every resistance proved futile. See Von der Thüsen, *Het verlangen*, 64.

35 K. Griewank, *Der neuzeitliche Revolutionsbegriff. Entstehung und Entwicklung*, 2nd ed. (Frankfurt a.M.: Europäische Verlagsanstalt, 1969), 184 (orig. ed. Weimar: Böhlau, 1955).

36 Cf. Jäger, *Politische Metaphorik*, 29-32 and Von der Thüsen, *Het verlangen*, 61 (in combination with the volcano, inundation, and earthquake metaphor).

37 Jäger, *Politische Metaphorik*, 30. For Hölderlin the French Revolution was a monumental shock, comparable to a thunderstorm, see B. Böschenstein, "Quatre sommets de la poésie allemande des orages: Klopstock – Goethe – Kleist – Hölderlin," in Le Roy Ladurie et al., *L'Événement climatique*, 307. A more detailed account is given in B. Böschenstein, *"Frucht des Gewitters". Hölderlins Dionysos als Gott der Revolution* (Frankfurt a.M.: Insel Verlag, 1989).

38 C. Rogge, *Tafereel van de geschiedenis der jongste omwenteling in de Vereenigde Nederlanden* (Amsterdam: J. Allart, 1796), 1-2. Cf. also S. Vuyk, *De verdraagzame gemeente van vrije christenen. Remonstranten op de bres voor de Bataafse Republiek (1780-1800)* (Amsterdam: De Bataafsche Leeuw, 1995), 58.

39 A. Principato, "Tourmente ou déluge: métaphores révolutionnaires chez Chateaubriand," in Le Roy Ladurie et al., *L'Événement climatique*, 474.

40 This idea can be found, e.g., with the German author C.L. Wienbarg (1802-1872), see C.L. Wienbarg, *Ästhetische Feldzüge*, ed. by W. Dietze (Berlin-Weimar: Aufbau-Verlag, 1964), 76, cited by Jäger, *Politische Metaphorik*, 136 n. 122.

41 Goubier-Robert, "De la fulgurisation sadienne," 425.

42 "een gloênde bliksemstraal ging uit van 's Dwinglands troon | (...) Wat Godheid, riep men, zal 't vernielende onweêr keeren? | Toen trad de Vryheid uit; Myn Burgers, sprak zy: "staa" | En keert die donderbui met uwe zwaanenveeren." *Hollandsche historische courant*, 8 March 1785. Cf. also Altena, "Die holländische Himmelsstürmer," 193 and Altena, *Gerrit Paape*, 192.

43 Cf., for instance, the anonymous, passionately Patriotic pamphlet written by the Amsterdam bookseller Cornelis Lorié, *Heuglyk vooruitzigt, of vaderlandsche droom* [n.p., n.p., 1784], 5-6.

44 Of course, the lightning conductor did not stop the thunder rolling, but as that noise is innocuous in itself, to conclude that the thunder metaphor was used by preference to indicate uncontrollable, incisive political changes seems something of an overinterpretation. Cf. Briese, *Macht der Metaphern*, 70. However, the precise relationship between thunder and lightning was still not very clear in early electricity research.

45 Cf. Jäger, *Politische Metaphorik*, 136 n. 122.

46 Cf. Altena, *Gerrit Paape*, 195.

47 P.W. van Wissing, *Stokebrand Janus 1787. Opkomst en ondergang van een achttiende-eeuws satirisch politiek-literair weekblad* (Nijmegen: Vantilt, 2003), 313-16.

48 Delbourgo, *Amazing Scene*, 160. Especially in revolutionary circles the metaphor of an electrical chain was sometimes used as a symbol of conspiracy, see ibid., 160, 162-63.

49 Jäger, *Politische Metaphorik*, 32 (a defence of the sudden and incisive phenomenon of revolution against those who believed only in age-long natural development) and 214. Cf. also Möhring, *Geschichte des Blitzableiters*, 195 n. 563.

50 The metaphor of bottled-up energy can be found, e.g., with the German writer Heinrich von Kleist (1777-1811) in his description of Mirabeau (1749-1791), see Gamper, *Elektropoetologie*, 208.

51 Streng, *'Het is thans zeer briljant'*, 159 and Leemans and Johannes, *Worm en donder*, 379.

52 Von der Thüsen, *Het verlangen*, 63. Cf. also J. von der Thüsen, "'Die Lava der Revolution fließt majestätisch'. Vulkanische Metaphorik zur Zeit der französischen Revolution," *Francia* 23, no. 2 (1996): 128.

53 Demandt, *Metaphern für Geschichte*, 135 (with the inclusion of a reference to Homer's *Iliad*). Conversely, Seneca had already tried to clarify the hard noise of thunder by comparing it to the roar of military missiles being launched, see Engammare, "Tonnerre de Dieu," 174. In the eighteenth century, too, thunderstorms were sometimes depicted in military metaphors, see Mitchell, "The Politics of Experiment," 312.

54 Demandt, *Metaphern für Geschichte*, 135. The military metaphor can be found in the Bible as well, e.g., in Nahum 2: 4: "The chariots rage in the streets, they rush to and fro through the squares; they gleam like torches, they dart like lightning."

55 See the references to Vondel in the entry "kartouw" in the *Woordenboek der Nederlandsche taal*: https://gtb.ivdnt.org/search/?owner=wnt (accessed 9 June 2021).

56 J.D. Macquet, *De vier jaergetyden* (Leiden: Johannes van Zyp, 1767), 33 and 42 (orig. ed. Leiden: Johannes le Mair, 1761).

57 "De donder gromt met forsch geklater | En rolt langs dek en boorden heen; | Tot bergen zwelt het vliegend water | En barst tot op den boôm van een." H. Tollens, "Tafereel van den Vierdaagschen Zeeslag," in idem, *Gezamenlijke dichtwerken* (3 vols.; Leeuwarden: Suringar, 1855), 1: 13. Cf. G.W. Huygens, *Hendrik Tollens. De dichter van de burgerij. Een biografie en een tijdsbeeld* (Rotterdam-The Hague: Nijgh & Van Ditmar, 1972), 115-16.

58 "Seit jenem Tage | Grollt über dieser Ebne unverrückt | Die Schlacht, mit immer reger Wuth, wie ein | Gewitter, zwischen waldgekrönter Felsen Gipfeln | Geklemmt. Als ich mit den Ätoliern gestern | Erschien, der unsern Reihen zu verstärken, | Schlug sie mit Donnerkrachen eben ein, | Als wollte sie den ganzen Griechenstamm | Bis auf den Grund, die Wüthende, zerspalten." H. von Kleist, *Penthesilea. Ein Trauerspiel* (Tübingen: Im Verlage der Cottaischen Buchhandlung, 1808), 8. On the role played by thunder and lightning in Kleist's plays, see also M.M. Tatar, *Spellbound. Studies on Mesmerism and Literature* (Princeton: Princeton University Press, 1978), 82-120 and Böschenstein, "Quatre sommets," 302-03. In Kleist's works lightning is the attribute of the war god. The English translation is taken from Heinrich von Kleist, *Penthesilea: A Tragic Drama* (trans. Joel Agee). 2nd ed. (New York: Harper Perennial, 2000) (orig. ed. 1998).

59 P. van Hemert, *Redevoering over het verhevene* (Amsterdam: M. Schalekamp, 1804), 5-6.

60 A short poem by H.D. van Hoorn in the *Liber amicorum* of Jacob Kantelaar (Royal Library The Hague, inventory no. 133 L 24, 30 October 1796). Cited by S. Vuyk, *Jacob Kantelaar. Veelzijdig verlicht verliezer 1759-1821* (Zwolle: Waanders, 2005), 135.

61 J.J.A. Goeverneur, *Gezamenlijke gedichten en rijmen* (Groningen: J.B. Wolters, 1874), 330.

62 Cf. Böschenstein, "Quatre sommets," 304-07.

63 Cf. Gamper, *Elektropoetologie*, 62.

64 B. Specht, "Die Elektrisierung der Aufklärung. Kontexte, Metaphorisierungen und Funktionen der Elektrizität im Wissenssystem des späten 18. Jahrhunderts," in *Die Sachen der Aufklärung. Beiträge zur DGEJ-Jahrestagung 2010 in Halle a.d. Saale*, ed. F. Berndt and D. Fulda (Hamburg: F. Meiner, 2012), 443.

65 G.P.M. Knuvelder, *Handboek tot de geschiedenis der Nederlandse letterkunde*, 5th ed. (4 vols.; Den Bosch: Malmberg, 1970-1976), 3: 36.

66 Of the many studies on the subject I mention here only a few: Specht, *Physik als Kunst*, Chapters 4 and 5; J. Barkhoff, *Magnetische Fiktionen. Literarisierung des Mesmerismus in der Romantik* (Stuttgart-Weimar: J.B. Metzler, 1995), passim and *The Oxford Handbook of Samuel Taylor Coleridge*, ed. F. Burwick (Oxford-New York: Oxford University Press, 2009), 649-50. Byron too, often made an intentional use of the ambiguity of the concept of electricity both in its literal, physiological, and its metaphorical meanings, see Fairclough, "Romantic Electricity," 10.

67 On the complex backdrop of this concept, especially in the works of Novalis and Friedrich Schlegel (1772-1829), see L. Pikulik, *Frühromantik. Epoche – Werke – Wirkung* (Munich: Beck, 1992), 98-100.

68 "Der Witz ist geistige Elektrizität – dazu sind feste Körper nötig." Novalis, *Schriften. Die Werke Friedrich von Hardenbergs*, ed. P. Kluckhohn and R. Samuel, 2nd ed. (6 vols.; Stuttgart: Kohlhammer, 1960-2006), 2: 621.

69 "wer ahndet hier nicht Seelenmagnetism?" Ibid. and Gamper, *Elektropoetologie*, 139. As observed in n. 30 above, Mary Fairclough in particular has pointed to the "contagious" and very rapid communication implied in the electricity metaphor, see Fairclough, *Literature, Electricity and Politics*, 26 and passim.

70 Cf. Specht, *Physik als Kunst*, 261 n. 125.

71 Jean Paul, *Ächten Samlung meiner besten Bonmots* (1784), cited by Gamper, *Elektropoetologie*, 243.

72 *The Nightingale*, 11 June 1796: "At his birth, the electric Spark, Nature's first and most attenuated Agent, sympathising with the fine aether of his Intellect, vibrated and coalesced, and unfolding the Laboratory of her Works, taught him whence her Lightnings flash, and why her Thunders roll," cited by Delbourgo, *Amazing Scene*, 143-44.

73 "als ein zweiter Prometheus den elektrischen Funken vom Himmel holen", cited by Specht, "Elektrisierung der Aufklärung," 443. On the points of comparison between electricity and artistic genius, see also Specht, *Physik als Kunst*, 60-61. Similarly, the English writer Mary Wollstonecraft (1759-1797) spoke of "the electric spark of genius," see Fairclough, *Literature, Electricity and Politics*, 153.

74 P. Frantz, "Diderot, Dorval, tempête et création," in Le Roy Ladurie et al., *L'Événement climatique*, 435.

75 Cf. P.B. Shelley, "A Defence of Poetry" (1821), in *The Complete Works of Percy Bisshe Shelley*, ed. R. Ingpen and W.E. Peck (10 vols.; New York: Scribner's Sons, 1926-1930), 7: 122, cited by Gilmore, *Aesthetic Materialism*, 72.

76 "Doch uns gebührt es, unter Gottes Gewittern, | Ihr Dichter! mit entblößtem Haupte zu stehen, | Des Vaters Strahl, ihn selbst, mit eigner Hand | Zu fassen und dem Volk ins Lied | Gehüllt die himmlische Gabe zu reichen." F. Hölderlin, *Sämtliche Werke*, ed. by J. Schmidt, vol. 1 *Gedichte* (Frankfurt a. M.: Deutscher Klassiker Verlag, 1992), 240, cited by Böschenstein, "Quatre sommets," 307. The English translation is taken from *Poems of Friedrich Hölderlin* (trans. James Mitchell) (San Francisco: Ithuriel's Spear, 2007). The passage is from "Wie wenn am Feiertage" (written 1799-1800). Cf. also Briese, *Macht der Metaphern*, 51. It is a translation of a fragment of Pindar's First Pythian Ode, see Beller, *Jupiter tonans*, 165-66.

77 Demandt, *Metaphern für Geschichte*, 9.

78 https://www.dwds.de/r/plot/?view=1&corpus=dta%2Bdwds&norm=date%2Bclass&smooth=spline&-genres=0&grand=1&slice=10&prune=0.05&window=3&wbase=0&logavg=0&logscale=0&xrange=1-750%3A1830&q1=Gedankenblitz (accessed 23 February 2026).

79 This applies to the inner struggle of both writers and the characters they created, in the Netherlands, Germany, and elsewhere. Cf. A. Vasak, "*Héloïse* et *Werther, Sturm und Drang*: comment la tempête, en entrant dans nos cœurs, nous a donné le monde," *Ethnologie française* 39(2009), 677-85. In some cases echoes of a pietistic past can be heard, especially when the ongoing battle between God and devil for the human soul was presented as a thunderstorm in the conscience, cf. Wodianka, *Betrachtungen des Todes*, 135.

80 Cf. Specht, *Physik als Kunst*, 282ff., 412.

81 Ibid., 357-58, 410.

82 Cf. L. Pikulik, *Romantik als Ungenügen an der Normalität. Am Beispiel Tiecks, Hoffmanns, Eichendorffs* (Frankfurt a. M.: Suhrkamp, 1979), 353 and 361.

83 Cf. ibid., 255.

84 "In trüber Schwüle liegt die Welt gefangen." Ibid., 351 (with a reference to Eichendorff's Collected Works). Cf. also ibid., 218. The quotation is taken from Eichendorff's novel *Dichter und ihre Gesellen. Novelle*, published at Christmas 1833.

85 F. Jakob, *Die Orgel und das Donnergrollen* (Männedorf: Orgelbau Kuhn, 1976), 16.

86 F.D. Zeiler, "Johann Carl Röhner (1774-1837). Een muzikant met hartstocht," *Zwols historisch tijdschrift* 11 (1994): 17.

87 "'t Gezis des blikzems scheen te zuizen; | 't Geknap van het elektriesch vuur; | Scheen door 't geschok der orgelbuizen, | Te vonken aan den tempelmuur." J. le Francq van Berkhey, *Dicht-verrukking op den alom beroemden organist den heer abt M. Vogler [...]. By het nabootsen van donder en onweder, op het orgel van St. Pieterskerk der stad Leyden, en de Groote Kerk te Haarlem* (Leiden: T. Koet, 1789), n.p. In a note, Berkhey added the announcement that Vogler would imitate a sea battle on the organ of Amsterdam's Old Church.

88 Bosch, *En nooit meer oude Psalmen zingen*, 143.

89 Jakob, *Die Orgel*, 24.

90 H. de la Motte-Haber, *Musik und Natur. Naturanschauung und musikalische Poetik* (Laaber: Laaber Verlag, 2000), 131. In this context, Strasbourg historian Hartmann rightly characterises Baroque music as "musique imitative et descriptive," see P. Hartmann, "Du baroque au classicisme: orages et tempêtes dans l'œuvre de Joseph Haydn," in Le Roy Ladurie et al., *L'Événement climatique*, 310-22, there 316. On the initial absence of temporal perspective, cf. ibid., 313.

91 On this development, cf. J. Neubauer, *The Emancipation of Music from Language. Departure from Mimesis in Eighteenth-Century Aesthetics* (New Haven-London: Yale University Press, 1986), 74-75 (without mention of Vogler). Vogler is a transitional figure, whose connections with the imitative tradition are stressed with some justification by some scholars, see, e.g., Schleuning's book, which is also important for sketching the wider context: P. Schleuning, *Die Sprache der Natur: Natur in der Musik des 18. Jahrhunderts* (Stuttgart-Weimar: J.B. Metzler, 1998), 195.

92 Jakob, *Die Orgel*, 5 and 15.

93 Zeiler, "Johann Carl Röhner," 15-16.

94 Beethoven finished this symphony in Vienna in 1807-1808. The opening performance was on 22 December 1808. This piece of music is analysed and contextualised at length (and somewhat speculatively) by R. Schmenner, *Die Pastorale. Beethoven, das Gewitter und der Blitzableiter* (Kassel: Bärenreiter, 1998). Previously, in 1801-1802, Beethoven had composed a sonata entitled "Der Sturm" afterwards. Cf. R. Kleinertz, "Beethovens 'Sturm'-Sonate op. 31,2 d-moll im Kontext von Sturm- und Gewittermusiken des 18. und frühen 19. Jahrhunderts," in Braungart and Büttner, *Wind und Wetter*, 69-86.

95 Cf. above, Chapter 4, n. 147. On Ruppe D. van Heuvel and F.D. Zeiler, see "Biografieën," in *Hef aan! Bataaf! Beschouwingen over muziek en muziekleven in Nederland omstreeks 1795*, ed. P. van Reijen (Alphen a.d. Rijn: Canaletto, [1997]), 169-71.

96 Hartmann, "Du baroque au classicisme," 320. Haydn influenced Beethoven here, see ibid., 321. Beethoven's *Pastoral* was heralded by a piece by Justin Heinrich Knecht (1752-1817), an unknown composer who dedicated his symphony *Le portrait musical de la nature* (1784) to Vogler and already practised such a succession of scenes, see C. Bockmaier, *Entfesselte Natur in der Musik des achtzehnten Jahrhunderts* (Tutzing: H. Schneider, 1992), 267-91, esp. 267-75.

97 Cf. the description of Elector Carl Theodor's innovating Mannheim orchestra as a force of nature by musician C.F.D. Schubart (1739-1791): "Its forte is like thunder, its crescendo a cataract." Though he added: "its diminuendo a clear brook babbling in the distance, its piano a spring breeze." See C.F.D. Schubart, *Ideen zu einer Ästhetik der Tonkunst*, ed. Ludwig Schubart (Leipzig: Reclam, 1977), 122 (written ca. 1785, but first published posthumously in Vienna: J.V. Degen, 1806).

98 Beethoven's composition is considered bombastic by many a biographer, e.g., by W. Kinderman, *Beethoven* (Berkeley-Los Angeles: University of California Press, 1995), 172.

99 What is meant here is a recitative from the First Part, sung by the angel Raphael, based on Genesis 1: 7.

100 The classic study on the sublime (which has its origins in rhetoric) is S.H. Monk, *The Sublime. A Study of Critical Theories in XVIIIth-Century England* (Ann Arbor: University of Michigan Press, 1960) (orig. ed. New York: Modern Language Association of America, 1935). On the musical sublime, see C.A.W. Brillenburg Wurth, *The Musically Sublime. Indeterminacy, Infinity, Irresolvability* (New York: Fordham University Press, 2009) and M. Stanyon, *Resounding the Sublime. Music in English and German Literature and Aesthetic Theory, 1670-1850* (Philadelphia: University of Pennsylvania Press, 2021).

101 "Aandoeningen van verhevenheid zyn, eenigermaate, onafscheidelyk van ongemeen sterk, of geweldig, geluid. Het slaan eens trommels, het gebalder eens kanons, het geraas des donders, het geweldig uitbarsten eens vuurbraakenden bergs, heeft iets verheevens: (...) eenzaamheid vermeerdert die aandoening grootlyks." "Aanmerkingen over de bronnen van grootsche denkbeelden," *Algemeene vaderlandsche letter-oefeningen* 5, pt. 2 (1783): 165 (translation from an unidentified English source).

102 E. Burke, *A Philosophical Enquiry into the Origin of Our Ideas of the Sublime and Beautiful*, 5th ed. (London: J. Dodsley, 1767), 151.

103 Cf. Von der Thüsen, *Het verlangen*, 22. As early as 1688 the sensation of "delightful horror" was already described by the English author John Dennis (1657-1734), see R. Doran, *The Theory of the Sublime from Longinus to Kant* (Cambridge: Cambridge University Press, 2015), 125. Joseph Addison (1672-1719), too, noted that "hideous objects" were sometimes regarded "at the same time as dreadful and harmless," see *Spectator* no. 418 (30 June 1712), cited by H. Conrad, *Die literarische Angst. Das schreckliche in Schauerromantik und Detektivgeschichte* (Düsseldorf: Bertelsmann, 1974), 45. On the seventeenth-century sublime, cf. below, Chapter 8, n. 24.

104 During the same period the simple, rigid contrast between the beautiful and the ugly was further complicated by the introduction of the "picturesque", see H. Dieckmann, "Das Abscheuliche und das Schreckliche in der Kunsttheorie des 18. Jahrhunderts," in *Die nicht mehr schönen Künste. Grenzphänomene des Ästhetischen*, ed. H.R. Jauß (Munich: Fink, 1968), 292-93.

105 Characteristically, the German philosopher F.H. Jacobi (1743-1819) blamed Enlightenment science for "extirpating all miracles" by focusing on intermediary causes only and consequently losing sight of final causes. Hence, in his opinion any numinous sense of the sublime in nature was eroded. See Weigl, "Entzauberung der Natur," 33 (with inclusion of references to Jacobi's writings).

106 R. Alewyn, "Die Lust an der Angst," in idem, *Probleme und Gestalten. Essays* (Frankfurt a.M.: Insel, 1974), 316. Alewyn's ideas are discussed below, pp. 229ff. On the impact of the lightning rod on the aestheticising of thunderstorms, cf. E. Weigl, *Instrumente der Neuzeit. Die Entdeckung der modernen Wirklichkeit* (Stuttgart: J.B. Metzler, 1990), 186-91.

107 Thalmann, *Der Trivialroman*, 218. Here I pass over the categories of the ugly and the grotesque, see Von der Thüsen, *Het verlangen*, 68.

108 A short comparison of the eerie (*Unheimliche*) and the sublime is made by Von der Thüsen, *Het verlangen*, 70-71.

109 With the notable exceptions of the recent publications cited in Chapter 4, n. 14 and n. 100 of the present Chapter, the theme of the musical sublime still seems to be a rather understudied phenomenon. For instance, the valuable doctoral dissertation by Chr. Madelein, *Juigchen in den adel der menschlijke natuur. Het verhevene in de Nederlanden (1770-1830)* (Ghent: Academia Press, 2010) ignores the musical aspect of the sublime, whereas Neubauer, *Emancipation of Music*, pays no attention to the category of the sublime itself.

110 Pliny the Elder, *Naturalis historia*, book 35, cited by L. Marin, "Sur une tour de Babel dans un tableau de Poussin," in *Du sublime*, ed. J.-F. Courtine et al. (Paris: Belin, 1988), 239 (cf. also 257). According to Pliny, only the famous Greek painter Apelles (4th century B.C.) had succeeded in adequately depicting this "unpaintable" subject.

111 Cf. Thalmann, *Der Trivialroman*, 218. See also G. Zacharias-Langhans, *Der unheimliche Roman um 1800* (Bonn: R. Schwarzbold, 1968), 53-54. The restless, flickering light is characteristic of other contemporary poems as well, such as those of Ludwig Hölty (1748-1776), see E. Albert, *Das Naturgefühl L.H.Chr. Höltys* (Dortmund: Ruhfus, 1910), 108 (repr. Hildesheim: Gerstenberg Verlag, 1978). On this theme, see also J. Christen, "L'orage nocturne, entre sublime et frénésie," *Revue des Sciences Humaines* 281, no. 1 (January-March 2006): 67-80.

112 C. Reichler, "Air, orages et météores au tournant du XVIIIe siècle," in Le Roy Ladurie et al., *L'Événement climatique*, 153.

113 W. Trautwein, *Erlesene Angst. Schauerliteratur im 18. und 19. Jahrhundert* (Munich: C. Hanser, 1980), 38.

114 Post, *Het land*, 260.

115 Ibid., 260-61. Comparable dialectics between light and darkness, as well as between loud noise and silence are found in Post's description of a thunderstorm at the Rhine, included in her book *Voor eenzamen* (Amsterdam: J. Allart, 1789), 39-40.

116 At any rate no translation of Burke's treatise into Dutch was published during the eighteenth century, see Chr. Madelein, "Kant in de fout? Het verhevene in de Nederlanden," *De achttiende eeuw* 39 (2007): 82. Post may have read the translation into German: Edmund Burke, *Philosophische Untersuchungen über den Ursprung unsrer Begriffe vom Erhabnen und Schönen. Nach der fünften Englischen Ausgabe* (Riga: J.F. Hartknoch, 1773).

117 Interestingly, Burke also considered a sudden transition from light to darkness a source of sublime experience, mentioning lightning as an example. See Burke, *A Philosophical Enquiry*, 145. Cf. also Madelein, *Juigchen*, 77.

118 The difference between literature and painting does not seem to be absolute here. This nuance should be made to the valuable contribution of M. Mathijsen, "Geschreven schilderijen. Toenadering tussen de Nederlandse literatuur en schilderkunst van de Romantiek," in *Meesters van de Romantiek. Nederlandse kunstenaars 1800-1850*, ed. R. de Leeuw, J. Reynaerts, and B. Tempel (Zwolle: Waanders, 2005), 196.

119 B. Tempel, "Nederlandse Romantiek. Een plaatsbepaling," in De Leeuw et al., *Meesters van de Romantiek*, 74. Surprisingly, here Tempel mentions only the Hegelian inspiration, ignoring its Burkean equivalent which it may have superseded. Cf. ibid., n. 8 (p. 308).

120 Reichler, "Air, orages et météores," 151.
121 Juel is to be considered a mentor to the German Romantic painter Caspar David Friedrich (1774-1840), who lived in Copenhagen from 1794 to 1798, see K. Monrad et al., *The Golden Age of Danish Painting* (New York: Hudson Hills Press, 1993), 138.
122 As a keen hunter, though, Tavenraat had a predilection for painting animals, see W. Busch, D. Jackson, and J. Reynaerts, *De Romantiek in het noorden. Van Friedrich tot Turner* (Zwolle: WBooks, 2017), 194 (entry by R. Schenk).
123 Thacker, *The Wildness Pleases*, 133.
124 Obviously this kind of painting has a much longer history. It may even be regarded as the very cradle of thunderstorm painting in the Netherlands, see R. Falkenburg, "Onweer bij Jan van Goyen. Artistieke wedijver en de markt voor het Hollandse landschap in de 17de eeuw," in *Natuur en landschap in de Nederlandse kunst 1500-1850*, ed. R. Falkenburg et al. (Zwolle: Waanders, 1998), 117-61, esp. 141 [*Nederlands kunsthistorisch jaarboek* no. 48, 1998].
125 Cf. M. Bertsch, "Zwischen Erhabenheitsästhetik und dokumentarischem Anspruch. Das Motiv des Schiffbruchs im ausgehenden 18. Jahrhundert," in *Entfesselte Natur. Das Bild der Katastrophe seit 1600*, ed. M. Bertsch and J. Trempler (Petersberg: Imhof, 2018), 60-62.
126 Generally speaking, all weather changes constitute a popular theme in Romantic painting, in the Netherlands as well as elsewhere, see J. Reynaerts, "De hand van God. Romantiek in de Nederlandse landschapschilderkunst," in De Leeuw et al., *Meesters van de Romantiek*, 140.
127 On this painting, see Monrad et al., *Golden Age*, 138 and Busch et al., *De Romantiek in het noorden*, 130-33 (entry by J. Cox).
128 According to M. Pinault-Sørensen the rainbow which was frequently depicted in such paintings is to be considered the symbol of serenity, see M. Pinault- Sørensen, "Orages et tempêtes. Peinture et dessin, deuxième moitié du XVIIIe siècle et début du xixe siècle," in Le Roy Ladurie et al., *L'Événement climatique*, 266.
129 I disregard here the ancient harmonious tradition of the pastoral and the idyll, which also remained in vogue during the entire eighteenth century, cf. Thacker, *The Wildness Pleases*, 192.
130 Brillenburg Wurth, *The Musically Sublime*, 5.
131 "Selbstbehauptung der Vernunft," see H. Böhme, *Natur und Subjekt* (Frankfurt a. M.: Suhrkamp, 1988), 394. The sublime is interpreted by Böhme as a sort of secular substitute of the apocalyptical, see ibid., 392-93.
132 Van Hemert, *Redevoering*, 41 and Th. van Swinderen, "Redevoering over het verhevene in de natuur," *De recensent ook der recensenten* 4 (1809): 252. Van Hemert's treatise was the result of a lecture given before the learned society *Felix Meritis* in Amsterdam on 1 February 1804. He emphasised the ethical aspect of the sublime even more than Kant did.
133 Van Hemert, *Redevoering*, 9.
134 Ibid., 28-29.
135 Ibid., 46. Cf. also Van Swinderen, "Redevoering," 239.
136 Van Hemert, *Redevoering*, 31. Cf. the title of Madelein's doctoral dissertation, see above, n. 109. See also S. Contarini, "In de spiegel van de natuur. Het verhevene bij Paulus van Hemert," in *Esthetica tussen klassiek en romantiek*, ed. H. Krop and P. Sonderen (Rotterdam: Faculteit der Wijsbegeerte van de Erasmus Universiteit Rotterdam, 1992), 97-106.
137 Schröder used the same metaphor and identical phrasings in "Redevoering over het gevoel van het verhevene," *Algemeene vaderlandsche letter-oefeningen* (1810), pt. 2: 133.
138 Van Hemert, *Redevoering*, 22.

By Way of Conclusion

1 Cf. above, Chapter 4, p. 83.
2 Briese, *Macht der Metaphern*, 8ff.
3 Von der Thüsen, *Het verlangen*, 8.
4 Ibid., 18-19.

5 Begemann, *Furcht und Angst*, 6ff. and passim.

6 "ritualisierte Wiederholung historisch ausgeständener Ängste," cited by J. Schönert, "Behaglicher Schauer und distanzierter Schrecken. Zur Situation von Schauerroman und Schauererzählung im literarischen Leben der Biedermeierzeit," in *Literatur in der sozialen Bewegung. Aufsätze und Forschungsberichte zum 19. Jahrhundert*, ed. A. Martino (Tübingen: Niemeyer, 1977), 33 n. 18.

7 A.H. Lemaire, *Filosofie van het landschap*, 6th ed. (Baarn: Ambo, 1996), 41 (cf. 37) (orig. ed. Bilthoven: Ambo, 1970).

8 Von der Thüsen, *Het verlangen*, 8 and 18.

9 Cf. Begemann, *Furcht und Angst*, 108.

10 In 1790 Kant observed that the view of bold, overhanging, and seemingly threatening rocks, thunderstorms, lightning, volcanoes and so on, became more attractive as it was more terrifying, provided one was in a position of safety, see Kant, *Kritik der Urteilskraft*. In: *Werke* 10: par. 28. As early as the seventeenth century the learned Jesuit D. Bouhours (1628-1702) also realised that the (relative) absence of danger was an essential precondition for aesthetic pleasure, see M. Labussière, "Naufrages avec sauveteur chez Prévost: théories esthétiques et tempêtes romanesques à l'ombre du *Suave mari magno*," in Le Roy Ladurie et al., *L'Événement climatique*, 403.

11 So, following in the footsteps of Kant, e.g., Van Swinderen, "Redevoering," 247.

12 Hochadel and Heering, "Introduction: Revisiting an Invisible Technology," 8.

13 Heering and Hochadel, "Epilogue," 271 and Schmenner, *Die Pastorale*.

14 Alewyn, "Die Lust an der Angst," in idem, *Probleme und Gestalten*, 307-30.

15 Ibid., 329.

16 Cf. C. Zelle, *'Angenehmes Grauen'. Literaturhistorische Beiträge zur Ästhetik des Schrecklichen im achtzehnten Jahrhundert* (Hamburg: F. Meiner, 1987).

17 Ibid., xxv and Begemann, *Furcht und Angst*, 11.

18 Begemann, *Furcht und Angst*, 11.

19 Cf. Zelle, *'Angenehmes Grauen,'* 347.

20 Cf. Delbourgo, *Amazing Scene*, 62.

21 Von der Thüsen, *Het verlangen*, 36. Here I disagree with McCallam, who thinks volcanoes lost a lot of their awesome quality in the eighteenth century because by then "the industrial forge had (...) replaced the mythological forge" (McCallam, *Volcanoes in Eighteenth-Century Europe*, 116). Not fortuitously, the reflections on the theatrically sublime which could be occasioned by this natural violence, sought their way to experimental science rather than to dramatic art, partly owing to the relative safety provided by precautions and scale models, cf. above, Chapter 7, p. 201 and below, n. 22.

22 J. von der Thüsen, *Schönheit und Schrecken der Vulkane. Zur Kulturgeschichte des Vulkanismus* (Darmstadt: Wissenschaftliche Buchgesellschaft, 2008), 111-26 and McCallam, *Volcanoes in Eighteenth-Century Europe*, 110-12. The artificial volcano in the English garden at Wörlitz in eastern Germany was built at the behest of Leopold III Friedrich Franz of Anhalt-Dessau (1740-1817) next to the Villa Hamilton on the Insel Stein (Stone Island). The curious fake Vesuvius may be considered an exemplar of contemporary efforts to elicit a sublime experience in a safe manner, reaching a degree of theatricality impossible to attain in using small objects such as thunderhouses, perhaps with the exception of Cuthbertson's "electrical" kites. On a much smaller volcano scale model, lit by electricity, see F. Fraunberger, *Elektrische Spielereien im Barock und Rokoko* (Munich: Oldenbourg; Düsseldorf: VDI-Verlag, 1967), 36.

23 Möhring, *Geschichte des Blitzableiters*, 302.

24 See Falkenburg, "Onweer bij Jan van Goyen," 117-61 (cf. above, Chapter 7, n. 124) and S. Bussels and B. van Oostveld, *The Sublime in the Visual Culture of the Seventeenth-Century Dutch Republic* (London-New York: Routledge 2023). Seventeenth-century representations of thunderstorms differed from medieval depictions because early modern painters tried to reproduce lightning faithfully rather than schematically, see Winter-Fritzsche, *'Naturgetreue' Blitzdarstellungen*, 5. On the other hand the Romantic pictorial portrayal of thunderstorms had its limits as well, because generally only a few side flashes were depicted, see M. Stromp et al., "How Realistic are Painted Lightnings? Quantitative Comparison of the Morphology of Painted and Real Lightnings: a Psychophysical Approach," *Proceedings of the Royal Society* A, vol. 474 (June 2018), Issue 2214. According to the Scottish astronomer James Nasmyth (1808-1890),

J.M.W. Turner (1775-1851) was the only painter who really depicted lightning in all its capriciousness, see A. Drahos, *Orages et tempêtes, volcans et glaciers. Les peintres et les sciences de la terre aux XVIIIe et XIXe siècles* (Paris: Hazan, 2014), 37.

25 Begemann, *Furcht und Angst*, 91. Cf. above, p. 13, 87.

26 Buisman, *Tussen vroomheid en Verlichting* 1: 17ff., 262-63.

27 Hochadel, *Öffentliche Wissenschaft*, 141.

28 Delbourgo, *Amazing Scene*, 83. See above, Chapter 5, n. 188.

29 Cf. Weigl, *Instrumente der Neuzeit*, 200.

30 Pikulik, *Romantik als Ungenügen an der Normalität*, 83.

31 Cf. the analysis of Vasak and others, see above Chapter 7, n. 79.

32 Cf. A. Le Brun, *Les châteaux de la subversion* (Paris: Pauvert, 1982), 103-04.

33 H. Walpole, *Essay on Modern Gardening*, included in vol. 4 of his *Anecdotes of Painting in England* (4 vols.; Strawberry Hill: T. Farmer, 1762-1780). Cited by Helmut J. Schneider, "Naturerfahrung und Idylle in der deutschen Aufklärung," in *Erforschung der deutschen Aufklärung*, ed. P. Pütz (Königstein: Athenäum, 1980), 301.

34 A.N. Paasman, *J.F. Martinet. Een Zutphens filosoof in de achttiende eeuw* (Zutphen: Van Someren, 1971), 55. The lightning rod's breakthrough in the Netherlands in the late 1770s is described above, Chapter 2, pp. 48ff.

35 M. Batt, *The Treatment of Nature in German Literature from Günther to the Appearance of Goethe's Werther* (Port Washington, NY-London: Kennikat, 1969), 104 (repr. of the orig. ed. Chicago: University of Chicago Press, 1902).

36 Thacker, *The Wildness Pleases*, 77.

37 Schneider, "Naturerfahrung und Idylle," 293-94.

38 Cf. ibid., 295.

39 Ibid., 297. See on Brockes above, Chapter 4, p. 106.

40 P. Van Tieghem, *Le sentiment de la nature dans le préromantisme européen* (Paris: A.G. Nizet, 1960), 62.

41 Ibid., 79.

42 Cf. Schneider, "Naturerfahrung und Idylle," 300 and the still useful book by W. Flemming, *Der Wandel des deutschen Naturgefühls vom 15. zum 18. Jahrhundert* (Halle a.d. Saale: Niemeyer, 1931), 87. As a matter of course this close attention to detail brought out the sublime in full relief.

43 Batt, *The Treatment of Nature*, 105.

44 Voltaire, *Poème sur le désastre de Lisbonne ou examen de cet axiome: 'tout est bien'*. In: idem, *Mélanges*, ed. J. Van den Heuvel (Paris: Gallimard, 1961), 307.

45 Cf. Flemming, *Der Wandel des deutschen Naturgefühls*, 93 and Lemaire, *Filosofie van het landschap*, 38.

46 According to Lemaire, *Filosofie van het landschap*, 95, this identification is not even possible, since science has disenchanted nature. Literary historian Brandt Corstius, however, does not go beyond a description here and only points to the chaos of emotions and inner unrest which is characteristic of the new experience of nature, see J.C. Brandt Corstius, "Verschijnselen in onze literatuur aan het einde van de 18de eeuw in verband met veranderende opvattingen omtrent geloof en natuur," *De nieuwe taalgids* 44 (1951): 248.

47 Even French historian J. Christen, who – erroneously, in my opinion – supposes that the sublime gave way to the secular despair of the eerie because the lightning had become an immanent phenomenon, implicitly confirms the congeniality of the sublime and the religious experiences. Cf. Christen, "L'orage nocturne," 76. See also K. Richter, *Literatur und Naturwissenschaft. Eine Studie zur Lyrik der Aufklärung* (Munich: Fink, 1972), 147 (the example of Klopstock), Weber, *Vom Gottesgericht*, 306, and (shortly) above, p. 209.

48 J.J. Hemmer, *Kurzer begriff und nuzen der Wetterleiter* (Düsseldorf: Stahl, 1782), 3, cited by Dross, "Gottes elektrischer Wille?," 281.

49 Van Swinderen, "Redevoering," 342.

50 See above, p. 106.

51 Cf. the section "Het bos, het graf, de worm en de ratelende donder" [The Forest, the Grave, and the Rolling Peal of Thunder] in Leemans en Johannes, *Worm en donder*, 469-71. The authors do not discuss the impact of the invention of the lightning rod, however.

52 See, e.g., A. Heumakers, *De esthetische revolutie. Hoe Verlichting en Romantiek de Kunst uitvonden* (Amsterdam: Boom, 2015), 73-74 and Vuyk, *Jacob Kantelaar*, 44.

53 Cf. D.G. Charlton, *New Images of the Natural in France. A Study in European Cultural History 1750-1800* (Cambridge: Cambridge University Press, 1984), 62. Cf. tevens Weber, *Vom Gottesgericht*, 219.
54 Charlton, *New Images*, 43.
55 Thacker, *The Wildness Pleases*, 12.
56 Lemaire, *Filosofie van het landschap*, 44.
57 Ibid., 99.
58 This dissimilarity can be seen clearly, e.g., in Oetinger's opinions mentioned in Chapter 6. Though he came close to some kind of pantheist emanationism, the Württemberg pietist finally maintained a distinction between Creator and creature, see Benz, *Theologie der Elektrizität*, 52 and 61.
59 Cf. Von der Thüsen, *Het verlangen*, 39 (on Hölderlin's unfinished drama *The Death of Empedocles*).
60 Cf. Flemming, *Der Wandel des deutschen Naturgefühls*, 85-86.
61 Van Tieghem, *Le sentiment de la nature*, 226.
62 Cf. Flemming, *Der Wandel des deutschen Naturgefühls*, 96.
63 Ibid., 91.
64 Thalmann, *Der Trivialroman*, 191-93.
65 Reynaerts, "De hand van God," 131.
66 Knuvelder, *Handboek* 3: 39.
67 When the notion of "electrical fluid" gradually fell into disuse after 1830 – in the wake of the scientific discoveries of Michael Faraday (1791-1867) –, the character of "electrical" metaphors changed accordingly, see Fairclough, *Literature, Electricity and Politics*, 223-34. Here I pass over the important issue of the historicising of (natural) disasters, which resulted in the genesis of the new artistic genre of depicting catastrophes, see J. Trempler, "Entfesselte Natur und Geschichte. Zur Interpretation und Darstellung von Katastrophen im 18. Jahrhundert," in Bertsch and Trempler, *Entfesselte Natur*, 11-19, as well as Mercier-Faivre and Thomas, *L'invention de la catastrophe*, préface.
68 Heumakers, *Esthetische revolutie*, 353-54.
69 "seelische Ausnahmezustände," term borrowed from Specht, *Physik als Kunst*, 419.
70 Cf. Reed, *Romantic Weather*, 58-59.
71 Cf. Thalmann, *Der Trivialroman*, 13.
72 It seems wise to avoid extremes here. On the one hand, e.g., music historian Bockmaier totally ignores the impact of the lightning rod, which is curious given all his attention to the numerous eighteenth-century thunder-music compositions, cf. Bockmaier, *Entfesselte Natur*, 354. On the other, French historian Julien Christen overestimates the discovery of the electric character of thunderstorms (and the invention of the lightning rod as a consequence) by claiming nothing less than a transformation of the concept of the universe and making every artistic reference to the numinous a void symbol, see Christen, "L'orage nocturne," 76.
73 See above, Introduction, p. 11.
74 "Ich weiß nichts Rührenderes als solche Szenen," Wieland to publisher and natural scientist Johann Heinrich Merck (1741-1791), letter dated 12 April 1778, cited by Flemming, *Der Wandel des deutschen Naturgefühls*, 85.

Bibliography

Manuscripts and Archives

Royal Library, The Hague, inv. no. 133 L 24, 30 October 1796 (Liber amicorum of Jacob Kantelaar, short poem by H.D. van Hoorn).

National Archives, The Hague, Archief Hof van Holland, no. 5527.

National Archives, The Hague, VOC-archief, entry 1.04.02, inv. no. 64, 25 March-18 October 1771.

Primary Literature

"Aanmerkingen over de bronnen van grootsche denkbeelden," *Algemeene vaderlandsche letter- oefeningen* 5, pt. 2 (1783): 164-66 (translation of an unidentified English source).

Abbildung und Beschreibung der gemeinnützlichen Hauptstände. Regensburg: C. Weigel, 1698.

Aepinus, F.U.T., *Tentamen theoriae electricitatis et magnetismi*. Saint Petersburg: Typis Academiae Scientiarum, 1759.

Ahlwardt, P., *Bronto-theologie, oder vernünftige und theologische Betrachtungen über den Blitz und Donner*. Greifswald-Leipzig: J.J. Weitbrecht, 1745.

Ahlwardt, P., *Bronto-theologie, of redelyke en theologische bespiegelingen over het onweder*. 2nd ed. Rotterdam: Ph. Losel, 1760 (orig. ed. Delft: A. Voorstad, 1750).

Algemeene vaderlandsche letter-oefeningen.

Almanach der musen voor het jaar 1795. Amsterdam: W. Brave, n.d.

Almanach der natuur, voor iedere maand van het jaar. Amsterdam: J. de Jongh, 1789.

Alphen, H. van, "Het onweder," in *Kleine gedigten voor kinderen*, 2e vervolg. Utrecht: J. van Terveen, 1782.

(Alphen, H. van), *De christelyke spectator. In twaalf vertoogen*. The Hague: J. Thierrij and C. Mensing, 1799.

Angelis, Augustinus de, *Lectiones meteorologicae*. Rome: Corbi, 1663.

Arnim, A. von, *Hollins Liebeleben*. Göttingen: Dieterich, 1802.

Arnim, A. von, "Wetterprophet, " in A. von Arnim and C. Brentano, *Des Knaben Wunderhorn: alte deutsche Lieder* (3 vols.; Stuttgart: Reclam, 1987), 3: Chapter 201 (orig. ed. Heidelberg: Mohr and Zimmer, 1806).

Balthasar, Augustin von, *Programma quo rector Academiae Gryphiswaldensis Augustinus Balthasar,... sacra pentecostalia rite ac pie celebranda civibus academicis indicit et in hunc usum pias quasdam meditationes de electrisatione spirituali cum iisdem communicat*. Greifswald: Struck, 1745.

Bapst von Rochlitz, M., *Wetterspiegel, oder nützliche unnd eigentliche Erzelung*. Leipzig: Berwaldt, 1589.

Barneveld, W. van, *Het onweersvuur in zyne richting en uitwerksels nagespoort*. Amsterdam: P. Hayman, 1780.

Barton, R., *The Analogy of Divine Wisdom, in the Material, Sensitive, Moral, Civil, and Spiritual System of Things, in Eight Parts*. Dublin: Printed for the author, and sold by George and Alexander Ewing, 1750.

"Bedenkingen over de vrees die veele menschen voor onweders hebben," *Hedendaagsche vaderlandsche letter-oefeningen* 4, pt. 2 (1775): 241-44 (by E.).

Bellamy, J., *Proeven voor het verstand, den smaak en het hart*. Dordrecht: De Leeuw and Krap, 1790.

Bellamy, J., "Het onweeder," in idem, *Proeven voor het verstand, den smaak en het hart*, 174-76.

Bertholon, P., *De l'électricité des végétaux*. Lyon: Bernuset, 1783.

"Beschouwing der natuur na eene zwaare dondervlaag," *De menschenvriend* 9 (1796): 249-56.

Bilderdijk, W., *Dichtwerken*, ed. I. da Costa. 16 vols., Haarlem: A.C. Kruseman, 1856-1859.

Bilderdijk, W., "Op het huwelijksbed van een bruid en bruidegom, in hun eersten huwlijksnacht door den bliksem getroffen" (1796), in W. Bilderdijk, *Dichtwerken*, ed. I. da Costa (16 vols.; Haarlem: A.C. Kruseman, 1856-1859), 13: 66.

Bilderdijk, W., "Grafschrift op twee gelieven, door den donder getroffen" (1808-1809), in idem, *Dichtwerken* 13: 186.

Bilderdijk, W., "Bruiloftsbrand" (1822), in idem, *Dichtwerken* 1: 303-08.

Blussé, A., *De beste opvoeding der jeugd, ten nutte van dit gemeenebest; en bespiegelingen bij een onweder*. Dordrecht: Blussé, 1776.

Boddaert, P. Czn., *Stichtelyke gedichten*. 5th ed. 2 vols., Leiden: Luchtmans, 1765 (orig. ed. 1726).

Bossuet, Jacques-Bénigne, *Œuvres complètes*, ed. F. Lachat. 31 vols., Paris: Vivès, 1862–1879.

"Brief over den elektrikalen vlieger, geschreven door B. Franklin, October 19, 1752," *Uitgezogte verhandelingen uit de nieuwste werken van de sociëteiten der wetenschappen in Europa en van andere geleerde mannen*, vol. 1 (1757): 617-18.

"Brief van den heer Franklin, betreffende het voorkomen van ongelukken, die het weerlicht veroorzaakt," *Vaderlandsche letter-oefeningen* 6, pt. 2 (1766): 29-31.

Brockes, B.H., *Rahts-Herrn der Stadt Hamburg, verdeutschte Grund-Sätze der Welt-Weisheit, des Herrn Abts Genest : nebst verschiedenen eigenen theils Physicalischen theils Moralischen Gedichten, als des Irdischen Vergnügens in Gott Dritter Theil*. Hamburg: König and Richter, 1728.

Burke, E., *A Philosophical Enquiry into the Origin of Our Ideas of the Sublime and Beautiful*. 5th ed. London: J. Dodsley, 1767.

Burke, E., *Philosophische Untersuchungen über den Ursprung unsrer Begriffe vom Erhabnen und Schönen. Nach der fünften englischen Ausgabe*. Riga: J.F. Hartknoch, 1773.

Buys, E., *Nieuw en volkomen woordenboek van konsten en weetenschappen*. 10 vols., Amsterdam: S.J. Baalde, 1769-1778.

Captain Cook's Journal during his First Voyage round the World, made in H.M. Bark Endeavour, 1768-71: a Literal Transcription of the Original Mss., ed. W.J.L. Wharton. Cambridge: Cambridge University Press, 2014.

Chomel, N., *Algemeen huishoudelijk-, natuur-, zedekundig- en konstwoordenboek*. 2e druk geheel verbetert en meer als de helfte vermeerdert door J.A. de Chalmot. 7 vols., Leiden-Leeuwarden: J. le Mair and J.A. de Chalmot, 1778 and *Vervolg*. 9 vols., Kampen-Amsterdam: J.A. de Chalmot and J. Yntema, 1786-1793.

"De christen bij het onweder," *Nieuwe vaderlandsche bibliotheek van wetenschap, kunst en smaak* 2, pt. 2 (1798): 432.

Cuthbertson, J., *Algemeene eigenschappen van de electriciteit, onderrichting van de werktuigen en het neemen van proeven in dezelve.* 2nd ed. 2 vols., Amsterdam: Pieter Hayman 1782 (orig. ed. Amsterdam 1776).

Delprat, G.H.M., "Het dierlijk magnetisme toegepast op de wonderen en voorspellingen in de H. Schrift voorkomende," *Bijdrage tot de beoefening en geschiedenis der godgeleerde wetenschappen* 5 (1817): 323-48.

Description de l'Égypte, ou recueil des observations et des recherches qui ont été faites en Egypte pendant l'expédition de l'armée française. 23 vols., Paris: Imprimerie impériale, 1809-1828.

Derham, W., *Physico-Theology, or a Demonstration of the Being and Attributes of God, from His Works of Creation.* 2 vols., London: A. Strahan; T. Cadell Jr. and W. Davies, 1798 (orig. ed. 1713).

Dichtoefeningen van het genootschap onder de zinspreuk: Kunstliefde spaart geen vlyt. The Hague: Van Cleef, 1813.

Dichtstukken van het Haagsch genootschap onder de spreuk: Kunstliefde spaart geen vlijt, en bekroonde keurstoffen. 3 vols., The Hague: J. Thierry and C. Mensing, 1791-1799.

Eichendorff, Joseph von, *Dichter und ihre Gesellen. Novelle.* Berlin: Duncker and Humblot, 1833.

Eijk, J. van, *Liederen voor den landman.* 3 pieces, Amsterdam: J. Allart, 1794-1796.

Eijk, J. van, *Het dorp Eerenhoef, uit het diepst verval tot voorige welvaard hersteld.* Rotterdam: Cornel, 1816.

Elikink, B., "Bij gelegenheid van een onweer," in idem, *Stichtelijke gezangen* (Amsterdam: Johannes Covens Jr, 1769), 171-74.

Engelberts, E.M., *Bespiegelingen over de vier getijden des jaars.* Amsterdam: Yntema and Tieboel, 1769.

Engelman, J., "Proeven over den aart der werkinge van de electriciteit," in *Verhandelingen van de Hollandsche Maatschappy van Wetenschappen*, vol. 1 (1754): 267-328.

Feenstra, P.W., *Leerrede en gebed, uitgesproken op den dank- en bededag 22 februari 1809, in de kerk der doopsgezinden op het Cingel te Sneek.* Sneek: C. van Gorcum, 1809.

Feith, R., *Julia*, ed. J.J. Kloek and A.N. Paasman. The Hague: Nijhoff, 1982 (orig. ed. Leiden: C. van Hoogeveen, 1783).

R. Feith, "Het onweder. Eene cantate," in *Oden en gedichten* (3 vols.; Zwolle: H. Doyer, 1824), 2: 48 and 51 (originally published in 1787 in *Tael- en dichtlievende oefeningen van 'Kunst wordt door arbeid verkreegen'* [7 vols.; Leiden: C. van Hoogeveen Jr., 1775-1790], 5: 167-68).

Francq van Berkhey, J. le, *Natuurlyke historie van Holland.* 9 vols., Amsterdam: Yntema and Tieboel, 1769-1811.

Francq van Berkhey, J. le, *Dicht-verrukking op den alom beroemden organist den heer abt M. Vogler [...]. By het nabootsen van donder en onweder, op het orgel van St. Pieterskerk der stad Leyden, en de Groote Kerk te Haarlem.* Leiden: T. Koet, 1789.

Franklin, B., *New Experiments and Observations.* London: D. Henry and R. Cave, 1754.

Franklin, B., *The Papers of Benjamin Franklin*, ed. L.W. Labaree et al. 42 vols., New Haven: Yale University Press, 1959- .

Gallitzin, D.A. von, *Sendschreiben an die Kaiserliche Akademie der Wissenschaften zu St. Petersburg über einige Gegenstände der Electricität.* Münster-Leipzig: Ph.H. Perrenon, 1780.

"Gedagten by het eindigen eens zomer-onweders," *Hedendaagsche vaderlandsche letter-oefeningen* 7, pt. 2 (1778): 314-18 (by J.V.P.).

Gerricken, J.P. (= J.P. Gericke), *De Heere in 't vuur of een leerreden gedaan op palm sondag 1750. Na het afbranden van (...) de kerke St. Michaels tot Hamburg (...).* Uit het Hoogduits vertaald. Amsterdam: D. Weege, 1750.

Gijsen, M., "The House by the Leaning Tree" (strophe VII), in G. Komrij, *De Nederlandse poëzie van de negentiende en twintigste eeuw in duizend en enige gedichten* (Amsterdam: Bert Bakker, 1996), 660.

"Gods grootheid bij het onweder," in *Almanach der musen voor het jaar 1795* (Amsterdam: W. Brave, n.d.): 12-14 (by Hn.).

"Godvrugtige bespiegelingen over de wolken," *De denker* 10 (1773): 241-48.

Goethe, J.W. von, *Maximen und Reflexionen* no. 9. Hamburg Edition, vol. 12 (Hamburg: C. Wegner, 1953), 365ff.

Goeverneur, J.J.A., *Gezamenlijke gedichten en rijmen.* Groningen: J.B. Wolters, 1874.

Goeze, J.A.E., "Beredeneerde, stichtlijke en proefondervindlijke bedenkingen over de vrees voor het onweder van donder en bliksem," *Nieuwe vaderlandsche bibliotheek van wetenschap, kunst en smaak* 5, pt. 2 (1801): 189-98.

Goslin, J.S., "Bij een onweder," in *Dichtoefeningen van het genootschap onder de zinspreuk: Kunstliefde spaart geen vlyt* (The Hague: Van Cleef, 1813), 176-77.

Gütle, J.C., *Allgemeine Sicherheitsregeln bei Gewittern.* Nuremberg: Schneider and Weigel, 1805.

Haastert, I. van, "Ode aan de wolken," in *Proeven van poëtische mengelstoffen, door het dichtlievend kunstgenootschap onder de spreuk: Kunstliefde spaart geen vlijt* (13 vols.; Leiden: C. van Hoogeveen Jr., 1774-1789), 11: 38-41.

Haller, A. von, "Gedanken über Vernunft, Aberglauben und Unglauben," (1729) in idem, *Die Alpen und andere Gedichte* (Stuttgart: Reclam, 1994), 3-22.

Haller, A. von, "An Historical Account of the Wonderful Discoveries Made in Germany etc. Concerning Electricity," *The Gentleman's Magazine* 15 (1745): 193-97.

Helmers, J.F., "De nacht. Ode," in *Kleine dichterlijke handschriften* (20 vols.; Amsterdam: Uylenbroek, 1788-1809), 1: 23-32.

Hemert, P. van, *Redevoering over het verhevene.* Amsterdam: M. Schalekamp, 1804.

Hemmer, J.J., *Kurzer begriff und nuzen der Wetterleiter.* Düsseldorf: Stahl, 1782.

's Hertogenbossche courant, 11 August 1772.

Hesselink, G.J., *Dissertatio physica inauguralis de montium ignivomorum, terraeque motuum origine, et cognatione, ex stupendis illorum effectibus colligenda.* Groningen: Jacob Dikema, 1778.

Hesselink, G.J., "Bericht wegens het onweder van 24 february 1804," *Algemeene vaderlandsche letter-oefeningen* (1804), pt. 2: 168-73.

Hoek, H., *Het heilzaam oogmerk van Godt in zyne oordeel-gerichten door vuur (...). Uitgesproken op den plegtigen boet- vast- en beededag den 19 Maart 1750.* Uit het Hoogduits vertaald. Haarlem: J. Nieuwenhuyzen and J. Huytink, 1750.

Hölderlin, F., *Sämtliche Werke*, ed. by J. Schmidt. Vol. 1 *Gedichte.* Frankfurt a.M.: Deutscher Klassiker Verlag, 1992.

Hollandsche historische courant, 8 March 1785.

Hottinger, S., *Brontologia physico-sacra.* Basel: n.p., 1698.

Hunolt, F., *Christliche Sitten-Lehr über die Evangelische Wahrheiten.* 6 vols., Cologne: Huisch, 1740-1748.

Immerzeel jr., J., "Bij een onweder," in *Nederlandsche muzen-almanak 1819* (Rotterdam: J. Immerzeel jr., n.d.): 46.

Ingenhousz, J., *Verzameling van verhandelingen over verschillende natuurkundige onderwerpen.* Vertaald door J. van Breda. 2 vols., The Hague: Isaac van Cleef, 1785.

Jacobi, J.G., *Taschenbuch von J.G. Jacobi und seinen Freunden*. Königsberg-Leipzig: F. Nicolovius, 1795.

Jacquet, L.S., *Proef eener elektrische natuurkunde*, na 't Fransch van den abt Jacquet. Vertaald en uitgebreid door C.R.Th. Kraijenhoff (Utrecht-Harderwijk: B. Wild and J. van Kasteel, 1783), 297-310 (appendix no. 3).

Jean Paul (Johann Paul Friedrich Richter), *Ächten Samlung meiner besten Bonmots*. n.p., 1784.

Kant, I., *Kritik der Urteilskraft* (1790). In: *Werke*, ed. W. Weischedel. 12 vols., Frankfurt a.M.: Suhrkamp, 1974-1977.

Kate, J.J.L. ten, *De planeeten: een gedicht*. 2nd ed. The Hague: Nijhoff, 1873 (orig. ed. 1869).

Kircher, A., *Sphinx mystagoga, sive diatribe hieroglyphica, qua Mumiae, ex Memphiticis Pyramidum adytis erutae, et non ita pridem in Galliam transmissae*. Amsterdam: Johannes Janssonius van Waesberghe, 1676.

Kleist, H. von, *Penthesilea. Ein Trauerspiel*. Tübingen: Im Verlage der Cottaischen Buchhandlung, 1808.

Klopstock, F.G., "Die Frühlingsfeier" (1759), in: idem, *Oden* (Stuttgart: Reclam, 1994), 59-67.

Komrij, G., *De Nederlandse poëzie van de negentiende en twintigste eeuw in duizend en enige gedichten*. Amsterdam: Bert Bakker, 1996.

Koopmans, R.K., *Hulde aan Gerrit Hesselink*. Amsterdam: J.W. Yntema, 1812.

"Kort begrip der natuurkunde," *Vaderlandsche letter-oefeningen* 7, pt. 2 (1767): 561-662.

Kraijenhoff, C.R.Th., "Onzydige beproeving en verdediging der metaalen blixem-afleideren," *Algemeen magazyn van wetenschap, konst en smaak* 1, pt. 2 (1785): 937-77.

Kraijenhoff, C.R.Th., *Voorschriften wegens het plaatsen van bliksemafleiders aan 's rijks gebouwen, behoorende onder de administratie van het Departement van Oorlog, en wel bijzonder de buskruid-magazijnen*. The Hague: s.n., 1819.

Kraijenhoff, C.R.Th., *Handleiding tot het stellen van bliksem-afleiders*. Nijmegen: Vieweg, 1836.

Kraus, J., *Gewitterkatechismus, oder Unterricht über Blitz und Donner*. Augsburg: N. Doll, 1797 (orig. ed. 1794).

Kruythoff, W.A.T., *Geschiedverhaal van de vuur en watersnood, de stad Edam meermaals, dog byzonder op den 24 february 1602 overgekomen. Nevens eene korte beschryving van dezelve stad. Met aantekeningen en byvoegsels vermeerderd en verbeterd door Mr. Jan Leoninus, oud schepen der gemelde stad*. Edam: H. Rolff, 1781.

Lier, J. van, *Oudheidkundige brieven*. The Hague: Van Thol, 1760.

Loo, J. van, *Leerredenen*. 4 vols., Utrecht: A. van Paddenburg, 1784-1789.

Lorié, C., *Heuglyk vooruitzigt, of vaderlandsche droom*. n.p., n.p. [1784].

Louwerse, P., *Jan Pietersz. Sweelinck of de toonkunst in Nederland in de 17e eeuw*. Schiedam: Roelants, 1879.

Lugt, H., *Onderwijs in de eerste beginsels der electriciteit*. Westzaandam: H. van Aken, 1797.

Lugt, H., *De theorie der electriciteit, rustende op proefondervindlyke waarheden*. Westzaandam: H. van Aken, 1797.

Lugt, H., *Korte beschrijving eener electriseer-machiene met geïsoleerde wrijvers*. 2 vols., Rotterdam: n.p., 1801-1803.

Macquet, J.D., *De vier jaergetyden*. Leiden: Johannes van Zyp, 1767 (orig. ed. Leiden: Johannes le Mair, 1761).

Martinet, J.F., "Verbaazende uitwerksels van een onweder," in *Verhandelingen uitgegeeven door de Hollandsche Maatschappye der Weetenschappen*, vol. 14, section Communications (*Berichten*) (1773): 33-45.

Martinet, J.F., *Katechismus der natuur.* 4 vols., Amsterdam: J. Allart, 1777-1779.

Martinet, J.F., *Kleine katechismus der natuur voor kinderen.* Amsterdam: J. Allart, 1779.

Martinet, J.F., *Huisboek voor vaderlandsche huisgezinnen.* Amsterdam: J. Allart, 1793.

Marum, M. van, *Verhandeling over het elektrizeeren.* Groningen: s.n., 1776.

Marum, M. van, "Verhandeling over de electrische luchtverhevelingen," *Verhandelingen van het Bataafsch Genootschap der Proefondervindelyke Wysbegeerte*, vol. 6 (1781): 1-74.

Marum, M. van, *Eerste vervolg der proefneemingen met Teyler's electrizeer-machine in 't werk gesteld.* Haarlem: Joh. Enschedé and Jan van Walré, 1787 (*Verhandelingen uitgegeeven door Teyler's tweede genootschap*, vol. 4).

Marum, M. van, "Bijvoegsel over de bliksem-afleiders, door den secretaris der maatschappij," in *Natuurkundige verhandelingen van de Hollandsche Maatschappij der Wetenschappen*, vol. 9 (1819): 319-44.

May, W., "Verhaal der uitwerkingen van eenen enkelden blixem slag, voorgevallen op één van 's lands oorlogschepen," in *Verhandelingen uitgegeeven door de Hollandsche Maatschappye der Weetenschappen*, vol. 12 (1770): 391-412.

Mémoires de l'Académie impériale et royale des sciences et belles-lettres de Bruxelles, vol. 3, Brussels: Imprimerie académique, 1780.

Mengels, W., "Chronijk van het dorp Opcanne bij Maastricht, 1740-1778," ed. by V. de Stuers, in *Publications de la Société historique et archéologique dans le Duché de Limbourg*, vol. 24, n.s. vol. 4 (Roermond: Romen, 1887): 167-297.

Moll, G., "Over afleiders en het onweder van 12 februarij 1836," *Algemeene konst- en letterbode* 1836, no. 21 (6 May): 322-26 and no. 22 (13 May): 339-45.

Montagu, M.W., *The Letters and Works of Lady Mary Wortley Montagu*, ed. Lord Wharncliffe. 2 vols., Paris: A. and W. Galignani, 1837.

"Nieuwe waarneemingen, gedaan op het afleiden van de blixemstoffe door yzeren staaven, ter voorkominge van ongelukken; benevens eenige aanmerkingen over den koers dien de blixem houdt," *Nieuwe vaderlandsche letter-oefeningen* 3, pt. 2 (1770): 20-25.

Nieuwenhuyzen, M., *Lettergeschenk voor de Nederlandsche jeugd.* 3rd ed. Haarlem: A. Loosjes, 1814 (orig. ed. 1790).

The Nightingale, 11 June 1796.

Nollet, J.-A., *Brieven over de elektrisiteit.* 3 vols., Utrecht-Amsterdam: S. de Waal and G. Warnars, 1773-1783.

Novalis, *Schriften. Die Werke Friedrich von Hardenbergs*, ed. by P. Kluckhohn and R. Samuel. 2nd ed. 6 vols., Stuttgart: Kohlhammer, 1960-2006.

Nuys Klinkenberg, J. van, *Onderwys in den godsdienst.* 11 vols., Amsterdam: J. Allart, 1780-1794.

Onderdewyngaart, C., "Het onweder," in *Proeven van poëtische mengelstoffen, door het dichtlievend kunstgenootschap onder de spreuk: Kunstliefde spaart geen vlijt* (13 vols.; Leiden-The Hague: C. van Hoogeveen Jr, 1774-1789), 9: 129-35.

Onderrigtingen omtrent de wyze, op welke men zich te gedragen heeft by het opkomen van onweder, benevens de middelen, om zich voor de nadeelige uitwerkselen van den blixem te beveiligen. Uitgegeeven tot onderrigting van onkundigen. Na den derden vermeerderden druk uit het Hoogduitsch vertaald. The Hague: I. van Cleef, 1779.

"De onmaatige vrees voor 't onweder tegengegaan," *De vaderlander* 1(1776): 233-48.

"Het onweder," *Nieuwe vaderlandsche bibliotheek van wetenschap, kunst en smaak* 4, pt. 2 (1800): 318-20.

Het onweêr. Eene gemeenzame zamenspraak. Amsterdam: n.p., 1805.

Oprechte Haarlemse courant, 4 October 1800.

Os, J. van, "Bij een onweder," in *Dichtstukken van het Haagsch genootschap onder de spreuk: Kunstliefde spaart geen vlijt, en bekroonde keurstoffen* (3 vols.; The Hague: J. Thierry and C. Mensing, 1791-1799), 2: 136-38.

"Over de vrees by een onweder," *De menschenvriend* 4 (1791): 257-64.

"Over den donder en den blixem, en hunnen oorsprong. Weinig daarvan te vrezen. Voorzorgen evenwel daartegen," *De opmerker* 3 (1774-1775): 337-44.

Paape, G., "Het onweder," in *Proeven van poëtische mengelstoffen, door het dichtlievend kunstgenootschap onder de spreuk: Kunstliefde spaart geen vlijt* (13 vols.; Leiden: C. van Hoogeveen Jr., 1774-1789), 5: 163-66.

(Paape, G.), *Vrolijke reis van een Engelschman, door Holland*. The Hague: J.C. Leeuwestyn, 1796.

Paets van Troostwijk, Adriaan, and Cornelis Kraijenhoff, *De l'application de l'électricité à la physique et à la medicine*. Amsterdam: D.J. Changuion, 1788.

Perponcher Sedlnitzky, W.E. de, *De wijsgeer der natuur*. 2 vols., Utrecht: J. van Schoonhoven, 1797.

Perponcher Sedlnitzky, W.E. de, *De Zuid-Bevelandse dorps-leeraar*. Utrecht: J. van Schoonhoven, 1799.

Perponcher Sedlnitzky, W.E. de, "Het onweer," in idem, *Gedichten* (3 vols.; Utrecht: J. van Schoonhoven, 1800-1802), 1: 116-22.

Perponcher Sedlnitzky,, W.E. de, *Wijsgeerige gronden*. Utrecht: J. van Schoonhoven, 1817.

Picart, Bernard, et al., *Cérémonies et coutumes religieuses de tous les peuples du monde*. 9 vols., Amsterdam: J.F. Bernard, 1723-1743.

Pluche, N.-A., *Le spectacle de la nature*. 9 vols., Paris: Estienne, 1732-1742.

Pluche, N.-A., *Schouwtoneel der natuur of samenspraaken over de bysonderheden der natuurlyke historie*. 16 vols., The Hague-Amsterdam: J. Neaulme et al., 1737-1749.

Poot, H.K., "Zomeronweer," in idem, *Gedichten*, 2nd ed. (3 vols.; Delft: R. Boitet, 1726-1735), 3: 99-105 (orig. ed. Rotterdam: A. Willis, 1716-1722).

Post, E.M., *Het land, in brieven*. Amsterdam: J. Allart, 1788.

Post, E.M., *Voor eenzamen*. Amsterdam: J. Allart, 1789.

Priestley, J., *Geschiedenis en tegenwoordige staat der proefondervindelijke wijsbegeerte*. 2 vols., Amsterdam: M. Magerus, 1773.

Procopii Diviš Theologiae Doctoris et Pastoris zu Prendiz bey Znaim in Mähren Längst verlangte Theorie von der Meteorologischen Electricité, welche er selbst Magiam naturalem benamt. Sammt einem Anhang vom Gebrauch der elektrischen Gründe zur Chemie. Zum Druck befördert durch Vorsorge des Württembergischen Superintendenten in Herrenberg, Fridr. Christoph Oetinger. Tübingen: Schramm, 1765.

Proeven van poëtische mengelstoffen, door het dichtlievend kunstgenootschap onder de spreuk: Kunstliefde spaart geen vlijt. 13 vols., Leiden: C. van Hoogeveen Jr., 1774-1789.

Radermacher, D., 'Berigt wegens de uitwerkingen van den donder op een Oost-Indisch schip'. In: *Verhandelingen van het Bataafsch Genootschap der Proefondervindelyke Wysbegeerte*, deel VIII (Rotterdam 1787) 173-180.

Rapport over den staat en de middelen ter verbetering der laagere schoolen; door eene daar toe benoemde commissie, bij den raad der gemeente van Leyden ingeleverd 28 Junij 1796. Leiden: Herdingh and Du Mortier, 1796.

Rechter, J., "Beschryving van een toestel, geschikt, om met een vlieger, veilig, electrische proeven te neemen," *Algemeene vaderlandsche letter-oefeningen* 1, pt. 2 (1779): 69-81.

Rogge, C., *Tafereel van de geschiedenis der jongste omwenteling in de Vereenigde Nederlanden*. Amsterdam: J. Allart, 1796.

Rossyn, J.Th., *De tonitru et fulmine ex nova electricitatis theoria deducendis*. Franeker: Gulielmus Coulon, 1762.

Saussure, H.-B. de, *Exposition abrégée de l'utilité des conducteurs électriques*. Geneva: n.p., 1771.

Schopff, J., *Wetterglöcklin*. Tübingen: Erhardo Cellio, 1599.

Schröder, J.F.L., "Redevoering over het gevoel van het verhevene," *Algemeene vaderlandsche letter-oefeningen* (1810), pt. 2: 129-47.

Schubart, C.F.D., *Ideen zu einer Ästhetik der Tonkunst*, ed. Ludwig Schubart. Leipzig: Reclam, 1977 (written ca. 1785, but published only posthumously in Vienna: J.V. Degen, 1806).

Shelley, Mary, *Frankenstein, or the Modern Prometheus*. 1818.

Shelley, P.B., *The Complete Works of Percy Bisshe Shelley*, ed. R. Ingpen and W.E. Peck. 10 vols., New York: Scribner's Sons, 1926-1930.

Spectator no. 418, 30 June 1712.

Stapfer, J.F., *Een dertig-tal van predikatien*. Uit het Hoogduits vertaalt en met eenige aantekeningen uitgegeven door Antonius de Stoppelaar. 2 vols., Utrecht: G.T. and A. van Paddenburg, 1768-1771.

Stöltzlin, Bonifacius, *Geistliches Donner- und Wetter-Büchlein : Das ist: Einfältige Erinnerung vom Donner, Blitz, Straal, Hagel und schädlichen Wettern*. Ulm: Vogt, 1650.

Supprian, F.L., *Vernünftige Gedanken von den Ursachen des Blitzes und dessen wunderbaren Wirkungen*. Potsdam: C.F. Voss, 1746.

Swinden, J.H. van, *Recueil des mémoires sur l'analogie de l'électricité et du magnétisme*. 3 vols., The Hague: Libraires Associés, 1784.

Swinderen, Th. van, "Redevoering over het verhevene in de natuur," *De recensent ook der recensenten* 3 (1808): 181-92 and 4 (1809): 237-53 and 340-51.

Tieck, L., *Franz Sternbalds Wanderungen. Eine altdeutsche Geschichte*, ed. K.-M. Guth. Berlin: Contumax, 2016 (orig. ed. 2 vols., Berlin: Unger, 1798).

Toaldo, J., *Verhandeling over het weeder met betrekking tot den landbouw (...)*. Vertaald en daarenboven voorzien met eene voorrede van M. van Geuns. Amsterdam-Utrecht: Esveldt and Holtrop, 1778.

Tollens, H., "Tafereel van den Vierdaagschen Zeeslag," in idem, *Gezamenlijke dichtwerken* (3 vols.; Leeuwarden: Suringar, 1855), 1: 5-27.

Tollens, H., *Gezamenlijke dichtwerken*. Leeuwarden: Suringar, 1855.

Uilkens, J.A., *Verhandeling over de vooroordeelen omtrent het onweder*. Groningen 1808 (now lost).

Ulber, C.S., *Der christliche Creutzträger*. Liegnitz-Hamburg: R. Beneke, 1760.

Ulber, C.S., *De christelyke kruis-drager*. Uit het Hoogduitsch vertaald door Antoni Fredrik Klenke. 2 vols., Amsterdam: A. Eichhorn, 1772-1780.

Valmont de Bomare, J.-C., *Algemeen en beredenerent woordenboek der natuurlyke historie*. 3 vols., Dordrecht: Abraham Blussé, 1767-1770.

Veldhoven, A., *Dichtkundige bespiegelingen over de aarde, den mensch en den starrenhemel*. Dordrecht: H. de Haas, 1774.

Ver Huell, Alexander, *Het dagboek van Alexander Ver Huell 1860-1865*, ed. Jan Bervoets and Rody Chamuleau. Zutphen: Walburg Pers, 1985.

Volledige verzameling van alle stukken, betreffende de afscheiding der kerk van den staat, door de Nationaale Vergadering, representeerende het volk van Nederland. Met eene voorrede van C. Rogge. Leiden: Du Mortier, 1796.

Volta, A., *Meteorologische Briefe nebst Beschreibung seines Eudiometers.* Leipzig: J.G. Müller, 1793.
Voltaire, *Poème sur le désastre de Lisbonne ou examen de cet axiome: 'tout est bien'.* In: idem, *Mélanges*, ed. J. Van den Heuvel, Paris: Gallimard, 1961.
"De vyanden des lichts. Een vertelzel," *Algemeene vaderlandsche letter-oefeningen* (1791), pt. 2: 534-37.
Walpole, H., *Anecdotes of Painting in England.* 4 vols., Strawberry Hill: T. Farmer, 1762-1780.
Watson, W., "Raadgeeving tot voorkoming van de ongelukken, die de schepen dikwyl door het weerlicht overkomen," *Vaderlandsche letter-oefeningen* 4, pt. 2 (1764): 455-58.
Watts, Isaac, *Hymns and Spiritual Songs. In Three Books.* London: printed by J. Humfreys, for John Lawrence, 1707.
Wieland, C.M., *Sämtliche Werke.* 39 + 6 vols., Leipzig: Göschen, 1794-1811.
Wienbarg, C.L., *Ästhetische Feldzüge*, ed. by W. Dietze. Berlin-Weimar: Aufbau Verlag, 1964 (orig. ed. Hamburg: Hoffmann and Campe, 1834).
Wijnpersse, S.J. van de, *Het loeijend onweder godvruchtig beschouwd, of opwekking tot een godvereerend gemoeds-bestaan, onder het rollen der donderslagen en het klateren der bliksemen, in eene leerrede over Psalm XXIX.* Amsterdam: H. Diederiks and H. Larde, 1799.
Winkler, J.H., *Nieuwe natuurkundige ontdekkingen, aangaande de eigenschappen, werkingen, en oorzaaken der electriciteyt, benevens eene beschryvinge van twee nieuwe electrische werktuigen. Mitsgaders de aftekening der zelve.* Amsterdam: Hendrik Vieroot, 1745.
Winkler, J.H., *De eigenschappen der electrische stoffe, en van het electrische vuur, uit verscheidene nieuwe proeven verklaart. Benevens eenige in 't koper gebragte nieuwe werktuigen om te electriseeren. Strekkende tot een tweede deel der Gedagten over de electriciteyt.* Amsterdam: Hendrik Vieroot, 1746.
Winkler, J.H., *Die Stärke der electrischen Kraft des Wassers in gläsernen Gefässen.* Leipzig: Breitkopf, 1746.
Woodward, J., *An Essay toward a Natural History of the Earth and Terrestrial Bodies.* London: R. Wilkin, 1695.
Zamenspraak tusschen een landheer en zijn' tuinman over het onweder. Amsterdam: De Vries, Van Munster, and Van der Hey, 1805.
Zamenspraken over de oorzaken van het onweder en het nut der afleiders. Amsterdam: De Vries, Van Munster, and Van der Hey, 1805.

Secondary Literature

Aaftink, H., *Leven met bliksem.* 4th ed. Amsterdam: Winterthur Verzekeringen, 1991.
Achermann, H., *Die Katakombenheiligen und ihre Translationen in der schweizerischen Quart des Bistums Konstanz.* Stans: Historischer Verein Nidwalden, 1979.
Adrian, K., "Wind und Wetter im Glauben und Brauchtum unseres Volkes," in *Mitteilungen der Gesellschaft für Salzburger Landeskunde* 84-85 (1944-1945): 1-48.
Albert, E., *Das Naturgefühl L.H.Chr. Höltys.* Dortmund: Ruhfus, 1910 (repr. Hildesheim: Gerstenberg Verlag, 1978).
Aldridge, A.O., *Franklin and his Contemporaries.* New York: New York University Press, 1957.
Alewyn, R., "Die Lust an der Angst," in idem, *Probleme und Gestalten. Essays*, 307-30.
Alewyn, R., *Probleme und Gestalten. Essays.* Frankfurt a.M.: Insel, 1974.

Altena, P., "Die holländische Himmelsstürmer. Sturm und Drang in der niederländischen Literatur 1770-1800," in *Sturm und Drang. Geistiger Aufbruch 1770-1790 im Spiegel der Literatur*, ed. B. Plachta and W. Woesler (Tübingen: Niemeyer, 1997), 183-98.

Altena, P., *Gerrit Paape (1752-1803). Levens en werken*. Nijmegen: Vantilt, 2012.

Andel, M.A. van, "De ceraunia of dondersteen in de volksgeneeskunst," *Volkskunde* 38 (1933): 3-13.

Anderson, R., *Lightning Conductors: Their History, Nature and Mode of Application*. London-New York: E. & F.N. Spon, 1880.

Aspalter, Chr., et al., eds., *Paradoxien der Romantik. Gesellschaft, Kultur und Wissenschaft in Wien im frühen 19. Jahrhundert*. Vienna: WUV, 2006.

Aspalter, Chr., and A. Tantner, "Ironieverlust und verleugnete Rezeption: Kontroversen um Romantik in Wiener Zeitschriften," in *Paradoxien der Romantik. Gesellschaft, Kultur und Wissenschaft in Wien im frühen 19. Jahrhundert*, ed. Chr. Aspalter et al. (Vienna: WUV, 2006), 47-120.

B., Pauline, *De vrouw om en in de pastorie*. Amsterdam: Johannes van der Hey, 1853.

Bächtold-Stäubli, H., ed., *Handwörterbuch des deutschen Aberglaubens*. 10 vols., Berlin-Leipzig: De Gruyter, 1927-1942.

Barkhoff, J., *Magnetische Fiktionen. Literarisierung des Mesmerismus in der Romantik*. Stuttgart-Weimar: J.B. Metzler, 1995.

Barnett, Cynthia, *Rain. A Natural and Cultural History*. New York: Crown, 2015.

Batt, M., *The Treatment of Nature in German Literature from Günther to the Appearance of Goethe's Werther*. Port Washington, NY-London: Kennikat, 1969 (repr. of the orig. ed. Chicago: University of Chicago Press, 1902).

Bausinger, H., "Aufklärung und Aberglaube," *Deutsches Vierteljahrsschrift für Literaturwissenschaft und Geistesgeschichte* 37 (1963): 345-62.

Beek, A. van de, et al., eds., *Waar is God in deze tijd? De betekenis van de geschiedenis in de theologie van Dr. H. Berkhof*. Nijkerk: Callenbach, 1994.

Begemann, Chr., *Furcht und Angst im Prozeß der Aufklärung. Zu Literatur und Bewußtseinsgeschichte des 18. Jahrhunderts*. Frankfurt a.M.: Athenäum, 1987.

Belhoste, B., and N. Edelman, eds., *Mesmer et les mesmérismes. Le magnétisme animal en contexte*. Paris: Montreuil Omniscience, 2015.

Beller, M., *Jupiter tonans. Studien zur Darstellung der Macht in der Poesie*. Heidelberg: Winter, 1979.

Bensaude-Vincent, B., and Chr. Blondel, eds., *Science and Spectacle in the European Enlightenment*. Abingdon-New York: Routledge, 2008.

Benz, E., *Theologie der Elektrizität. Zur Begegnung und Auseinandersetzung von Theologie und Naturwissenschaft im 17. und 18. Jahrhundert*. Mainz-Wiesbaden: Steiner, 1971.

Berger, G., and S.A. Amar, "The Noteworthy Involvement of Jacques de Romas in the Experiments on the Electric Nature of Lightning," *Journal of Electrostatics* 67 (2009): 531-35.

Bernardi, W., "The Controversy on Animal Electricity in Eighteenth-Century Italy: Galvani, Volta and Others," *Revue d'histoire des sciences* 54 (2001): 53-70.

Bernd-Brinkmann, A., "Wetterlieder im 17. und 18. Jahrhundert," *Lied und populäre Kultur* 45 (2000): 89-108.

Berndt, F., and D. Fulda, eds., *Die Sachen der Aufklärung. Beiträge zur DGEJ-Jahrestagung 2010 in Halle a.d. Saale*. Hamburg: F. Meiner, 2012.

Bertsch, M., "Zwischen Erhabenheitsästhetik und dokumentarischem Anspruch. Das Motiv des Schiffbruchs im ausgehenden 18. Jahrhundert," in *Entfesselte Natur. Das Bild der Katastrophe seit 1600*, ed. M. Bertsch and J. Trempler (Petersberg: Imhof, 2018), 59-69.

Bertsch, M., and J. Trempler, eds., *Entfesselte Natur. Das Bild der Katastrophe seit 1600*. Petersberg: Imhof, 2018.

Bertucci, P., "Revealing Sparks: John Wesley and the Religious Utility of Electrical Healing," *The British Journal for the History of Science* 39 (2006): 341-62.

Bertucci, P., "Domestic Spectacles: Electrical Instruments between Business and Conversation," in *Science and Spectacle in the European Enlightenment*, ed. B. Bensaude-Vincent and Chr. Blondel (Abingdon-New York: Routledge, 2008), 75-87.

Bertucci, P., "Enlightening Towers. Public Opinion, Local Authorities, and the Reformation of Meteorology," in *Playing with Fire. Histories of the Lightning Rod*, ed. P. Heering et al. (Philadelphia: American Philosophical Society, 2009), 25-44.

Beutel, A., and M. Nooke, eds., *Religion und Aufklärung. Akten der Ersten Internationalen Kongresses zur Erforschung der Aufklärungstheologie (Münster, 30. März bis 2. April 2014)*. Tübingen: Mohr Siebeck, 2016.

Bevilacqua, F., and L. Fregonese, eds., *Nuova Voltiana. Studies on Volta and his Times*. 5 vols., Milan: Hoepli, 2000-2003.

Bierens de Haan, J.C., *Rosendael, Groen Hemeltjen op Aerd. Kasteel, tuinen en bewoners sedert 1579*. Zutphen: Walburg Pers, 1994.

Bijker, W.E., *Of Bicycles, Bakelites, and Bulbs. Toward a Theory of Sociotechnical Change*. Cambridge, MA-London: MIT Press, 1995.

Blair, A., and K. von Greyerz, eds., *Physico-Theology. Religion and Science in Europe, 1650-1750*. Baltimore: Johns Hopkins University Press, 2020.

Blécourt, W. de, *Volksverhalen uit Noord-Brabant*. Utrecht-Antwerp: Spectrum, 1980.

Blinkenberg, Chr., *The Thunderweapon in Religion and Folklore. A Study in Comparative Archaeology*. Cambridge: The University Press, 1911.

Bockmaier, C., *Entfesselte Natur in der Musik des achtzehnten Jahrhunderts*. Tutzing: H. Schneider, 1992.

Böhme, H., *Natur und Subjekt*. Frankfurt a.M.: Suhrkamp, 1988.

Boia, L., *The Weather in the Imagination*. London: Reaktion Books, 2005.

Bosch, R.A., *En nooit meer oude Psalmen zingen. Zingend geloven in een nieuwe tijd 1760-1810*. Zoetermeer: Boekencentrum, 1996.

Böschenstein, B., *"Frucht des Gewitters". Hölderlins Dionysos als Gott der Revolution*. Frankfurt a.M.: Insel Verlag, 1989.

Böschenstein, B., "Quatre sommets de la poésie allemande des orages: Klopstock – Goethe – Kleist – Hölderlin," in *L'Événement climatique et ses représentations (XVIIe-XIXe siècle)*, ed. E. Le Roy Ladurie et al. (Paris: Desjonquères, 2007), 297-309.

Botke, IJ., *Jacobus Albertus Uilkens, 1772-1825*. Groningen: Universiteitsmuseum, 1984.

Bots, J., *Tussen Descartes en Darwin. Geloof en natuurwetenschap in de achttiende eeuw in Nederland*. Assen: Van Gorcum, 1972.

Bouchard, G., *Le village immobile. Sennely-en-Sologne au XVIIIe siècle*. Paris: Plon, 1972.

Brandt Corstius, J.C., "Verschijnselen in onze literatuur aan het einde van de 18de eeuw in verband met veranderende opvattingen omtrent geloof en natuur," *De nieuwe taalgids* 44 (1951): 241-53.

Braungart, G., and U. Büttner, eds., *Wind und Wetter. Kultur-Wissen-Ästhetik*. Paderborn: Fink, 2018.

Breidert, W., "Das Erdbeben von Lissabon und die Erschütterung seiner Zeitgenossen," in *Lichtenberg Jahrbuch* (1994): 56-67.

Brewer, J., and R. Porter, eds., *Consumption and the World of Goods*. London-New York: Routledge, 1993.

Briese, O., "Der abgeleitete Blitz. Metapherngeschichte als Mentalitätengeschichte," *Euphorion. Zeitschrift für Literaturgeschichte* 92 (1998): 413-35.

Briese, O., *Die Macht der Metaphern. Blitz, Erdbeben und Kometen im Gefüge der Aufklärung*. Stuttgart-Weimar: J.B. Metzler, 1998.

Briese, O., "Blitze: Klopstocks Ode Die Frühlingsfeyer," in *Phänomene der Atmosphäre. Ein Kompendium literarischer Meteorologie*, ed. U. Büttner and I. Theilen (Stuttgart: J.B. Metzler, 2017), 133-42.

Brillenburg Wurth, C.A.W., *The Musically Sublime. Indeterminacy, Infinity, Irresolvability*. New York: Fordham University Press, 2009.

Brise, C., and A. Romano, eds., *Sciences et religions de Copernic à Galilée (1540-1610)*. Rome: École Française de Rome, 1999.

Brittinger, A., *Die bayerische Verwaltung und das volksfromme Brauchtum im Zeitalter der Aufklärung*. Munich: Mössl, 1938.

Broeckhove, M., "Devotie in het Vlaamssprekend gedeelte van België tot de H. Donatus, biezondere patroon tegen schadelijke natuur- en vernielende kunstmatige luchtelementen," *Oost- vlaamsche zanten* 34 (1959) and 35 (1960).

Bruijn, J.G. de, *Inventaris van de prijsvragen uitgeschreven door de Hollandsche Maatschappij der Wetenschappen 1753-1917*. Haarlem-Groningen: H.D. Tjeenk Willink, 1977.

Buisman, J., *Duizend jaar weer, wind en water in de Lage Landen*. 7 vols., Franeker: Van Wijnen, 1995-2019.

Buisman, J.W., *Tussen vroomheid en Verlichting. Een cultuurhistorisch en -sociologisch onderzoek naar enkele aspecten van de Verlichting in Nederland (1755-1810)*. 2 vols., Zwolle: Waanders, 1992.

Buisman, J.W., "Opvattingen over het onweer in de overgang van Verlichting naar Romantiek," *Zeeland* 3 (1994): 70-74.

Buisman, J.W., "Het onweer in Nederland in godsdienst, wetenschap en kunst, 1752-1830," *Nederlands archief voor kerkgeschiedenis* 78 (1998): 228-42.

Buisman, J.W., "Inleiding: vreemde Verlichting," In: *Verlichting in Nederland 1650-1850. Vrede tussen rede en religie?*, ed. J.W. Buisman (Nijmegen: Vantilt, 2013), 9-13.

Buisman, J.W., "Op zoek naar het juiste midden: C.R.Th. Kraijenhoff en S.J. van de Wijnpersse over het onweer," In: *Verlichting in Nederland 1650-1850. Vrede tussen rede en religie?*, ed. J.W. Buisman (Nijmegen: Vantilt, 2013), 191-206.

Buisman, J.W., ed., *Verlichting in Nederland 1650-1850. Vrede tussen rede en religie?* Nijmegen: Vantilt, 2013.

Buisman, J.W., "Thunderstorms: the Devil's Work?," in *Religion und Aufklärung*, ed. A. Beutel and M. Nooke (Tübingen: Mohr Siebeck, 2016), 639-46.

Bulhof, F., *Ma patrie est au ciel. Leven en werk van Willem Emmery de Perponcher Sedlnitzky (1741-1819)*. Hilversum: Verloren, 1993.

Burwick, F., ed., *The Oxford Handbook of Samuel Taylor Coleridge*. Oxford-New York: Oxford University Press, 2009.

Busch, W., D. Jackson, and J. Reynaerts, *De Romantiek in het noorden. Van Friedrich tot Turner*. Zwolle: W Books, 2017.

Bussels, S., and B. van Oostveld, *The Sublime in the Visual Culture of the Seventeenth-Century Dutch Republic*. London-New York: Routledge 2023.

Büttner, M., "Theologie und Klimatologie im 18. Jahrhundert," *Neue Zeitschrift für systematische Theologie und Religionsphilosophie* 6 (1964): 154-92.

Büttner, U., and I. Theilen, eds., *Phänomene der Atmosphäre. Ein Kompendium literarischer Meteorologie.* Stuttgart: J.B. Metzler, 2017.

Cantor, G.N., "The Theological Significance of Ethers," in *Conceptions of Ether. Studies in the History of Ether Theories 1740-1900*, ed. G.N. Cantor and M.J.S. Hodge (Cambridge: Cambridge University Press, 1981), 135-55.

Cantor, G.N., and M.J.S. Hodge, eds., *Conceptions of Ether. Studies in the History of Ether Theories 1740-1900*. Cambridge: Cambridge University Press, 1981.

Catalogus van de bibliotheken van J. J. L. ten Kate en van een rechtsgeleerde. Amsterdam: Van Looy, 1890.

Charlton, D.G., *New Images of the Natural in France. A Study in European Cultural History 1750-1800*. Cambridge: Cambridge University Press, 1984.

Christen, J., "L'orage nocturne, entre sublime et frénésie," *Revue des Sciences Humaines* 281, no. 1 (January-March 2006): 67-80.

Clerc, A., "Tempêtes libertins et naufrages spirituels au XVIIe siècle," in *L'Événement climatique et ses représentations (XVIIe-XIXe siècle)*, ed. E. Le Roy Ladurie, J. Berchtold, and J.-P. Sermain (Paris: Desjonquères, 2007), 347-62.

Cock, A. de, "Eerbied voor de natuur en haar verschijnsels," *Volkskunde* 17 (1905): 45-53.

Cohen, I.B., "Prejudice against the Introduction of Lightning Rods," *Journal of the Franklin Institute* 253 (1952): 393-440.

Cohen, I.B., and R. Schofield, "Did Diviš erect the First European Protective Lightning Rod, and was his Invention Independent?," *Isis* 43 (1952): 358-64.

Cohen, I.B., *Benjamin Franklin's Science*. Cambridge, MA-London: Harvard University Press, 1990.

Conrad, H., *Die literarische Angst. Das schreckliche in Schauerromantik und Detektivgeschichte*. Düsseldorf: Bertelsmann, 1974.

Contarini, S., "In de spiegel van de natuur. Het verhevene bij Paulus van Hemert," in *Esthetica tussen klassiek en romantiek*, ed. H. Krop and P. Sonderen (Rotterdam: Faculteit der Wijsbegeerte van de Erasmus Universiteit Rotterdam), 97-106.

Corbin, A., *Les cloches de la terre. Paysage sonore et culture sensible dans les campagnes au XIXe siècle*. Paris: Flammarion, 1994.

Courtine, J.F., et al., *Du sublime*. Paris: Belin, 1988.

Darnton, R., *Mesmerism and the End of the Enlightenment in France*. 4th ed. Cambridge, MA: Harvard University Press, 2009 (orig. ed. 1968).

Deinse, J.J. van, *Uit het land van katoen en heide*. 2nd ed. Enschede: Van der Loeff, 1925 (orig. ed. 1922).

Davies, O., and W. de Blécourt, eds., *Beyond the Witch Trials. Witchcraft and Magic in Enlightenment Europe*. Manchester-New York: Manchester University Press, 2004.

Delbourgo, J., *A Most Amazing Scene of Wonders. Electricity and Enlightenment in Early America*. Cambridge, MA-London: Harvard University Press, 2006.

Delon, M., "Electriser, un mot d'ordre au siècle des Lumières," *Revue des sciences humaines* 281 (2006): 39-51.

Delumeau, J., *La peur en Occident (XIVe-XVIIIe siècles). Une cité assiégée*. Paris: Fayard, 1978.

Delumeau, J., *Rassurer et protéger. Le sentiment de sécurité dans l'Occident d'autrefois*. Paris: Fayard, 1989.

Demandt, A., *Metaphern für Geschichte. Sprachbilder und Gleichnisse im historisch-politischen Denken*. Munich: Beck, 1978.

Demarée, G.R., "'De grote droge nevel' van 1783 in de Zuidelijke Nederlanden: een historisch- klimatologische studie," *Tijdschrift voor ecologische geschiedenis* 2 (1997): 27-36.

Dieckmann, H., "Das Abscheuliche und das Schreckliche in der Kunsttheorie des 18. Jahrhunderts," in *Die nicht mehr schönen Künste. Grenzphänomene des Ästhetischen*, ed. H.R. Jauß (Munich: Fink, 1968), 271-317.

Diederen, R., and Chr. Kürzeder, eds., *Mit Leib und Seele. Münchner Rokoko von Asam bis Günther*. Munich: Sieveking, 2014.

Dijkhuizen, J.F. van, et al., *Living in Posterity. Essays in Honour of Bart Westerweel*. Hilversum: Verloren, 2004.

H. Donatus Martelaar: 200-jarige viering St. Donatusbroederschap Reek. s.l., s.n., 1979.

Dongelmans, B.P.M., *Johannes Immerzeel junior (1776-1841). Het bedrijf van een uitgever-boekhandelaar in de eerste helft van de negentiende eeuw*. Amstelveen: Ernst, 1992.

Doran, R., *The Theory of the Sublime from Longinus to Kant*. Cambridge: Cambridge University Press, 2015.

Drahos, A., *Orages et tempêtes, volcans et glaciers. Les peintres et les sciences de la terre aux XVIIIe et XIXe siècles*. Paris: Hazan, 2014.

Dross, F., "Gottes elektrischer Wille? Zum Düsseldorfer 'Blitzableiter-Aufruhr' 1782/83," in *Landes- und Reichsgeschichte*, ed. J. Engelbrecht and S. Laux (Bielefeld: Verlag für Regionalgeschichte), 281-302.

Dülmen, R. van, *Kultur und Alltag in der Frühen Neuzeit. Dritter Band: Religion, Magie, Aufklärung 16.-18. Jahrhundert*. Munich: Beck, 1994.

Edgerton, D., *The Shock of the Old. Technology and Global History since 1900*. 3rd ed. London: Profile Books, 2011 (orig. ed. 2006).

Egelie, G.C.M., *Gietijzeren wegkruisen in Limburg*. Zutphen: Walburg Pers, 1983.

Ehrard, J., *L'idée de nature en France dans la première moitié du XVIIIe siècle*. 2 vols., Paris: S.E.V.P.E.N., 1963.

Eijnatten, J. van, *Hogere sferen. De ideeënwereld van Willem Bilderdijk (1756-1831)*. Hilversum: Verloren, 1998.

Eliade, M., *The Forge and the Crucible*. 2nd ed. Chicago: University of Chicago Press, 1978 (orig. French ed. Paris: Flammarion, 1956).

Elsom, D.M., *Lightning. Nature and Culture*. London: Reaktion Books, 2015.

Engammare, M., "Tonnerre de Dieu et 'courses d'exhalations encloses es nuées'. Controverses autour de la foudre et du tonnerre au soir de la Renaissance," in *Sciences et religions de Copernic à Galilée (1540-1610)*, ed. C. Brise and A. Romano (Rome: École Française de Rome, 1999), 161-81.

Engelbrecht, J., and S. Laux, eds., *Landes- und Reichsgeschichte. Festschrift für Hansgeorg Molitor zum 65. Geburtstag*. Bielefeld: Verlag für Regionalgeschichte, 2004.

English, M., "De donderheilige Sint-Donatus en zijn verering in West-Vlaanderen," *Biekorf* 60 (1959).

Fabich, F., *Bauernmedizin*. Rosenheim: Rosenheimer, 1991.

Fairclough, M., *Literature, Electricity and Politics 1740–1840: 'Electrick Communication Every Where'*. London: Palgrave, 2017.

Fairclough, M., "Romantic Electricity: Fields and Currents in Recent Research," *Literature Compass* 15 (2018): 1-15.

Falkenburg, R., et al., eds., *Natuur en landschap in de Nederlandse kunst 1500-1850*. Zwolle: Waanders, 1998 (*Nederlands kunsthistorisch jaarboek* no. 48, 1998).

Falkenburg, R., "Onweer bij Jan van Goyen. Artistieke wedijver en de markt voor het Hollandse landschap in de 17de eeuw," in *Natuur en landschap in de Nederlandse kunst 1500-1850*, ed. R. Falkenburg et al. (Zwolle: Waanders, 1998), 117-61.

Fara, P., "An Attractive Therapy: Animal Magnetism in Eighteenth-Century England," *History of Science* 33 (1995): 127-77.

Fara, P., *An Entertainment for Angels. Electricity in the Enlightenment*. 4th ed. London: Icon Books, 2017 (orig. ed. New York: Columbia University Press, 2002).

Feldmann, A., "'Donnerkeil, Feuer, und ein entsetzlicher Hagel....' Der Heilige Donatus von Münstereifel," in *Brand. Heimatkundliche Blätter* (2002): 56-91.

Ferré, F., *Hellfire and Lightning Rods. Liberating Science, Technology, and Religion*. Maryknoll: Orbis Books, 1993.

Filipetti H., and J. Trotereau, *Symboles et pratiques rituelles dans la maison paysanne traditionnelle*. Paris: Berger-Levrault, 1978.

Figuier, L., *Les merveilles de la science*. Paris: Furne, Jouvet et Cie, 1870.

Flemming, W., *Der Wandel des deutschen Naturgefühls vom 15. zum 18. Jahrhundert*. Halle a.d. Saale: Niemeyer, 1931.

Fontane, Th., *Wanderungen durch die Mark Brandenburg*. Berlin: Hertz, 1862-1882.

Fonvielle, W. de, *Éclairs et tonnerre*. 2nd ed. Paris: Hachette, 1869 (orig. ed. 1867).

Frantz, P., "Diderot, Dorval, tempête et creation," in *L'Événement climatique et ses représentations (XVIIe-XIXe siècle)*, ed. E. Le Roy Ladurie, J. Berchtold, and J.-P. Sermain (Paris: Desjonquères, 2007), 430-38.

Fraunberger, F., *Elektrische Spielereien im Barock und Rokoko*. Munich: Oldenbourg; Düsseldorf: VDI-Verlag, 1967.

Frazer, J.G., *The Golden Bough*. Abr. ed. Mineola, N.Y.: Dover, 2002 (orig. ed. 12 vols., London: Macmillan, 1906-1915).

Freudenthal, H., *Das Feuer im deutschen Glauben und Brauch*. Berlin: De Gruyter, 1931.

Friedman, J.S., *Out of the Blue. A History of Lightning: Science, Superstition, and Amazing Stories of Survival*. New York: Delacorte Press, 2008.

Frijhoff, W., "Vraagtekens bij het vroegmoderne kersteningsoffensief," in *Religieuze volkscultuur. De spanning tussen de voorgeschreven orde en de geleefde praktijk*, ed. G. Rooijakkers and Th. van der Zee (Nijmegen: SUN, 1986), 71-98.

Frisinger, H.H., "Early Theories on the Cause of Thunder and Lightning," *Bulletin of the American Meteorological Society* 46 (1965): 785-87.

Fuhrmeister, Chr., "'Eripuit Caelo Fulmen Sceptrumque Tyrannis.' The Political Iconography of Lightning in Europe and North America, 1750-1800," in *Playing with Fire. Histories of the Lightning Rod*, ed. P. Heering et al. (Philadelphia: American Philosophical Society, 2009), 144-63.

Gaderer, R., *Poetik der Technik. Elektrizität und Optik bei E.T.A. Hoffmann*. Freiburg: Rombach, 2009.

Gamper, M., *Elektropoetologie. Fiktionen der Elektrizität, 1740-1870*. Göttingen: Wallstein, 2009.

Gari, M., *Le vinaigre et le fiel*. Paris: Plon, 1983.

Geertz, C., *Local Knowledge. Further Essays in Interpretative Anthropology*. New York: Basic Books, 1983.

Gessel, J. van, *Muziek beleven in het negentiende-eeuwse Nederland*. Utrecht: IJzer, 2020.

Geudeke, E., *De classis Edam, 1572-1650. Opbouw van een nieuwe kerk in een verdeelde samenleving*. Gouda: Vereniging voor Nederlandse Kerkgeschiedenis, 2010.

Geurts, H., "De uitzonderlijke zomer van 1783 met de warmste juli sinds het begin der waarnemingen," *Weerspiegel* 7 (1982): 355-59.

Gilmore, P., *Aesthetic Materialism. Electricity and American Romanticism*. Stanford: Stanford University Press, 2009.

Gleichmann, E., *Die Glocke im Volksglauben, Sage und Dichtung*. Kulmbach: R. Rehm, 1930.

Golinski, J., *British Weather and the Climate of Enlightenment*. Chicago: The University of Chicago Press, 2007.

Golinski, J., "Enlightenment Science," in *The Oxford Illustrated History of Science*, ed. I. Rhys Morus (Oxford: Oxford University Press, 2017), 180-210.

Goodrick-Clarke, N., "The Esoteric Uses of Electricity: Theologies of Electricity from Swabian Pietism to Ariosophy," *Aries* 4 (2004): 69-90.

Gou, L. de, *Biografische bijdragen over achttiende-eeuwers*. Haarlem: Joh. Enschedé, 1988.

Goubier-Robert,G., "De la fulgurisation sadienne aux foudres républicaines," in *L'Événement climatique et ses représentations (XVIIe-XIXe siècle)*, ed. E. Le Roy Ladurie, J. Berchtold, and J.-P. Sermain (Paris: Desjonquères, 2007), 415-29.

Govers, N., *Conferenties over de heiliging van het huisgezin*. Roermond-Maaseik: Romen, 1946.

Graaf, I. de, "Als de bliksem...," *Nijvertijd* 7, no. 12 (December 2003).

Grabo, C., *A Newton among Poets. Shelley's Use of Science in* Prometheus Unbound. Chapel Hill: University of North Carolina Press, 1930 (repr. 1968).

Grattan, J., and M. Brayshay, "An Amazing and Portentous Summer. Environmental and Social Responses in Britain to the 1783 Eruption of an Iceland Volcano," *The Geographical Journal* 161 (1995): 125-34.

Griewank, K., *Der neuzeitliche Revolutionsbegriff. Entstehung und Entwicklung*. 2nd ed. Frankfurt a.M.: Europäische Verlagsanstalt, 1969 (orig. ed. Weimar: Böhlau, 1955).

Grijzenhout, F., N. van Sas en W. Velema, eds., *Het Bataafse experiment. Politiek en cultuur rond 1800*. Nijmegen: Vantilt, 2013.

Grimminger, R., ed., *Deutsche Aufklärung bis zur Französischen Revolution 1680-1789*. Munich: Hanser, 1980.

Gruys, J.A., and C. de Wolf, *Thesaurus 1473-1800*. Nieuwkoop: B. de Graaf, 1989.

Hache, S., "Thunder or Celestial Harmony: French Theological Debates on the Sonorous Sublime," in *Music and the Sonorous Sublime in European Culture, 1680-1880*, ed. S. Hibberd and M. Stanyon (Cambridge-New York: Cambridge University Press, 2020), 26-43.

Hackmann, W.D., *John and Jonathan Cuthbertson. The Invention and Development of the Eighteenth Century Plate Electrical Machine*. Leiden: Rijksmuseum voor de geschiedenis der natuurwetenschappen, 1973.

Hackmann, W.D., "Instruments and Experiments. The Case of Atmospheric Electricity in Eighteenth-Century Holland," *Tijdschrift voor de geschiedenis der geneeskunde, natuurwetenschappen, wiskunde en techniek* 10 (1987): 190-207.

Hackmann, W.D., "The Lightning Rod. A Case Study of Eighteenth Century Model Experiments" in *Playing with Fire. Histories of the Lightning Rod*, ed. P. Heering et al. (Philadelphia: American Philosophical Society, 2009), 209-29.

Hagen, E., "Een zaal van staatsmannen, niet van godgeleerden. Godsdienstige sentimenten in de Nationale Vergadering," in *Het Bataafse experiment. Politiek en cultuur rond 1800*, ed. F. Grijzenhout et al. (Nijmegen: Vantilt, 2013), 124-53.

Hager, F., and H. Heyn, *Drudenhax und Allelujawasser. Volksbrauch im Jahreslauf*. Rosenheim: Förg, 1975.

Hankins, Th.L., *Science and the Enlightenment*. Cambridge: Cambridge University Press, 1985.

Hardenberg, H., "Benjamin Franklin en Nederland," *Bijdragen voor de geschiedenis der Nederlanden* 5 (1950): 213-30.

Hardwick, J., and R.J. Stephens, 'Acts of God: Continuities and Change in Christian Responses to Extreme Weather Events from Early Modernity to the Present,' *WIREs Climate Change* 11, no. 2 (2020): 1-16.

Hart, Darryl G., *Benjamin Franklin. Cultural Protestant*. Oxford-New York: Oxford University Press, 2021.

Hartmann, P., "Du baroque au classicisme: orages et tempêtes dans l'œuvre de Joseph Haydn," in *L'Événement climatique et ses représentations (XVIIe-XIXe siècle)*, ed. E. Le Roy Ladurie, J. Berchtold, and J.-P. Sermain (Paris: Desjonquères, 2007), 310-22.

Hattum, M. van, *Lezingen en verhandelingen in 'Concordia et libertate' (1769-1806) en 'Felix meritis' (Dep. letterk.) (1779-1808, 1810-1832, 1865-1873)*. Amstelveen: s.n., 1983 (typoscript).

Haubelt, J., "Václav Prokop Diviš und Johann Ludwig Fricker," in *Mathesis, Naturphilosophie und Arkanwissenschaft im Umkreis Friedrich Christoph Oetingers (1702-1782)*, ed. S. Holtz et al. (Stuttgart: F. Steiner, 2005), 153-64.

Heckscher, K., *Die Volkskunde des germanischen Kulturkreises*. Hamburg: M. Riegel, 1925.

Heering, P., "Styles of Experimentation and the Attempts to Establish the Lightning Rod in Pre-Revolutionary Paris," in *Playing with Fire. Histories of the Lightning Rod*, ed. P. Heering et al. (Philadelphia: American Philosophical Society, 2009), 121-43.

Heering, P., O. Hochadel, and D.J. Rhees, eds., *Playing with Fire. Histories of the Lightning Rod*. Philadelphia: American Philosophical Society, 2009.

Heering, P., and O. Hochadel, "Epilogue. An Invisible Technology – What Remains to be Seen," in *Playing with Fire. Histories of the Lightning Rod*, ed. P. Heering et al. (Philadelphia: American Philosophical Society, 2009), 269-76.

Heilbron, J.L., *Electricity in the 17th and 18th Centuries. A Study of Early Modern Physics*. Berkeley and Los Angeles-London: University of California Press, 1979.

Hempler, F., *Das Gewitter im Volksglauben unserer Heimat*. Danzig: A.W. Kafemann, 1928 (*Heimatblätter des deutschen Heimatbundes Danzig* 5 [1928], Issue 5/6).

Hentschel, K.-H., "Man erhitze sich daher zur Zeit der Ungewitter nicht. Dunst und Dämpfe, Kult und Kämpfe: Menschen, Blitz und Donner," *Hierzuland* 8, no. 15 (1993): 6-23.

Heumakers, A., *De esthetische revolutie. Hoe Verlichting en Romantiek de Kunst uitvonden*. Amsterdam: Boom, 2015.

Heuvel, D. van, and F.D. Zeiler, "Biografieën," in *Hef aan! Bataaf! Beschouwingen over muziek en muziekleven in Nederland omstreeks 1795*, ed. P. van Reijen (Alphen a.d. Rijn: Canaletto, [1997]), 149-80.

Hibberd, S., and M. Stanyon, eds., *Music and the Sonorous Sublime in European Culture, 1680-1880* Cambridge-New York: Cambridge University Press, 2020.

Hielkema, H.W., "It kloklieden as folksbrûkme yn Fryslân," in *Frysk studinte-almenak* 1937: 36-45.

Hildebrand, P., *De Kapucijnen in de Nederlanden en het prinsbisdom Luik*. 10 vols., Antwerp: Archief der Kapucijnen, 1945-1956.

Hill, Chr., *Society and Puritanism in Pre-Revolutionary England*. New York: Schocken Books, 1964.

Hochadel, O., *Öffentliche Wissenschaft. Elektrizität in der deutschen Aufklärung*. Göttingen: Wallstein, 2003.

Hochadel, O., “The Sale of Shocks and Sparks: Itinerant Electricians in the German Enlightenment,” in *Science and Spectacle in the European Enlightenment*, ed. B. Bensaude-Vincent and Chr. Blondel (Abingdon-New York: Routledge, 2008), 89-101.

Hochadel, O., “In Nebula Nebulorum. The Dry Fog of the Summer of 1783 and the Introduction of Lightning Rods in the German Empire,” in *Playing with Fire. Histories of the Lightning Rod*, ed. P. Heering et al. (Philadelphia: American Philosophical Society, 2009), 45-70.

Hochadel, O., and P. Heering, “Introduction: Revisiting an Invisible Technology,” in *Playing with Fire. Histories of the Lightning Rod*, ed. P. Heering et al. (Philadelphia: American Philosophical Society, 2009), 1-24.

Hoftijzer, P.G., “An Eighteenth-Century Amateur of Books and Science in Warmond,” in *Living in Posterity. Essays in Honour of Bart Westerweel*, ed. J.F. van Dijkhuizen et al. (Hilversum: Verloren, 2004), 147-56.

Hoggard, B., “The Archaeology of Counter-Witchcraft and Popular Magic,” in *Beyond the Witch Trials. Witchcraft and Magic in Enlightenment Europe*, ed. O. Davies and W. de Blécourt (Manchester-New York: Manchester University Press, 2004), 167-86.

Holmes, F.L., “Phlogiston in the Air,” in *Nuova Voltiana. Studies on Volta and his Times*, ed. F. Bevilacqua and L. Fregonese (5 vols.; Milan: Hoepli, 2000-2003), 2: 73-113.

Holtz, S., G. Betsch, and E. Zwink, eds., *Mathesis, Naturphilosophie und Arkanwissenschaft im Umkreis Friedrich Christoph Oetingers (1702-1782)*. Stuttgart: F. Steiner, 2005.

Home, R.W., “Points or Knobs. Lightning Rods and the Basis of Decision Making in Late Eighteenth Century British Science,” in *Playing with Fire. Histories of the Lightning Rod*, ed. P. Heering et al. (Philadelphia: American Philosophical Society, 2009), 97-120.

Huizinga, J., “Natuurbeeld en historiebeeld in de achttiende eeuw,” in idem, *Verzamelde werken* (9 vols.; Haarlem: H.D. Tjeenk Willink, 1948-1953), 4: 341-59.

Huygens, G.W., *Hendrik Tollens. De dichter van de burgerij. Een biografie en een tijdsbeeld*. Rotterdam-The Hague: Nijgh & Van Ditmar, 1972.

Israelowich, I., “The Rain Miracle of Marcus Aurelius: (Re-)Construction of Consensus,” *Greece and Rome*, 2nd ser. 55 (2008): 83-102.

Jäger, H.-W., *Politische Metaphorik im Jakobinismus und im Vormärz*. Stuttgart: J.B. Metzler, 1971.

Jakob, F., *Die Orgel und das Donnergrollen*. Männedorf: Orgelbau Kuhn, 1976.

Jakubowski-Tiessen, M., and H. Lehmann, eds., *Um Himmels Willen. Religion in Katastrophenzeiten*. Göttingen: Vandenhoeck & Ruprecht, 2003.

Janković, V., *Reading the Skies. A Cultural History of English Weather, 1650-1820*. Manchester: Manchester University Press, 2000.

Jauß, H.R., ed., *Die nicht mehr schönen Künste. Grenzphänomene des Ästhetischen*. Munich: Fink, 1968.

Jedin, H., ed., *Handbuch der Kirchengeschichte*. 10 vols., Freiburg-Basel-Vienna: Herder, 1962-1979.

Jensen, J., *Kirchliche Rituale als Waffen gegen Dämonenwirken und Zauberei*. Berlin: LIT, 2007.

L. Jensen, “Disaster upon Disaster Inflicted on the Dutch,” – *BMGN-Low Countries Historical Review* 134, no. 2 (2019): 45-70.

Kalmijn, D., *Abraham Capadose*. The Hague: Boekencentrum, 1955.

Keyser, M., “Het intekenboek van Benjamin Bosma – natuurwetenschappelijk en wijsgerig onderwijs te Amsterdam, 1752-1790. Een verkenning,” in *Jaarverslagen Koninklijk Oudheidkundig Genootschap, uitgebracht door de voorzitters in de 124e, de 125e, de 126e en de 127e jaarvergadering* (1986): 65-81.

Kielman, R.H., *In het laatste der dagen. Eindtijdverwachting in Nederland op de drempel van de moderne tijd (1790-1880)*. Delft: Eburon, 2017.

Kinderman, W., *Beethoven*. Berkeley-Los Angeles: University of California Press, 1995.

Kington, J.A., *The Weather of the 1780s over Europe*. Cambridge: Cambridge University Press, 1988.

Kittsteiner, H.D., "Das Gewissen im Gewitter," in *Jahrbuch für Volkskunde der Görres-Gesellschaft*, n.s. 10 (1987): 7-26.

Kittsteiner, H.D., *Die Entstehung des modernen Gewissens*. Frankfurt a.M.: Insel Verlag, 1991.

Kleemann, K., *A Mist Connection. An Environmental History of the Laki Eruption of 1783 and Its Legacy*. Berlin-Boston: De Gruyter, 2023.

Kleijn, H., *Planten en hun naam: een botanisch lexicon voor de Lage Landen*. 3rd ed. Amsterdam: Meulenhoff, 1980.

Kleinertz, R., "Beethovens 'Sturm'-Sonate op. 31,2 d-moll im Kontext von Sturm- und Gewittermusiken des 18. und frühen 19. Jahrhunderts," in *Wind und Wetter. Kultur-Wissen-Ästhetik*, ed. G. Braungart and U. Büttner (Paderborn: Fink, 2018), 69-86.

Kloek, J.J., and W.W. Mijnhardt, *1800: Blauwdrukken voor een samenleving*. The Hague: Sdu, 2001.

Knippenberg, W.H.Th., *Devotionalia. Religieuze voorwerpen uit het katholieke leven*. 2nd ed. 2 vols., Eindhoven: Bura, 1985 (orig. ed. 1980).

Knuvelder, G.P.M., *Handboek tot de geschiedenis der Nederlandse letterkunde*. 5th ed. 4 vols., Den Bosch: Malmberg, 1970-1976.

Kreissl, E., ed., *Aberglauben – Aberwissen. Welt ohne Zufall*. Graz: Universalmuseum Joanneum, 2014.

Kroonenberg, S., *Why Hell Stinks of Sulfur. Mythology and Geology of the Underworld*. London: Reaktion Books, 2013.

Krop, H., en P. Sonderen, eds., *Esthetica tussen klassiek en romantiek*. Rotterdam: Faculteit der Wijsbegeerte van de Erasmus Universiteit Rotterdam, 1992.

Kück, E., *Wetterglaube in der Lüneburger Heide*. Hamburg: R. Hermes, 1915.

Kuijt, G.H., *De schaduw van drie sterren*. Goes: Oosterbaan & Le Cointre, 1994.

Kürzeder, Chr., "Wetterläuten in Freising und Mr. Burneys Todesangst. Phänomene der Frömmigkeit im Rokoko," in *Mit Leib und Seele. Münchner Rokoko von Asam bis Günther*, ed. R. Diederen and Chr. Kürzeder (Munich: Sieveking, 2014), 211-19.

Kyll, N., "Die Glocke im Wetterglauben und Wetterbrauch des Trierer Landes," in *Rheinisches Jahrbuch für Volkskunde* 9 (1958): 130-79.

Kyll, N., "Die Hagelfeier im alten Erzbistum Trier und seinen Randgebieten," in *Rheinisches Jahrbuch für Volkskunde* 13-14 (1962-1963): 113-71.

Labussière, M., "Naufrages avec sauveteur chez Prévost: théories esthétiques et tempêtes romanesques à l'ombre du *Suave mari magno*," in *L'Événement climatique et ses représentations (XVIIe-XIXe siècle)*, ed. E. Le Roy Ladurie, J. Berchtold, and J.-P. Sermain (Paris: Desjonquères, 2007), 401-14.

Landon, E.H., *A Manual of Councils of the Holy Catholic Church*. Rev. ed., 2 vols., Edinburgh: Grant, 1909 (orig. ed. London: Rivington, 1846).

Le Brun, A., *Les châteaux de la subversion*. Paris: Pauvert, 1982.

Le Roy Ladurie, E., J. Berchtold and J.-P. Sermain, eds., *L'Événement climatique et ses représentations (XVIIe-XIXe siècle). Histoire, littérature, musique et peinture*. Paris: Desjonquères, 2007.

Leemans, I., and G.-J. Johannes, *Worm en donder. Geschiedenis van de Nederlandse literatuur 1700-1800: de Republiek*. Amsterdam: Bert Bakker, 2013.

Leeuw, R. de, J. Reynaerts, and B. Tempel, eds., *Meesters van de Romantiek. Nederlandse kunstenaars 1800-1850*. Zwolle: Waanders, 2005.

Lehner, U.L., *The Catholic Enlightenment. The Forgotten History of a Global Movement.* Oxford-New York: Oxford University Press, 2016.

Lehr, A., *Fulgura frango, ik breek de bliksem. Een middeleeuws klokopschrift.* Asten: Nationaal Beiaardmuseum, 1983.

Lehr, A., *Magie en klokken. Het occult gebruik van klokken en bellen, vroeger en nu – hier en elders. Een verkenning in de literatuur.* Asten: Nationaal Beiaardmuseum, 2000.

Lemaire, A.H., *Filosofie van het landschap.* 6th ed. Baarn: Ambo, 1996 (orig. ed. Bilthoven: Ambo, 1970).

Lemay, J.A.L., *Ebenezer Kinnersley, Franklin's Friend.* Philadelphia: University of Pennsylvania Press, 1964.

Lieburg, M.J. van, and H.A.M. Snelders, *'De bevordering en volmaking der proefondervindelijke wijsbegeerte'. De rol van het Bataafsch Genootschap te Rotterdam in de geschiedenis van de natuurwetenschappen, geneeskunde en techniek (1769-1988).* Amsterdam: Bataafsch Genootschap, 1989.

Madelein, Chr., "Kant in de fout? Het verhevene in de Nederlanden," *De achttiende eeuw* 39 (2007): 77-93.

Madelein, Chr., *Juigchen in den adel der menschlijke natuur. Het verhevene in de Nederlanden (1770-1830).* Ghent: Academia Press, 2010.

Mandos, H., "De strijkster van Middelbeers," *Brabants heem* 16 (1964): 12-15.

Margry, P.J., et al., eds., *Bedevaartplaatsen in Nederland.* 4 vols., Amsterdam-Hilversum: Meertens Instituut and Verloren, 1997-2004.

Marin, L., "Sur une tour de Babel dans un tableau de Poussin," in *Du sublime*, ed. J.F. Courtine et al. (Paris: Belin, 1988), 237-58.

Martin, Chr., "Femmes, orages, tempêtes dans le roman français du XVIIIe siècle," in *L'Événement climatique et ses représentations (XVIIe-XIXe siècle)*, ed. E. Le Roy Ladurie, J. Berchtold, and J.-P. Sermain (Paris: Desjonquères, 2007), 363-86.

Martino, A., ed., *Literatur in der sozialen Bewegung. Aufsätze und Forschungsberichte zum 19. Jahrhundert.* Tübingen: Niemeyer, 1977.

Mathijsen, M., "Geschreven schilderijen. Toenadering tussen de Nederlandse literatuur en schilderkunst van de Romantiek," in *Meesters van de Romantiek. Nederlandse kunstenaars 1800-1850*, ed. R. de Leeuw, J. Reynaerts, and B. Tempel (Zwolle: Waanders, 2005), 193-210.

McCallam, D., *Volcanoes in Eighteenth-Century Europe. An Essay in Environmental Humanities.* Oxford: Oxford University Press, 2019.

Meidinger, H., *Geschichte des Blitzableiters.* Karlsruhe: Braun, 1888.

Mercier-Faivre, A.-M. and C. Thomas, eds., *L'invention de la catastrophe au XVIIIe siècle. Du châtiment divin au désastre naturel.* Geneva: Droz, 2008.

Merkle, S., *Die kirchliche Aufklärung im katholischen Deutschland.* Berlin: Reichl, 1910.

Meurs, B. van, *Verhandelingen over neus, niezen, bliksem, anagram en slaapwandelen.* Culemborg: Blom & Olivierse, 1890.

Meya, J., and H.O. Sibum, *Das fünfte Element. Wirkungen und Deutungen der Elektrizität.* Reinbek: Rowohlt, 1987.

Michel, P., *Physikotheologie. Ursprünge, Leistung und Niedergang einer Denkform.* Zürich: Beer, 2008.

Midelfort, H.C.E., *Exorcism and Enlightenment. Johann Joseph Gassner and the Demons of Eighteenth-Century Germany.* New Haven: Yale University Press, 2005.

Miskotte, K.H., *Antwoord uit het onweer. Een verhandeling over het boek Job.* 2nd ed. Amsterdam: Uitgeversmaatschappij Holland, 1936 (orig. ed. Amsterdam 1929).

Mitchell, T.A., "The Politics of Experiment in the Eighteenth Century: the Pursuit of Audience and the Manipulation of Consensus in the Debate over Lightning Rods," *Eighteenth-Century Studies* 31 (1998): 307-31.

Möhring, C., *Eine Geschichte des Blitzableiters. Die Ableitung des Blitzes und die Neuordnung des Wissens um 1800*. Weimar 2005 (available online at: https://e-pub.uni-weimar.de/opus4/frontdoor/index/index/year/2009/docId/1374).

Mohun, A., "Lightning Rods and the Commodification of Risk in Nineteenth Century America," in *Playing with Fire. Histories of the Lightning Rod*, ed. P. Heering et al. (Philadelphia: American Philosophical Society, 2009), 167-80.

Monk, S.H., *The Sublime. A Study of Critical Theories in XVIIIth-Century England*. Ann Arbor: University of Michigan Press, 1960 (orig. ed. New York: Modern Language Association of America, 1935).

Monrad, K., et al., *The Golden Age of Danish Painting*. New York: Hudson Hills Press, 1993.

Monteiro, M., G. Rooijakkers, and J. Rosendaal, eds., *De dynamiek van religie en cultuur*. Kampen: Kok, 1993.

Moore, C.B., G.D. Aulich, and W. Rison, "A Modern Assessment of Benjamin Franklin's Lightning Rods," in *Playing with Fire. Histories of the Lightning Rod*, ed. P. Heering et al. (Philadelphia: American Philosophical Society, 2009), 256-68.

Motte-Haber, H. de la, *Musik und Natur. Naturanschauung und musikalische Poetik*. Laaber: Laaber Verlag, 2000.

Nauta, D., et al., *Biografisch lexicon voor de geschiedenis van het Nederlandse protestantisme*. 6 vols., Kampen: Kok, 1978-2006.

Neubauer, J., *The Emancipation of Music from Language. Departure from Mimesis in Eighteenth-Century Aesthetics*. New Haven-London: Yale University Press, 1986.

Oeser, E., *Historische Erdbebentheorien von der Antike bis zum Ende des 19. Jahrhunderts*. Vienna 2003 (online available at: https://docplayer.org/41470765-Abhandlungen-geologische-bundesanstalt-historische-erdbebentheorien-von-der-antike-bis-zum-ende-des-19-jahrhunderts-band-58-erhard-oeser.html).

Otte, H., *Glockenkunde*. 2nd ed. Leipzig: Weigel, 1884 (orig. ed. 1858).

Paasman, A.N., *J.F. Martinet. Een Zutphens filosoof in de achttiende eeuw*. Zutphen: Van Someren, 1971.

Pater, C. de, *Petrus van Musschenbroek (1692-1761), een newtoniaans natuuronderzoeker*. Utrecht: s.n., 1979.

Peperkamp, B., "'Bald wird komen das Feur'. Over de representatie van natuurwetenschappelijke en medische kennis in de klucht *De electriciteit, of Pefroen, met het schaepshoofd ge-electriseerd* (1746)," *Gewina* 29 (2006): 269-92.

Le Petit Journal, Supplément illustré, 11 September 1910.

Phayer, F.M., *Religion und das gewöhnliche Volk in Bayern in der Zeit von 1750 bis 1850*. Munich: Stadtarchiv, 1970.

Philipp, W., *Das Werden der Aufklärung in theologiegeschichtlicher Sicht*. Göttingen: Vandenhoeck & Ruprecht, 1957.

Pikulik, L., *Romantik als Ungenügen an der Normalität. Am Beispiel Tiecks, Hoffmanns, Eichendorffs*. Frankfurt a.M.: Suhrkamp, 1979.

Pikulik, L., *Frühromantik. Epoche – Werke – Wirkung*. Munich: Beck, 1992.

Pinault-Sørensen, M., "Orages et tempêtes. Peinture et dessin, deuxième moitié du XVIIIe siècle et début du XIXe siècle," in *L'Événement climatique et ses représentations (XVIIe-XIXe siècle)*, ed. E. Le Roy Ladurie, J. Berchtold, and J.-P. Sermain (Paris: Desjonquères, 2007), 254-70.

Plachta, B., and W. Woesler, eds., *Sturm und Drang. Geistiger Aufbruch 1770-1790 im Spiegel der Literatur*. Tübingen: Niemeyer, 1997.

Plück, B., Die Verehrung des hl. Donatus von Münstereifel in den Bistümern Köln und Trier 1652-1773. Unpublished Master Thesis, Bonn 1980.

Plück, B., "Der Kult des Katakombenheiligen Donatus von Münstereifel," in *Jahrbuch für Volkskunde*, n.s. 4 (1981): 112-26.

Pohle, F., *'Mera ossa et cadavera.' Studien zum Jesuitentheater in Jülich-Berg, Ravenstein und Aachen (1601-1817)*. Unpublished Ph.D. Thesis, Aachen 2006, online available at: https://publications.rwth-aachen.de/record/62507.

Poirier, J.-P., *L'abbé Bertholon. Un électricien des Lumières en province*. Paris: Hermann, 2008.

Poirier, J.-P., "Electrical Earthquakes: A Short-Lived Theory in the 18th Century," *Earth Sciences History* 35 (2016): 283-302.

Post, P.G.J., "'Het suizen van een zachte koelte'. Het weer en christelijke rituelen: terreinverkenning en pleidooi voor wetenschappelijk grensverkeer gethematiseerd via palmzondagriten," *Volkskundig bulletin* 15 (1989): 177-200.

Pott, M., *Aufklärung und Aberglaube. Die deutsche Frühaufklärung im Spiegel ihrer Aberglaubenskritik*. Tübingen: Niemeyer, 1992.

Praamsma, R.F.H., *Zacharias Webber (1644-1696), irenisch lutheraan – verlicht protestant. Kerk en theologie in het denken van een zeventiende-eeuwse kunstschilder*. Delft: Eburon, 2013.

Principato, A., "Tourmente ou déluge: métaphores révolutionnaires chez Chateaubriand," in *L'Événement climatique et ses représentations (XVIIe-XIXe siècle)*, ed. E. Le Roy Ladurie, J. Berchtold, and J.-P. Sermain (Paris: Desjonquères, 2007), 464-78.

Prinz, H., "Fulminantes über Wolkenelektrizität," *Bulletin des schweizerischen elektrotechnischen Vereins* 64, no. 1 (1973): 3-17.

Prinz, H., *Gewitterelektrizität*. Munich: Oldenbourg, 1979.

Pütz, P., ed., *Erforschung der deutschen Aufklärung*. Königstein: Athenäum, 1980.

Raven, J.J., *The Bells of England*. London: Methuen, 1906.

Reed, A., *Romantic Weather. The Climates of Coleridge and Baudelaire*. Hanover, NH-London: Brown University Press, 1983.

Reichler, C., "Air, orages et météores au tournant du XVIIIe siècle," in *L'Événement climatique et ses représentations (XVIIe-XIXe siècle)*, ed. E. Le Roy Ladurie, J. Berchtold, and J.-P. Sermain (Paris: Desjonquères, 2007), 143-56.

Reijen, P. van, ed., *Hef aan! Bataaf! Beschouwingen over muziek en muziekleven in Nederland omstreeks 1795*. Alphen a.d. Rijn: Canaletto, [1997].

Reynaerts, J., "De hand van God. Romantiek in de Nederlandse landschapschilderkunst," in *Meesters van de Romantiek. Nederlandse kunstenaars 1800-1850*, ed. R. de Leeuw, J. Reynaerts, and B. Tempel (Zwolle: Waanders, 2005), 129-52.

Rhys Morus, I., ed., *The Oxford Illustrated History of Science*. Oxford: Oxford University Press, 2017.

Richter, K., *Literatur und Naturwissenschaft. Eine Studie zur Lyrik der Aufklärung*. Munich: Fink, 1972.

Rieth, A., *Der Blitz in der bildenden Kunst*. Munich: Heimeran, 1953.

Riskin, J., "The Lawyer and the Lightning Rod," *Science in Context* 12 (1999): 61-99.

Riskin, J., *Science in an Age of Sensibility. The Sentimental Empiricists of the French Enlightenment*. Chicago: University of Chicago Press, 2002.

Roberts, L., "Science Becomes Electric: Dutch Interaction with the Electrical Machine during the Eighteenth Century," *Isis* 90 (1999): 680-714.

Robles, W.B., "Atlantic Disaster: Boston Responds to the Cape Ann Earthquake of 1755," *The New England Quarterly* 90 (2017): 7-35.

Rocholz, E.L., *Deutscher Glaube und Brauch im Spiegel der heidnischen Vorzeit.* 2 vols., Berlin: Dümmler, 1867.

Roloff, E., *Göttliche Geistesblitze. Pfarrer und Priester als Erfinder und Entdecker.* Weinheim: Wiley-VCH, 2010.

Rooijakkers, G., and Th. van der Zee, eds., *Religieuze volkscultuur. De spanning tussen de voorgeschreven orde en de geleefde praktijk.* Nijmegen: SUN, 1986.

Rooijakkers, G., "De dynamiek van devotionalia. De materiële cultuur van het geleefde geloof in oostelijk Noord-Brabant," in *De dynamiek van religie en cultuur*, ed. M. Monteiro, G. Rooijakkers, and J. Rosendaal (Kampen: Kok, 1993), 80-106.

Rooijakkers, G., *Rituele repertoires. Volkscultuur in oostelijk Noord-Brabant 1559-1853.* Nijmegen: SUN, 1994.

Rooijakkers, G., *Eer en schande. Volksgebruiken in het oude Brabant.* Nijmegen: SUN, 1995.

Rooijakkers, G., A. van Lierop, and R. van de Weijer, eds., *De musealisering van het platteland: de historie van een Brabants boerenhuis.* Nijmegen: SUN, [2002].

Rooseboom, M., *Bijdrage tot de geschiedenis der instrumentmakerskunst in de Noordelijke Nederlanden tot omstreeks 1840.* Leiden: s.n., 1950.

Rose, Amber J., "The Witch on the Wall; the Milk-Stealing Witch in Scandinavian Iconography," *Arv : Nordic Yearbook of Folklore* 71 (2015): 27-43.

Ruhe, S., *Visserstruien uit 40 Nederlandse vissersplaatsen.* Baarn: Forte, 2013.

Ruhe, S., *Visserstruien 2. 65 historische truien met breischema's uit 55 Nederlandse vissersplaatsen.* Baarn: Forte, 2015.

Ruppert, W., "Volksaufklärung im späten 18. Jahrhundert," in *Deutsche Aufklärung bis zur Französischen Revolution 1680-1789*, ed. R. Grimminger (Munich: Hanser, 1980), 341-61.

Santillana, G. de, "Volta," *Scientific American* 212 (January 1965): 82-91.

Sartori, P., *Das Buch von deutschen Glocken.* Berlin: De Gruyter, 1932.

Schaffer, S., "The Consuming Flame: Electrical Showmen and Tory Mystics in the World of Goods," in *Consumption and the World of Goods*, ed. J. Brewer and R. Porter (London-New York: Routledge, 1993), 489-526.

Scharfe, M., *Die Religion des Volkes. Kleine Kultur- und Sozialgeschichte des Pietismus.* Gütersloh: Mohn, 1980.

Scharfe, M., "Kruzifix mit Blitzableiter," *Österreichische Zeitschrift für Volkskunde* 102 (1999): 289-336.

Scharfe, M., *Menschenwerk. Erkundungen über Kultur.* Cologne-Weimar-Vienna: Böhlau, 2002.

Schiffer, M.B., *Draw the Lightning Down. Benjamin Franklin and Electrical Technology in the Age of Enlightenment.* Berkeley and Los Angeles-London: University of California Press, 2003.

Schimmelpenninck van der Oije, C.O.A., "Een reysje na Gelderland en de electrike vloeistof," *Gelders oudheidkundig contactbericht* 47 (1970): 12-17.

Schleuning, P., *Die Sprache der Natur: Natur in der Musik des 18. Jahrhunderts.* Stuttgart-Weimar: J.B. Metzler, 1998.

Schmenner, R., *Die Pastorale. Beethoven, das Gewitter und der Blitzableiter.* Kassel: Bärenreiter, 1998.

Schmidt, A., "Gewitter und Blitzableiter. Historische Deutungsmuster von Gewitter und deren Umschlag in Technik," in *Natur-Bilder. Wahrnehmungen von Natur und Umwelt in der Geschichte*, ed. R.P. Sieferle and H. Breuninger (Frankfurt a.M.-New York: Campus Verlag, 1999), 279-96.

Schneider, Helmut J., "Naturerfahrung und Idylle in der deutschen Aufklärung," in *Erforschung der deutschen Aufklärung*, ed. P. Pütz (Königstein: Athenäum, 1980), 289-315.

Schneider, J.-P., "De l'orage châtiment au chaos maîtrisé," in *L'Événement climatique et ses représentations (XVIIe-XIXe siècle)*, ed. E. Le Roy Ladurie, J. Berchtold, and J.-P. Sermain (Paris: Desjonquères, 2007), 126-42.

Schönert, J., "Behaglicher Schauer und distanzierter Schrecken. Zur Situation von Schauerroman und Schauererzählung im literarischen Leben der Biedermeierzeit," in *Literatur in der sozialen Bewegung. Aufsätze und Forschungsberichte zum 19. Jahrhundert*, ed. A. Martino (Tübingen: Niemeyer, 1977), 27-92.

Schonland, B.F.J., *The Flight of the Thunderbolts*. 2nd ed. Oxford: Oxford University Press, 1964 (orig. ed. Oxford 1950).

Schott, H., *Magie der Natur. Historische Variationen über ein Motiv der Heilkunst*. 2 vols., Aachen: Shaker, 2014.

Schulte, A.G., ed., *Arnhems Historisch Genootschap 'Prodesse Conamur' 1792-1992*. Zutphen: Walburg Pers, 1992.

Schuth, D., *Die Farbe Blau. Versuch einer Charakteristik*. Münster: LIT, 1995.

Schwarz, J., "Aachen- oder Wetterhörner," *Westdeutsche Zeitschrift für Volkskunde* 32 (1935), Appendix 'Heimatmuseum und Volkskunde': 1-20.

Sébillot, P., *Le folk-lore de France*. 4 vols., Paris: Librairie Orientale et Américaine, 1904-1907.

Sieferle, R.P., and H. Breuninger, eds., *Natur-Bilder. Wahrnehmungen von Natur und Umwelt in der Geschichte*. Frankfurt a.M.-New York: Campus Verlag, 1999.

Snelders, H.A.M., *Het gezelschap der Hollandsche scheikundigen. Amsterdamse chemici uit het einde van de achttiende eeuw*. Amsterdam: Rodopi, 1980.

Snelders, H.A.M., "The Amsterdam Pharmacist Willem van Barneveld (1747-1826) and the Discovery of Photosynthesis (1778)," *Janus* 68 (1981): 1-14.

Snelders, H.A.M., "Romantiek en natuurwetenschappers in Nederland (1800-1850)," *De negentiende eeuw* 8 (1984): 72-90.

Snelders, H.A.M., "Benjamin Franklin (1706-1790), lid-correspondent van het Bataafsch Genootschap der Proefondervindelijke Wijsbegeerte te Rotterdam (1771)," *Nieuwe verhandelingen van het Bataafsch genootschap der proefondervindelijke wijsbegeerte te Rotterdam*, 3rd ser., vol. 6 (1991): 5-12.

Specht, B., *Physik als Kunst. Die Poetisierung der Elektrizität um 1800*. Berlin-New York: De Gruyter, 2010.

Specht, B., "Die Elektrisierung der Aufklärung. Kontexte, Metaphorisierungen und Funktionen der Elektrizität im Wissenssystem des späten 18. Jahrhunderts," in *Die Sachen der Aufklärung. Beiträge zur DGEJ-Jahrestagung 2010 in Halle a.d. Saale*, ed. F. Berndt and D. Fulda (Hamburg: F. Meiner, 2012), 437-47.

Spinks, J., and C. Zika, eds., *Disaster, Death and the Emotions in the Shadow of the Apocalypse, 1400-1700*. London: Palgrave Macmillan, 2016.

Stanyon, M., *Resounding the Sublime. Music in English and German Literature and Aesthetic Theory, 1670-1850*. Philadelphia: University of Pennsylvania Press, 2021.

Steffan, F., "Wetterabweisende Heiltümer. Fund in der Turmkugel von Altenhohenau bei Wasserburg am Inn," *Volkskunst. Zeitschrift für volkstümliche Sachkultur*, Issue 4 (1985): 23-25.

Stichweh, R., *Zur Entstehung des modernen Systems wissenschaftlicher Disziplinen. Physik in Deutschland 1740-1890*. Frankfurt a.M.: Suhrkamp, 1984.

Strele, R. von, "Wetterläuten und Wetterschiessen. Eine culturgeschichtliche Studie," in *Jahrbuch des Deutschen und Österreichischen Alpenvereins* 29 (1898): 123-42.

Streng, J.C., *'Het is thans zeer briljant'. Aspecten van het Zwolse culturele leven tijdens de overgang van ancien régime naar moderne tijd*. Hilversum: Verloren, 1999.

Stromp, M., et al., "How Realistic are Painted Lightnings? Quantitative Comparison of the Morphology of Painted and Real Lightnings: a Psychophysical Approach," *Proceedings of the Royal Society* A, vol. 474 (June 2018), Issue 2214.

Sturkenboom, D., *De elektrieke kus. Over vrouwen, fysica en vriendschap in de 18de en 19de eeuw. Het verhaal van het Natuurkundig genootschap der dames in Middelburg*. Amsterdam: Augustus, 2004.

Sudduth, W.M., "Eighteenth-Century Identifications of Electricity with Phlogiston," *Ambix* 25 (1978): 131-47.

Tatar, M.M., *Spellbound. Studies on Mesmerism and Literature*. Princeton: Princeton University Press, 1978.

Tempel, B., "Nederlandse Romantiek. Een plaatsbepaling," in *Meesters van de Romantiek. Nederlandse kunstenaars 1800-1850*, ed. R. de Leeuw, J. Reynaerts, and B. Tempel (Zwolle: Waanders, 2005), 71-94.

Thacker, C., *The Wildness Pleases. The Origins of Romanticism*. London: Croom Helm, 1983.

Thalmann, M., *Der Trivialroman des 18. Jahrhunderts und der romantische Roman. Ein Beitrag zur Entwicklungsgeschichte der Geheimbundmystik*. Berlin: E. Ebering, 1923.

Thomas, K., *Religion and the Decline of Magic. Studies in Popular Beliefs in Sixteenth- and Seventeenth-Century England*. 2nd ed. Harmondsworth: Penguin, 1973 (orig. ed. London: Weidenfeld & Nicolson, 1971).

Thorel-Cailleteau, S., and C. Jamain, eds., *Revue des Sciences Humaines* 281, no. 1 (January-March 2006): 1-255, Special Issue entitled *L'Imaginaire de l'électricité dans les lettres et les arts*.

Thuesen, P.J., *Tornado God. American Religion and Violent Weather*. Oxford-New York: Oxford University Press, 2020.

Thüsen, J. von der, "'Die Lava der Revolution fließt majestätisch'. Vulkanische Metaphorik zur Zeit der französischen Revolution," *Francia* 23, no. 2 (1996): 113-43.

Thüsen, J. von der, *Het verlangen naar huivering. Over het sublieme, het wrede en het unheimliche. Essays*. Amsterdam: Querido, 1997.

Thüsen, J. von der, *Schönheit und Schrecken der Vulkane. Zur Kulturgeschichte des Vulkanismus*. Darmstadt: Wissenschaftliche Buchgesellschaft, 2008.

Torlais, J., *Un physicien au siècle des Lumières: l'abbé Nollet, 1700-1770*. Paris: Sipuco, 1954.

Trautwein, W., *Erlesene Angst. Schauerliteratur im 18. und 19. Jahrhundert*. Munich: C. Hanser, 1980.

Trempler, J., "Entfesselte Natur und Geschichte. Zur Interpretation und Darstellung von Katastrophen im 18. Jahrhundert," in *Entfesselte Natur. Das Bild der Katastrophe seit 1600*, ed. M. Bertsch and J. Trempler (Petersberg: Imhof, 2018), 11-19.

Trénard, L., *Lyon. De l'Encyclopédie au Préromantisme*. 2 vols., Paris: P.U.F., 1958.

Trousson, R., *Le thème de Prométhée dans la littérature européenne*. 2 vols., Geneva: Droz, 1964 (3rd rev. ed. Geneva: Droz, 2001).

Uitterhoeve, W., *Cornelis Kraijenhoff 1758-1840. Een loopbaan onder vijf regeervormen*. Nijmegen: Vantilt, 2009.

Van Tieghem, P., *Le sentiment de la nature dans le préromantisme européen*. Paris: A.G. Nizet, 1960.

Vasak, A., "*Héloïse* et *Werther, Sturm und Drang* : comment la tempête, en entrant dans nos cœurs, nous a donné le monde," *Ethnologie française* 39(2009), 677-85.

Vermeir, K., ed., *Centaurus* 57, no. 3 (2015): 129-228.

Vermij, R., "Erschütterung und Bewältigung. Erdbebenkatastrophen in der Frühen Neuzeit," in *Um Himmels Willen. Religion in Katastrophenzeiten*, ed. M. Jakubowski-Tiessen and H. Lehmann (Göttingen: Vandenhoeck & Ruprecht, 2003), 235-52.

Vermij, R., *Thinking on Earthquakes in Early Modern Europe*. London-New York: Routledge, 2021.

Vervoort, R., *'Vrouwen op den besem en derghelijck gespoock'. Pieter Bruegel en de traditie van hekserijvoorstellingen in de Nederlanden tussen 1450 en 1700*. Nijmegen: Vantilt, 2011.

Vijselaar, J., *De magnetische geest. Het dierlijk magnetisme 1770-1830*. Nijmegen: SUN, 2001.

Vogel, B., Transmission and Performance of Taiko in Edo Bayashi, Hachijo, and Modern Kumi-daiko Styles. Unpublished Ph.D. Thesis, Houston 2009 (available online at: https://scholarship.rice.edu/bitstream/handle/1911/61883/3362425.pdf?sequence=1&isAllowed=y).

Vuyk, S., *De verdraagzame gemeente van vrije christenen. Remonstranten op de bres voor de Bataafse Republiek (1780-1800)*. Amsterdam: De Bataafsche Leeuw, 1995.

Vuyk, S., *Jacob Kantelaar. Veelzijdig verlicht verliezer 1759-1821*. Zwolle: Waanders, 2005.

Walsham, A., "Deciphering Divine Wrath and Displaying Godly Sorrow: Providentialism and Emotion in Early Modern England," in *Disaster, Death and the Emotions in the Shadow of the Apocalypse, 1400-1700*, ed. J. Spinks and C. Zika (London: Palgrave Macmillan, 2016), 21-43.

Warburg, A.M., *Schlangenritual. Ein Reisebericht*. 5th ed. Berlin: K. Wagenbach, 2011 (a revision of a lecture held in 1923).

Weber, Chr., *Vom Gottesgericht zur verhängnisvollen Natur. Darstellung und Bewältigung von Naturkatastrophen im 18. Jahrhundert*. Hamburg: F. Meiner, 2015.

Weigl, E., "Entzauberung der Natur durch Wissenschaft – dargestellt am Beispiel der Erfindung des Blitzableiters," in *Jahrbuch der Jean-Paul-Gesellschaft* 22 (1987): 7-39.

Weigl, E., *Instrumente der Neuzeit. Die Entdeckung der modernen Wirklichkeit*. Stuttgart: J.B. Metzler, 1990.

Weijer, R. van de, "Een huis als herinneringsmachine. Van agrarisch bedrijf Sigmans naar museumboerderij," in *De musealisering van het platteland: de historie van een Brabants boerenhuis*, ed. G. Rooijakkers, A. van Lierop, and R. van de Weijer (Nijmegen: SUN, [2002]), 136-69.

Westmijse, G., "Protestantsche bede-, dank- en vastendagen in de Republiek der Zeven Vereenigde Nederlanden," in *Kerkhistorische opstellen van het gezelschap S.S.S. Nieuwe bundel*, ed. R. Riphaagen Jr. et al. (The Hague: Nijhoff, 1914): 96-114.

Westra, F., *Martinitoren*. Groningen: Passage, 2009.

White, A.D., *A History of the Warfare of Science with Theology in Christendom*. 2 vols., London: Macmillan, 1896.

Wiechmann, A., and L.C. Palm, eds., *Een elektriserend geleerde. Martinus van Marum 1750-1837*. Haarlem: Joh. Enschedé, 1987.

Wiel, C. van de, *Het recht van de Kerk met betrekking tot handelingen van goddelijke eredienst, gewijde plaatsen en heilige tijden*. Louvain: Peeters, 2000.

Wielema, M.R., *Ketters en verlichters. De invloed van het Spinozisme en het Wolffianisme op de Verlichting in gereformeerd Nederland*. [Amsterdam:] s.n., 1999.

Wiesner, J., *Jan Ingen-Housz. Sein Leben und sein Wirken als Naturforscher und Arzt*. Vienna: C. Konegen, 1905.

Wild, R., "Prometheus-Franklin: Die Gestalt Benjamin Franklins in der deutschen Literatur des 18. Jahrhunderts," *Amerikastudien/American Studies* 23 (1978): 30-39.

Wilson, S., *The Magical Universe. Everyday Ritual and Magic in Pre-Modern Europe*. London: Hambledon and London, 2000.

Winckel, Floris, "Medicalising Electricity in the Dutch Republic, 1745-1789," *BMGN- Low Countries Historical Review* 137, no. 3 (2022): 60-86.

Winter-Fritzsche, A., *'Naturgetreue' Blitzdarstellungen in der venezianischen Malerei des 16. Jahrhunderts mit einer Einführung in Deutung und Darstellung des Blitzes*. Frankfurt a.M.: Peter Lang, 1990.

Wissing, P.W. van, *Stokebrand Janus 1787. Opkomst en ondergang van een achttiende-eeuws satirisch politiek-literair weekblad*. Nijmegen: Vantilt, 2003.

Wodianka, S., *Betrachtungen des Todes. Formen und Funktionen der 'meditatio mortis' in der europäischen Literatur des 17. Jahrhunderts*. Tübingen: Niemeyer, 2004.

Woud, A. van der, *De Bataafse hut. Denken over het oudste Nederland (1750-1850)*. 2nd ed. Amsterdam-Antwerp: Contact, 1998 (orig. ed. Amsterdam: Meulenhoff, 1990).

Young, F., *A History of Exorcism in Catholic Christianity*. London-New York: Palgrave Macmillan, 2016.

Zacharias-Langhans, G., *Der unheimliche Roman um 1800*. Bonn: R. Schwarzbold, 1968.

Zak, S., *Musik als 'Ehr und Zier'*. Neuss: Päffgen, 1979.

Zanetti, F., *L'Électricité médicale dans la France des Lumières*. Oxford: Voltaire Foundation, 2017.

Zeiler, F.D., "Johann Carl Röhner (1774-1837). Een muzikant met hartstocht," *Zwols historisch tijdschrift* 11 (1994): 13-23.

Zelle, C., *'Angenehmes Grauen'. Literaturhistorische Beiträge zur Ästhetik des Schrecklichen im achtzehnten Jahrhundert*. Hamburg: F. Meiner, 1987.

Zwierlein, C., *Prometheus Tamed. Fire, Security, and Modernities, 1400 to 1900*. Leiden-Boston: Brill, 2021.

Websites

http://cnum.cnam.fr

https://www.dbnl.org/arch/louw005janp01_01/pag/louw005janp01_01.pdf

https://dhw-schultz.de

https://docplayer.org/41470765-Abhandlungen-geologische-bundesanstalt-historische-erdbebentheorien-von-der-antike-bis-zum-ende-des-19-jahrhunderts-band-58-erhard-oeser.html

https://www.dwds.de/r/plot/?view=1&corpus=dta%2Bdwds&norm=date%2B-class&smooth=spline&genres=0&grand=1&slice=10&prune=0.05&window=3&wbase=0&logavg=0&logscale=0&xrange=1750%3A1830&q1=Gedankenblitz

https://e-pub.uni-weimar.de/opus4/frontdoor/index/index/year/2009/docId/1374

https://founders.archives.gov/documents/Franklin/01-09-02-0153

https://gtb.ivdnt.org/search/

www.martinicarillon.nl/geschiedenis.html

https://publications.rwth-aachen.de/record/62507

https://scholarship.rice.edu/bitstream/handle/1911/61883/3362425.pdf?sequence=1&isAllowed=y

Illustration Credits

The bold figures in the following list refer to the pages on which the illustrations are printed
t=top, b=bottom, l=left, r=right
tl= top left, tr= top right, bl=bottom left, br=bottom right

AKG-images **2** (Liszt Collection), **13**r (Liszt Collection), **26** (World History Archive), **42** (World History Archive), **113, 123, 145, 200, 202, 212-13, 224-25, 226** (Erich Lessing), **242-43** (Erich Lessing)
Alamy Photo **85**
Atlas van Stolk, Rotterdam **73**r, **104**
Bibliothèque nationale de France, Paris **160, 165, 189, 191**
© BPK-Bildagentur, Berlin **30** (Hamburger Kunsthalle | Elke Walford), **76-77** (Bayerische Staatsgemäldesammlungen), **107**l (Hamburger Kunsthalle | Elke Walford), **121** (Hamburger Kunsthalle | Elke Walford), **193** (Staatsbibliothek zu Berlin | Ruth Schacht), **206-07, 221** (Staatsgalerie Stuttgart)
© Claude Bernard Lyon 1 University | Communication Department, Eric Le Roux **164**
Collection of the author **101**
Digitale Bibliotheek voor de Nederlandse Letteren **205** (https://www.dbnl.org/arch/louw005janp01_01/pag/louw005janp01_01.pdf)
James Delbourgo, *A Most Amazing Scene of Wonders. Electricity and Enlightenment in Early America* (Cambridge MA: 2006, p. 155) **187**
Gelders Archief, Arnhem **45**
Getty Images | Universal History Archives **44**
Gleimhaus – Museum der Deutschen Aufklärung, Halberstadt **66**
Granger | Bridgeman Images **178**
Herzog August Bibliothek Wolfenbüttel **59**
Katholiek Documentatie Centrum, Nijmegen **138**
Kunsthuis Secretarie, Meppel (photograph: Albert Brunsting) **158-59**
Kupferstich-Kabinett, Staatliche Kunstsammlungen Dresden (photograph: Herbert Boswank) **31, 144**
Laing Art Gallery, Newcastle-upon-Tyne, UK © Tyne & Wear Archives & Museums | Bridgeman Images **54-55**
Musée des Beaux-Arts, Bordeaux **80**
© Musée des Beaux-Arts, Rennes, Dist. RMN-Grand Palais | Patrick Merret **18-19**
Musées de la Ville de Poitiers et de la Société des Antiquaires de l'Ouest © Musées de Poitiers/Christian Vignaud **184-85**
Museon, The Hague **130**
Museum Catharijneconvent, Utrecht **134, 137**br, **141**l, **194**
Museum Klok & Peel, Asten **125**
Museum Krona, Uden **141**r
National Galleries of Scotland, Edinburgh **57, 237**
National Gallery of Art, Washington DC **218**
Nationalmuseum Sweden, Stockholm via Google Art Project **152**

Index of Names

G

H

MIX
Papier | Fördert
gute Waldnutzung
FSC® C083411

Zeitfracht Medien GmbH
Ferdinand-Jühlke-Straße 7
99095 Erfurt, Deutschland
produktsicherheit@kolibri360.de